W0079829

SUSTAINABLE INDUSTRIAL CHEMISTRY
Principles, Tools and Techniques

SUSTAINABLE INDUSTRIAL CHEMISTRY
Principles, Tools and Techniques

Contributors :
Shuling Chen,
Shourong Wang, *et al.*

AURIS REFERENCE LTD.
London, UK

Sustainable Industrial Chemistry : Principles, Tools and Techniques
Contributors : Shuling Chen *and* Shourong Wang, *et al.*

Auris Reference Ltd., UK

www.aurisreference.com

United Kingdom

Copyright 2016

Printed in 2017 for Sale in the Indian Subcontinent

The information in this book has been obtained from highly regarded resources. The copyrights for individual articles remain with the authors, as indicated. All chapters are distributed under the terms of the Creative Commons Attribution License, which permit unrestricted use, distribution, and reproduction in any medium, provided the original author and source are credited.

Notice

Contributors, whose names have been given on the book cover, are not associated with the Publisher. The editors and the Publisher have attempted to trace the copyright holders of all material reproduced in this publication and apologise to copyright holders if permission has not been obtained. If any copyright holder has not been acknowledged, please write to us so we may rectify.

Reasonable efforts have been made to publish reliable data. The views articulated in the chapters are those of the individual contributors, and not necessarily those of the editors or the Publisher. Editors and/or the Publisher are not responsible for the accuracy of the information in the published chapters or consequences from their use. The Publisher accepts no responsibility for any damage or grievance to individual(s) or property arising out of the use of any material(s), instruction(s), methods or thoughts in the book.

No part of this publication maybe reproduced, stored in a retrieval system or transmitted in any form or by any means, electronic, mechanical, photocopying, recording, scanning or otherwise without prior written permission of the publisher.

Sustainable Industrial Chemistry : Principles, Tools and Techniques

ISBN: 978-1-78154-518-8

British Library Cataloguing in Publication Data
A CIP record for this book is available from the British Library

Exclusively distributed by CBS Publishers & Distributors Pvt. Ltd.

Sales & Distribution Rights only for India, Pakistan, Bangladesh, Sri Lanka, Nepal and Bhutan.This book is not to be sold outside these territories.

PREFACE

Industrial chemistry is part of the long chain in the design and manufacturing process. Industrial chemists deal with the ideas, the design, the testing, and prototyping of new industrial products. In order to design something entirely new to help solve the major problems of the world their essential skills are, in-depth knowledge and application of chemistry and creativity with chemicals.

Green and Sustainable Chemistry covers subjects relating to reducing the environmental impact of chemicals and fuels by developing alternative and sustainable technologies that are non-toxic to living things and the environment. Unique feature the Green and Sustainable Chemistry projects can be in any field of Green or Sustainable Chemistry so long as they are applicable for use in developing countries. This will not only contribute to a more sustainable growth of industry but also to a more green economy and environment.

Sustainable Industrial Chemistry: Principles, Tools and Techniques present text is an effort to fill up this void in the Sustainable Industrial Chemistry literature. With the concept of the subject having undergone a tremendous change in recent years, the present volume is intended to provide an up-to-date knowledge of the subject according to the needs of the students. The purpose of the book is also to show as to how an understanding of the concept of Sustainable Industrial Chemistry can be usefully applied in thinking about and planning to answer related problems.

This book describes in detail some of the basic principles, methods and results of Industrial chemistry that lead to our understanding of Sustainable behaviour. It has been endeavour of author to present the students the fundamental Principles, Tools and Techniques of Sustainable Industrial Chemistry.

This page left intentionally blank.

CONTENTS

This page left intentionally blank.

LIST OF CONTRIBUTORS

Dunzhu Xia

School of Instrument Science and Engineering, Southeast University, Nanjing City, Jiangsu Province, 210096, China; E-Mails: chenshuling318@126.com (S.C.); srwang@seu.edu.cn (S.W.)

Shuling Chen

School of Instrument Science and Engineering, Southeast University, Nanjing City, Jiangsu Province, 210096, China; E-Mails: chenshuling318@126.com (S.C.); srwang@seu.edu.cn (S.W.)

Shourong Wang

School of Instrument Science and Engineering, Southeast University, Nanjing City, Jiangsu Province, 210096, China; E-Mails: chenshuling318@126.com (S.C.); srwang@seu.edu.cn (S.W.)

This page left intentionally blank.

Chapter 1

FROM GREEN TO SUSTAINABLE
INDUSTRIAL CHEMISTRY

GREEN CHEMISTRY

A traditional concept in process chemistry has been the optimization of the time-space yield. From our modern perspective, this limited viewpoint must be enlarged, as for example toxic wastes can destroy natural resources and especially the means of livelihood for future generations. In addition, many feedstocks for the production of chemicals are based on petroleum, which is not a renewable resource. The key question to address is: what alternatives can be developed and used? In addition, we must ensure that future generations can also use these new alternatives. "Sustainability" is a concept that is used to distinguish methods and processes that can ensure the long-term productivity of the environment, so that even subsequent generations of humans can live on this planet. Sustainability has environmental, economic, and social dimensions.

Paul Anastas of the Environmental Protection Agency formulated some simple rules of thumb for how sustainability can be achieved in the production of chemicals - the "Green chemical principles":

1. Waste prevention instead of remediation
2. Atom economy or efficiency
3. Use of less hazardous and toxic chemicals
4. Safer products by design
5. Innocuous solvents and auxiliaries
6. Energy efficiency by design
7. Preferred use of renewable raw materials
8. Shorter syntheses (avoid derivatization)
9. Catalytic rather than stoichiometric reagents

10. Design products to undergo degradation in the environment

11. Analytical methodologies for pollution prevention

12. Inherently safer processes

Implementing these Green Chemical Principles requires a certain investment, since the current, very inexpensive chemical processes must be redesigned. However, in times when certain raw materials become more expensive (for example, as the availability of transition metals becomes limited) and also the costs for energy increase, such an investment should be paid back as the optimized processes become less expensive than the unoptimized ones. The development of greener procedures can therefore be seen as an investment for the future, which also helps to ensure that the production complies with possible upcoming future legal regulations.

A typical chemical process generates products and wastes from raw materials such as substrates, solvents and reagents. If most of the reagents and the solvent can be recycled, the mass flow looks quite different:

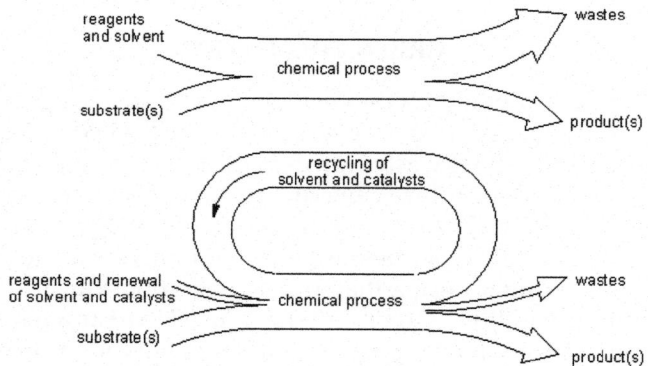

Thus, the prevention of waste can be achieved if most of the reagents and the solvent are recyclable. For example, catalysts and reagents such as acids and bases that are bound to a solid phase can be filtered off, and can be regenerated (if needed) and reused in a subsequent run. In the production of chemical products on very large scale, heterogeneous catalysts and reagents can be kept stationary while substrates are continuously added and pass through to yield a product that is continuously removed (for example by distillation).

The mass efficiency of such processes can be judged by the E factor (Environmental factor):

$$E \text{ factor} = \frac{\text{Mass of wastes}}{\text{Mass of product}}$$

Whereas the ideal E factor of 0 is almost achieved in petroleum refining, the production of bulk and fine chemicals gives E factors of between 1 and 50. Typical E factors for the production of pharmaceuticals lie between 25 and 100.

Note that water is not considered in this calculation, because this would lead to very high E factors. However, inorganic and organic wastes that are diluted in the aqueous stream must be included. Sometimes it is easier to calculate the E factor from a different viewpoint, since accounting for the losses and exact waste streams is difficult:

$$\text{E factor} = \frac{\text{Mass of materials - Mass of product}}{\text{Mass of product}}$$

In any event, the E factor and related factors do not account for any type of toxicity of the wastes. Such a correction factor (an "unfriendliness" quotient, Q) would be 1 if the waste has no impact on the environment, less than 1 if the waste can be recycled or used for another product, and greater than 1 if the wastes are toxic and hazardous. Such discussions are at a very preliminary stage, and E factors can be used directly for comparison purposes as this metric has already been widely adopted in the industry.

Another attempt to calculate the efficiency of chemical reactions that is also widely used is that of atom economy or efficiency. Here the value can be calculated from the chemical equation:

$$\text{atom efficiency} = \frac{\text{molecular weight of desired product}}{\text{molecular weight of all substances formed}}$$

examples:

$$\underset{Ph}{\overset{OH}{\diagup}} + 2\,CrO_3 + 3\,H_2SO_4 \longrightarrow 3 \underset{Ph}{\overset{O}{\diagup}} + Cr_2(SO_4)_3 + 6\,H_2O$$

120 g / mol • 3   392 g / mol   18 g / mol • 6

$$\text{atom efficiency} = \frac{3 \cdot 120}{3 \cdot 120 + 392 + 6 \cdot 18} = 42\%$$

$$\underset{Ph}{\overset{OH}{\diagup}} + 0.5\,O_2 \xrightarrow{\text{catalyst}} \underset{Ph}{\overset{O}{\diagup}} + H_2O$$

120 g / mol   18 g / mol

$$\text{atom efficiency} = \frac{120}{120 + 18} = 87\%$$

Atom efficiency is a highly theoretical value that does not incorporate any solvent, nor the actual chemical yield. An experimental atom efficiency can be calculated by multiplying the chemical yield with the theoretical atom efficiency. Anyway, the discussion remains more qualitative than quantitative, and does not yet quantify the type of toxicity of the products and reagents used. Still, atom economy as a term can readily be used for a direct qualitative description of reactions.

Considering specific reactions, the development of green methods is focused on two main aspects: choice of solvent, and the development of catalyzed reactions. By way of example, the development of catalyzed reactions for dihydroxylations have made possible the replacement of the Woodward Reaction in the manufacture of steroids, in which huge amounts of expensive silver salts were used and produced, and thus had become an economic factor:

Woodward Reaction

The Woodward reaction can be replaced through the use of stoichiometric quantities of OsO_4, but osmium tetroxide is both very toxic and very expensive, making its use on a commercial scale prohibitive. Only in its catalytic variant, which employs N-methylmorpholine-N-oxide as the stoichiometric oxidant and catalytic quantities of OsO_4, can this be considered a green reaction that can be used on industrial scale.

Upjohn Dihydroxylation

Some systems have already been reported in which H_2O_2 is used to reoxidize the N-methylmorpholine, allowing this material also to be used in catalytic amounts. Considering the atom efficiency using H_2O_2 as the terminal oxidant, H_2O as the stoichiometric byproduct is much better than N-methylmorpholine. Notably, catalytic systems are available in which the osmium catalyst is encapsulated in a polyurea matrix or bound to a resin, so that the catalyst can be more easily recovered and reused. An additional advantage of such polymer-bound catalysts is the avoidance of toxic transition metal impurities, for example in pharmaceutical products.

A key point is still the choice of solvent, as this is the main component of a reaction system by volume (approx. 90%). Chlorinated solvents should be avoided, as many of these solvents are toxic and volatile, and are implicated in the destruction of the ozone layer. Alternative solvents include ionic liquids, for example, which are non-volatile and can provide non-aqueous reaction media of varying polarity. Ionic liquids have significant potential, since if systems can be developed in which the products can be removed by extraction or distillation and the catalyst remains in the ionic liquid, theoretically both the solvent and the catalyst can be reused. The solvent of choice for green chemistry is water, which is a non-toxic liquid but with limited chemical compatibility. On the one hand, reactions such as the Diels-Alder Reaction are often even accelerated when run in an aqueous medium, while on the other hand, many reactants and reagents, including most organometallic compounds, are totally incompatible with water. There is thus a great need to develop newer methods and technologies that would make interesting products available through reactions in water or other aqueous media. For a short review of reactions in water, please check: S. Varma, Clean Chemical Synthesis in Water, *Org. Chem. Highlights* 2007, February 1. Chemical reactions run under neat conditions (no solvent) and in a supercritical CO_2 medium can also be considered as green choices. Other possible improvements can be considered, such as for example replacement of benzene by toluene (as a less toxic alternative), or use of solvents that can be rapidly degraded by microorganisms.

POLLUTION PREVENTION & EARLY ADVOCATES OF GREEN CHEMISTRY

In 1990, Congress passed the Pollution Prevention Act, a policy that states that pollution should be prevented or reduced at the source and recycled in an environmentally safe manner whenever feasible, and that unpreventable pollution should be treated and disposed of in an environmentally safe manner. The Act set a precedent of eliminating pollution from its source, and 1991, the Environmental Protection Agency (EPA)adopted this principle as one of its declared objectives.

The groundwork for the movement that became Green Chemistry emerged from these two events and developed further through the efforts and collaboration of several key advocates in the US Government. In the early 1990's, Kenneth Hancock, Director of the Division of Chemistry at the National Science Foundation (NSF), advocated the role of chemistry and chemists both in mitigating the environmental effects of past inventions and in preventing environmental problems in the future—all in an economically feasible way. Dr. Hancock died unexpectedly in 1993 while attending a conference in Eastern Europe. The Kenneth G. Hancock Memorial Award provides national recognition to outstanding student contributions to furthering the goals of green chemistry through research or education. Several early advocates of Green Chemistry were instrumental in the founding of the Green Chemistry Institute in 1997.

Joe Breen, whose twenty-year career at the EPA informed his understanding of the necessity of sustainable chemistry, was a pioneer and relentless early advocate of Green Chemistry. As the co-founder and first Director of the Green Chemistry Institute in 1997, he toured the world talking with students, teachers, and scientists about the urgency of promoting Green Chemistry. At his death in 1999, he was called the "Heart and Soul of Green Chemistry." In honor of his efforts, the Joe Breen Memorial Fellowship award sponsors the participation of a young international green chemistry scholar in a green chemistry technical meeting, conference or training program.

Dennis Hjeresen, a subsequent Director of the Green Chemistry Institute and currently at Los Alamos National Laboratory, also worked to make Green Chemistry a known entity in the chemistry world; through his efforts, the American Chemical Society, intending to focus on the role of chemistry in the environment, formed an alliance with—and began to provide core funding for—the Green Chemistry Institute.

The Founding of Green Chemistry:
A Collaboration of Government and Industry

Green Chemistry gained its current standing as a scientific discipline as well as practical means to pollution prevention as the result of collaboration between the US government, Industry, and Academia. In the early 1990's, Paul Anastas, who was then the chief of the Industrial Chemistry Branch at the US EPA, moved forward the concept of Green Chemistry. Despite the Pollution Prevention Act of 1990 and the early work of several tireless advocates, the focus of the US EPA at that time remained on end of pipe regulations, pollution clean-up, rather than preventing pollution. These discussions lead to Paul Anastas and John Warner to develop the12 Principles of Green Chemistry: a framework to help us think about how to prevent pollution when inventing new chemicals and materials, by the mid-1990's. Paul Anastas and John Warner's work as founders of a new field called Green Chemistry, based on the productive collaboration of government and industry, was just beginning.

In 1996, John Warner and Paul Anastas and others were stakeholders in the founding of Presidential Green Chemistry Challenge Award. This award increased awareness of Green Chemistry in industry and government by annually acknowledging individuals, groups, and organizations in academia, industry, and the government for their innovations in cleaner, cheaper, smarter chemistry. This remains the only award given by the President of the United States specifically for work in chemistry.

In 1998, John Warner and Paul Anastas published the seminal book Green Chemistry: Theory and Practice, which gave a precise definition to Green Chemistry and enumerated the Twelve Principles fundamental to the science. The definition and principles have become the generally-accepted guidelines for Green Chemistry. The book has achieved world-wide renown and has been re-printed in several languages.

Paul Anastas went on to become the director of the U.S. Green Chemistry Program at the EPA. He also held the role of Director of the Green Chemistry Institute, where he established twenty-four Green Chemistry chapters in countries around the world. He went on to found a Green Chemistry program at Yale University. In May 2009, President Barack Obama nominated Paul Anastas to lead the U.S. Environmental Protection Agency's (EPA's) Office of Research and Development. The nomination is a decisive achievement for the adoption and advancement of the principles of Green Chemistry.

John Warner turned his focus next on the need for educating a new generation of scientists in Green Chemistry principles. He founded the world's first Ph.D. program in Green Chemistry as well as a Center for Green Chemistry at the University of Massachusetts. In 2007, John Warner returned to industry to develop green technologies, partnering with Jim Babcock to found the first company completely dedicated to developing green chemistry technologies, the Warner Babcock Institute for Green Chemistry. The Institute was created with the mission to develop nontoxic, environmentally benign, and sustainable technological solutions for society. Simultaneously, John Warner founded a non-profit foundation, Beyond Benign, to promote K-12 science education and community outreach.

Green Chemistry Today:
World-Wide Initiatives in Government and Academia

Individuals and organizations are consistently becoming more aware of the growing need for moresustainable product and processes, Green Chemistry has emerged as a definitive answer to that need. As a result, Green Chemistry is becoming a focus for governments as well as academic institutions around the world.

Countries throughout the world are engaging and adopting Green Chemistry as a sustainable and economical development. The year 2000 marked the founding of the Green Chemistry Institute of Spain; the Green and Sustainable Chemistry Network in Japan; and the Centre of Green Chemistry at Monash University in Australia. In early 2009, PARTEQ Innovations founded GreenCentre Canada, North America's first all-inclusive commercialization center for Green Chemistry research innovations. It was founded with the mission of developing and commercializing early-stage Green Chemistry discoveries generated by academic researchers and industry.

An integral component to the growth of Green Chemistry is the proliferation of Green Chemistry education and research. To this end, several universities around the world have created specific departments focused on teaching, studying, and expanding Green Chemistry.

Since 1997, the University of Oregon has been developing undergraduate chemistry curricula that incorporates the principles and practices of Green Chemistry into both laboratory and lecture classes.Carnegie Mellon University, in Pittsburg, Pennsylvania, has established the Institute for Green Science, a center focused on research, education and development. Since the fall of 2001, the University of Massachusetts has offered a Green Chemistry Track in the Chemistry Ph.D. program; it was the first such program in the world. In England, the University of York houses the Green Chemistry Centre of Excellence, supporting Green Chemistry research, industrial collaboration, and development of educational materials, including Masters course in Green Chemistry and Sustainable Industrial Technology. The University of California at Berkeley, which has the US's #1 ranked chemistry program, has founded the The Berkeley Center for Green Chemistry, which aims to advance green chemistry principles and practice through interdisciplinary research, education, and novel collaborations among investigators working at the intersection of chemistry, toxicology, environmental health, business, law, and public policy.

Recent Developments in Green Chemistry: Legislation

In April 2007, the California Environmental Protection Agency (Cal/EPA) began a Green Chemistry Initiative to promote innovation, create new jobs, and keep people safe from harmful substances. After gathering input from over 600 participants, including industry leaders, community organizers, and scientists, Cal/EPA created a series of options of ways to reduce the effects of toxic chemicals on people and the environment. In September 2008, following advice from the Cal/EPA, California legislature passed two landmark bills focused on promoting Green Chemistry. California was the first state to pass a comprehensive Green Chemistry policy.

The bills set a precedent for Green Chemistry legislation, and states around the nation and countries around the world have been following suit. In 2008, for example, Michigan's Department of Environmental Quality began the first stage of a multi-step process entitled "Advancing Green Chemistry" in 2009, the Michigan legislature passed several bills promoting Green Chemistry innovation and development. Other states, including Minnesota, and other countries, including Canada, Spain, and China, have hosted large conferences to discuss the opportunities and applications of green chemistry on local, national, and international scales. With these initiatives and still others announced on almost a daily basis, Green Chemistry continues to grow in importance and impact on both national and international stages.

PRINCIPLES OF GREEN CHEMISTRY

The creators based their Green Chemistry on 12 principles, which were first published in 1998. Decoded Science spoke about this with Dr. Robert Pullar, research assistant from Aveiro University (Portugal). He tells us,

"One the key points of Green Chemistry is the use of benign, less toxic chemicals; this implies the development of alternative synthesis methods which could use less toxic solvents and/or starting materials. Ideally, processes should be based on water, not organic solvents.

Green Chemistry, however, is not just about toxicity and hazards, but also about energy efficiency/saving, reduction of wastes, and uses of renewable and sustainable sources. Basically green chemistry is about minimizing the impact on the environment in any possible way. An innovative non-toxic catalyst, for instance, which could allow a reaction to take place in milder conditions, is a good example of Green Chemistry. The use of wastes to make chemical compounds or materials we need is another example."

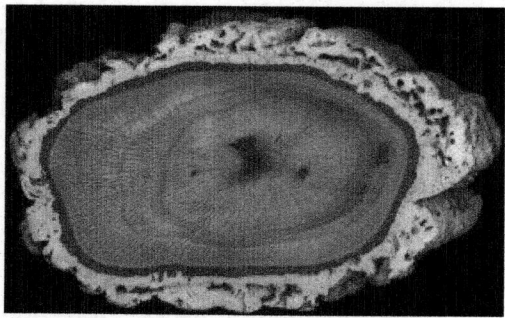

Fig. : Some natural materials show good energy efficiency.

Learning from Mother Nature

According to Dr. Pullar, we can learn about green chemistry from nature itself.

"If we consider energy efficiency, Mother Nature has a lot to teach us. Indeed, some natural materials, for example rocks and minerals, wood such as cork, and naturally occurring polymers have the best energy efficiency, and surely we should use them and/or consider as a model. We can also use natural structures as templates to create new forms of material, such as the porous or fibrous structure of wood."

The prevention of waste generation, and accidental spillages and leaks, are also crucial aspects of green chemistry.

Green Chemistry: Growing Interest

Since its introduction, Green Chemistry has attracted more and more interest. Considering scientific publications published, for instance, in 1991 less than 100 papers on this topic; in 2010, however, the publication number was higher than 600. Of these publications in 2010, the two countries producing the highest number were US and China, with 18.14 and 15.34 % shares, respectively.

"Both the Royal Society of Chemistry (UK) and the American Chemical Society have journals dedicated to this topic – Green Chemistry and ACS Sustainable Chemistry & Engineering. This clearly shows the interest of the scientific community."

Dr. Pullar said.

Barriers to the Chemistry Cleanup

Despite the progress made in recent years, however, there are still barriers which prevent a further diffusion and acceptance of Green Chemistry. Decoded Science talked about this with Dr. David Constable, director of the ACS Green Chemistry Institute. He explains,

"Like for any new technology innovation, there are always barriers and resistance to full acceptance.

In the case of Green Chemistry, it is sometimes difficult to find economically viable 'safer' alternatives to a certain product or process. In many cases, there is a lack of agreement of what is considered 'safer.'"

On the business supply chain side, it takes always time to switch from something old to something new; issues such as new product re-design, for instance, can be difficult and long.

Surely education is the key to the future; it is essential to have chemists and chemical engineers who understand how to incorporate Green Sustainable Chemistry into their work."

The First Steps to the Future

It is still early days for Green and Sustainable Chemistry, and it has yet to gain general acceptance globally as a viable alternative to current practices which govern how scientists perform chemistry and chemical engineering in industry and academia. However, there is growing interest in these ideas, both from the scientific and industrial communities. Political, societal and economic pressures should increase its importance in the future, driven by consumer demand and environmental needs.

SUSTAINABLE CHEMICAL PRODUCTION AND REACH

The chemical sector plays a significant role in the European economy, contributing 30% of the world's total chemical production, employing 1.2 million people and generating €537 billion in sales in 2007.

The EU registration, evaluation, authorisation and restriction of chemical substances (REACH) Regulation is at the centre of a shift towards more sustainable solutions. Aiming to protect human health and the environment, REACH makes companies responsible for registering information on the properties of their chemicals with the European Chemicals Agency (ECHA).

Since coming into force in June 2007, REACH has prompted the chemical industry to increase its innovative capabilities and competitiveness. Moreover, the Regulation has enshrined the substitution principle within the industry's regulations, ensuring the most dangerous chemicals will be replaced by more suitable alternatives.

By adopting measures to increase resource and energy efficiency, and developing improved production processes, the European chemical industry is lessening its overall environmental impact. In addition to addressing sustainability within the chemical industry, innovation and new technologies are enabling chemical companies to develop sustainable solutions for society as a whole.

As with all other sectors of society, the chemical industry has been forced to take stock of its environmental impact, and adopt measures to improve it, particularly as the industry accounts for 12% of total European energy demand.

A report by the High Level Group (HLG) on the competitiveness of the European chemical industry published in July 2009 indicated the sector must invest in eco-innovation to increase its competitiveness and ensure its sustainability. Furthermore, it called for the responsible use of natural resources and the overall improvement of energy efficiency.

The HLG, made up of representatives from the European Commission, Member States, industry and academia, capped 18 months of dialogue with 40 recommendations which aimed to help develop the sector's global competitiveness and boost its unrivalled innovative potential. The recommendations highlight the importance of innovation as essential for a sustainable and healthy European chemical industry.

Technology Platform Targets Sustainability

Sustainability in the chemical industry has long been a key topic for discussion. SusChem – the technology platform for sustainable chemistry – was launched in 2004 by the European Chemical Industry Council (Cefic) and the European Association for Bioindustries (EuropaBio), and endorsed by the European Commission, to work on boosting environmental innovation in the chemical industry. These actions can be viewed in two separate groups in terms of the measures being taken to increase sustainability: measures which increase sustainability within the industry, and those which are of benefit to the rest of society.

In terms of adopting innovative measures to increase sustainability within the industry, the European chemical sector has adopted behavioural changes to production processes, such as adjusting feedstock materials and introducing lifecycle thinking. Furthermore, through bodies such as SusChem and Cefic's long-range research initiative (LRI), innovative ideas have been developed.

One such example is the F3 Factory concept supported by SusChem. The F3 Factory is intended to represent the chemical factory of the future, which would not only have a lower environmental impact and higher energy efficiency, but would also require fewer materials. The combined result of such a design would potentially eliminate the environmental impact of new factories.

In addition, SusChem has also supported the development of integrated 'biofineries' that use biowaste to produce chemicals traditionally supplied by the petrochemical industry. Outside the sector, developments facilitated by innovation in the chemical industry are increasing the sustainability of the wider community. Through the application of nanotechnologies, the sector is providing solutions to the problems of environmental remediation, environmental monitoring and resource efficiency.

Furthermore, SusChem has also launched the 'Smart energy home' initiative which incorporates chemical innovations – many based on nanotechnology – to create a residential building that is a net energy generator.

Call for Improved Environmental Performance

Despite the progress made by the chemical industry in terms of helping increase sustainability both inside and outside the sector, the world's largest chemical firms have been criticised for not making adequate progress.

A report from the German-based environmental and social rating agency Oekom in October 2009 stated that the world's largest chemical companies have made 'scarcely any progress' in improving environmental performance in the last three years. Taking factors such as energy and water consumption, resource use and environmental reporting into account, Oekom found that only 3 of the 24 major listed chemical companies obtained a 'prime' best-in-class status. Overall ratings were given on a scale from A+ to D-; the average rating across the industry was a C.

"Chemical corporations are still hesitant in their approach to the social and environmental challenges we face," says Oekom research director Oliver Rüdel. "On climate protection, however, the companies do appear to be slowly rethinking their approach. However, while most companies do carry out risk analyse, problematic substances are still replaced by environmentally-friendly substitutes too seldom and too slowly."

The Oekom report went on to say that the same could be said of the industries application of new technologies. In particular, nanotechnology was highlighted as warranting greater use due to its potential to offer increased sustainability through the production of environmentally-friendly products and processes.

While Oekom indicated that the chemical industry could do more in terms of improving their environmental impact, it did acknowledge the achievements made in terms of lower energy consumption in the sector.

In particular, it highlighted the fact that while production increased by 56% in the sector between 1990 and 2004, through more efficient energy consumption, the industry was able to halve CO_2 emissions per unit of production of chemical product.

REACH CHEMICAL LAW 'WORTH THE MONEY IN THE END', SAYS BASF

Fig. : BASF, the German chemical giant, has distanced itself from the rest of the industry by saying Europe's REACH chemical safety law was worth the investment in the end. A review of the legislation is expected this month.

From the moment it was tabled until its eventual adoption in 2006, the REACH regulation gave rise to one of the most epic lobbying battles in the EU's history.

The bitter campaign saw chemical companies warn they could be forced to close factories and leave Europe because of the extra costs generated by the EU law, which sought to protect consumer health and the environment.

These included widely publicised industry studies which claimed that REACH would cost billions of euros to implement, causing millions of job losses in Germany.

BASF U-turn on REACH

Six years on, the European Commission is preparing to launch a review of the controversial legislation. And BASF has now radically changed its communications strategy.

Ronald Drews, vice president for chemical regulations and trade control at BASF, said the company has recruited 250 employees to prepare registration dossiers for submission to the European Chemicals Agency (ECHA) in Helsinki. Overall, Drews said the company expected to submit around 5,000 REACH dossiers over 10 years, costing the company between €500 million and €550 million, he told journalists at a June Brussels briefing.

"The costs are high," Drews said, citing an average of €50 million per year, which includes registration fees at ECHA, and other costs such as consultancy fees. But, "I think at the end, it is worth the money", he added in response to a question from EurActiv on whether REACH had helped promote innovation in the chemicals sector and bring safer products to the market.

Production did not Flee Europe

BASF's apparent endorsement of REACH may raise eyebrows among the wider chemicals industry because the cost of implementing the regulation may in fact have been under-estimated.

An interim evaluation by the European Commission, published in March, indicated that industry has already spent around €2.1 billion on REACH registration dossiers since the legislation was introduced, close to the €2.3 billion it had initially estimated would cost the industry until 2018. (A broader cost estimate could range between €1.1 billion and €4.1 billion, the Commission report added, pointing to a number of caveats in relation to these numbers).

"At this point, we see that costs [are] higher than originally estimated", said James Pieper, a spokesperson for the European Chemical Industry Council (CEFIC). "We expect further on that it will indeed benefit human health and the environment, but it is too early to see any impact on innovation", Pieper told EurActiv in an emailed statement.

BASF is equally dismissive about earlier industry claims that REACH would force chemical production abroad. The first phase of REACH dealt with substances produced or imported in large quantities (more than 1,000 tonnes per year). A plant of such production capacity "is not easily transferred to a place out of Europe," Drews explained.

"We have to produce where the market is. So for us, having a big downstream user market in Europe with the car manufacturers and so on, it makes sense to have our chemical production here."

In the long run, factory relocation "may happen in some cases", he said, explaining this could happen for chemicals produced in smaller quantities. "But

I don't see a big move out of Europe", he replied when asked about whether industry claims of de-industrialisation had materialised.

CEFIC too admits that the industry's early claims about cost had not materialised but prefers to remain cautious about the future. "The first deadline did not result in a dramatic situation for industry, but we must have a prudent approach for the upcoming deadlines", Pieper said.

SUSTAINABLE CHEMISTRY – INDUSTRIES – INNOVATION

The synthesis and manufacturing of molecules, substances and formulated chemicals primarily rely on the use of limited, non-renewable resources (oil) and often generate toxic and/or non selective molecules, ultimately leading to pollutions that modify or even damage the environment.

Solving this problem and finding renewable resources allowing the access to chemical intermediates (synthons) – in a nutshell, integrating the principles of ecodesign into chemistry – has become a vital necessity. The selection, scaling and control tools of a process must indeed:

(i) promote intensification and control reactions (reliability and security), the quality of the product(s) developed, the economy of reagents and energy,

(ii) minimizing negative impacts (the production of products that are harmful in terms of the quality of the product developed, of controlling the reaction and/or in terms of the environment).

Thus, the chemical industry must respond to human needs by offering products and processes efficient both in terms of health and environmental preservation. Molecules and products must be made in clean and safe conditions, with a constant concern for saving energyand raw materials, all the while taking the requirements of the European chemicals law (REACH) into account.

The CD2I programme aims to enable academic laboratories and industry to share their knowledge and expertise through collaborative projects. This involves getting researchers to think differently by incorporating the principles of "eco-design" into their synthetic methodologies, in their approach to improving or establishing new processes, and in their search of new renewable resources to replace fossil fuels. The CD2I programme also aims to contribute to the maintenance and development of the competitiveness of the chemical industry, offering, particularly to the numerous small and medium enterprises (SMEs) in this sector, means of improving their research relationships with the academic sphere. Under the requirements of REACH, this programme also aims to develop new tools and methods to be made available to manufacturers and expert bodies. Furthermore, it aims to come up with solutions to the substitution of products subject to authorization. Finally, the programme aims to increase the interaction between synthetic chemists and chemical engineers, and to make this community take the concepts of sustainable chemistry into account.

SUSTAINABLE CHEMISTRY

The **Organisation for Economic Cooperation and Development** (OECD) defines sustainable chemistry as:

"Sustainable chemistry ... seeks to improve the efficiency with which natural resources are used to meet human needs for chemical products and services. Sustainable chemistry encompasses the design, manufacture and use of efficient, effective, safe and more environmentally benign chemical products and processes.

Sustainable chemistry ... stimulates innovation across all sectors to design and discover new chemicals, production processes, and product stewardship practices that will provide increased performance and increased value while meeting the goals of protecting and enhancing human health and the environment."

Essentially sustainable chemistry is about doing more with less: reducing the environmental impact of processes and products, optimizing the use of finite resources and minimizing waste. Sustainable chemistry can ensure eco-efficiency in everything we do, both individually and as a society. Sustainable chemistry also means protecting and extending employment, expertise and quality of life. It provides a sustainable basis for the innovation needed to stimulate a competitive, knowledge-based, enterprise-led economy across Europe.

But solutions provided by sustainable chemistry must also be acceptable to society: they must be trusted and designed according to what society wants and needs and they must be economically sound. In practice, it means that SusChem projects and programmes should address clear societal needs, be environmentally sound and economically viable.

Sustainability and competitiveness are strategic priorities for SusChem. Progress on sustainability, competitiveness and environment protection are intimately linked; chemical products and chemistry-driven technological advances provide critical answers and ensure the sustainable development of modern societies.

The chemical industry, being a critical element of most value chains and supplying its products to all sectors of the economy, is at the forefront of the transition to a more sustainable development; the move towards increased sustainability requires developments that entail the knowledge and competences present in the European chemical sector and within its networks.

Chemistry for a Sustainable Future

Sustainable chemistry has a huge part to play in solving society's current and future challenges. These challenges include meeting future energy needs, reducing our energy and water consumption and emissions, favoring a circular economy (recycling, waste management, making best use of essential raw materials or finding substitutes, helping our urban areas to be better places to live.

These efforts are supported by enabling technologies, falling in three main categories in the Chemical Industry: Industrial Biotechnologies, Reaction and Process Design and Materials Technologies.

SusChem believes that sustainable chemistry can help inspire the change of pace and mindset that is needed to make the vision of a sustainable, smart and inclusive society a reality.

Sustainability and the Chemical Enterprise

Today's society is faced with limited resources, expanding population, economic and social welfare disparities, increasing consumption and pollution, climate change, and degradation of human and ecosystem health. These challenges require innovations and policies to find and follow a sustainable path, allowing humanity "to meet current environmental and human health, economic, and societal needs without compromising the progress and success of future generations".

The chemistry enterprise includes chemical and allied industries, their associated trade associations and professional societies, government laboratories, and the academic educational and research institutions that train the chemical workforce and advance scientific knowledge.

The chemistry enterprise has many roles in sustainability. It provides chemicals, materials, and technologies that improve the safe and efficient use of energy and natural resources and is responsible for delivering these in a way that protects human and environmental health. Chemistry — in labs, classrooms, and industry — is a central science for the development of sustainable technologies and innovations.

Chemical processes often require significant inputs of materials, water, and energy that are derived from limited resources. These processes also are potential sources of emissions and waste materials. However, the high-volume supply chains within the chemical industry provide opportunities to leverage and magnify improvements in material sourcing, process efficiency, waste reduction, co-product utilization, and the use and end-use of products.

In addition to reduced energy use in manufacture, chemical industry products also can improve sustainability in downstream uses. For example, the use of foam insulation in industrial, commercial, and home settings enables energy savings and associated drops in greenhouse gas emissions. A 2009 analysis found that every pound of CO_2 emitted by chemical industry operations was offset by more than two pounds of CO_2 emissions savings associated with the use the resulting products. The authors estimated that 2005 GHG emissions would have been 8-to-11 percent higher in a world without the chemical industry (ICCA).

A report by the National Research Council (NRC) in 2005 identified areas wherefocus would advance sustainability, including education; green and sustainable chemistry and engineering; life cycle assessment; toxicology; renewable fuels and chemical feedstocks; energy intensity of chemical processing; and separation, sequestration, and utilization of CO_2. While there has been progress in the ensuing years, those identified needs still remain relevant today. In addition, there is a need to address the professional system in which chemists and engineers work. These systems of incentives, regulations, markets, grant cycles, educational background, and corporate, academic, and government structures and climates has tremendous influence on the sustainable solutions that chemists and engineers can provide. More recent NRC reports have continued to examine these sustainability issues and largely reinforce those previous chemical industry-specific findings.

A 2008 ACS workshop (Satterfield, et. al.) recommended a framework that would address non-technical challenges to sustainability within the chemistry enterprise. NRC has noted that sustainability issues are complex and inter-linked, demanding a kind of holistic thinking and problem solving. NRC also laid out specific recommendations for federal agencies, noting that effectively managing sustainability can add value and efficiency across the government.

Chemists and engineers are embracing sustainability challenges through innovation and are using a life-cycle perspective to minimize or eliminate potential environmental and health implications of their technologies. In schools, curricula are including more instruction on green chemistry, green engineering and life cycle thinking; researchers are advancing the science and tools needed; and academic facilities are teaching by example in reducing the environmental footprint of their campus operations. However, the professional system in which scientists and engineers operate could be improved to further incentivize and support these efforts.

ACS believes that support for research to promote sustainability, green chemistry, and green engineering, combined with incentives for the adoption of sustainable technologies and new regulatory strategies that promote sustainable products and processes, will be instrumental in meeting the challenges of protecting human health and the environment, meeting our societal and energy needs, enhancing national and homeland security, and strengthening the economy.

Sustainable Chemistry, Smart Industry and #Digital4EU

ICT and digital innovation is an important technology area for the chemical industry. For decades the chemical industry has made extensive use of ICT systems throughout its value chain, from logistics, to modelling, design, control, monitoring and repair. In addition, the chemical industry is a key provider of materials and technologies that form the basis for many ICT and digital solutions.

Smart Chemical Processes

Within the total chemical industry value chain from product design to delivery to the customer, ICT plays a key role. ICT is key to a successful, efficient and competitive industry. As chemical products, process and plants become ever more complex and resource usage and performance requirements become tougher, ICT can deliver a large portion of the innovation needed to keep the European chemical industry competitive on the global stage.

Process Control is a critical factor for sustainability in the production process. Advanced process methods allow production units to run at optimal operating points under appropriate constraints. Monitoring is a related area of importance for the process industry where improved digital modelling can contribute to increased plant availability, reduced costs and improved product quality.

Modelling for innovation is also a key topic. ICT-enabled innovation can significantly reduce (20-40%) time lines for product and process developments and save costs. Overall ICT technologies can enable increased resource efficiency, will enable new process and product capabilities, and strengthen the chemical industry and European competitiveness.

Smart Materials for Smart Industry

Sustainable chemistry is all about developing 'Smart materials' – materials that will enable the development of important ICT such as nanoelectronics and haptic devices. Sustainable chemistry also provides the specialty polymers and other materials that will be required for new 3D printing technologies to produce components with demanding specifications.

Sustainable chemistry is looking to develop polymers that enable nano-structured self-organisation for use as templates to support advanced nano-lithography or other nanoelectronic fabrication techniques for the fast prototyping and production of complex electronic devices. Such advanced fabrication techniques can reduce development time for microelectronic devices and boost the capability and competitiveness of the European ICT sector.

Polymers and polymer-based ink formulations are also essential for printed fabrication techniques, such as roll-to-roll lithography that allow mass production of low-cost microelectronic circuits for a wide range of applications including RFID tags, flexible displays and OLED lighting.

Future chemical developments include improved conductive polymers, piezoelectric and electro-active polymers that can inspire new and emerging end-use applications including wearable electronics.

Additive Manufacturing Aka 3D Printing

3D printing will change the way society manufactures and its development heralds an era of mass-customisation. 3D printing or Additive Manufacturing produces a three-dimensional object from an electronic data set through an additive process making material layers in successive steps under computer control – truly digital manufacturing.

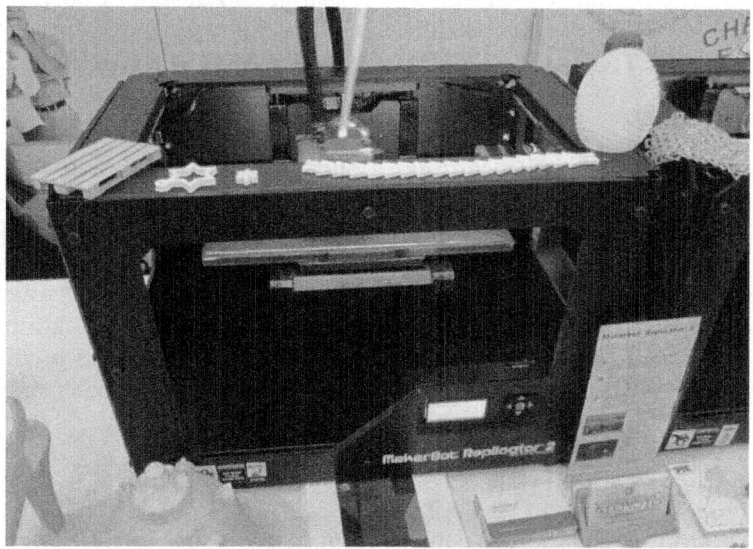

The global market for materials and services for 3D printing (excluding printer equipment) was estimated to be US$ 1.8 billion in 2013 and is projected to grow to US$ 10.8 billion by 2018.

The ability to produce small lot sizes and highly specialised added value products makes 3D printing technology a key technology for the next generation of industry: Industry 4.0. Innovation and pre-industrialisation, competitive small series production, improved time-to-market, custom made parts for personalised products, manufacturing of complex structures and geometries are all drivers for the development of additive manufacturing technologies. 3D printing also contributes to lower energy and resource use.

Polymers with appropriate end-use performances and adapted to specific 3D printing technologies are needed along with suitable metallic or ceramic materials. The European chemical industry already delivers many of these materials, but research is needed to widen the range of materials and mechanical properties of polymers available for 3D printing. Development of new electrically and thermally conductive materials will provide new opportunities for the development of additive manufacturing. Solutions to improve the surface finish of manufactured parts are also required.

Sustainable Chemistry Is Key

Additive manufacturing is a key technology for fostering the European innovation and manufacturing industries. And its full development requires key inputs from sustainable chemistry.

Digital technologies, such as 3D printing technologies, can reduce the gap between innovation and manufacturing, stimulate the renewal of European manufacturing industry and boost industrial research and design opportunities too.

INTERNATIONAL CHEMICALS POLICY AND SUSTAINABILITY

Chemical Policy

A broad chemicals policy addresses a health care organization's commitment to and support of changes in national (and potentially international) chemicals policies. In the United States, many credible groups are working together to support policies that protect public health. To this end, the Centers for Disease Control and Prevention (CDC), the Agency for Toxic Substances and Disease Registry (ATSDR), the EPA, and others have created the National Conversation on Public Health and Chemical Exposures. The National Conversation is a collaborative initiative intended to strengthen the nation's approach to protecting the public from harmful chemical exposures.

Lab Chemical Management

The bottom line is there are many compliance issues to consider when buying, using, and disposing of hazardous materials and waste. If, after undertaking the process of looking for less hazardous or non-hazardous alternative products or services, it is found that use of a chemical is necessary, the EPA suggests using this sustainable chemistry hierarchy to select products and processes, with the goal of purchasing products and using processes that represent the highest level possible of these green actions:

1. Practice green chemistry (source reduction/prevention of chemical hazards).
    o   Design chemical products to be less hazardous to human health and the environment.
    o   Use feedstocks and reagents that are less hazardous to human health and the environment.
    o   Design syntheses and other processes to use less energy and fewer materials (high-atom economy, low-E-factor).

o   Use feedstocks derived from annually renewable resources or from abundant waste.

o   Design chemical products for increased, more facile reuse or recycling.

2.   Reuse or recycle chemicals.

3.   Treat chemicals to render them less hazardous.

4.   Dispose of chemicals properly.

5.   *Chemicals that are less hazardous to human health and the environment are:

o   Less toxic to organisms and ecosystems.

o   Not persistent or bioaccumulative in organisms or the environment.

o   Inherently safer with respect to handling and use.

The EPA and DOE's Labs 21 program focus typical laboratories because in addition to chemicals, they use far moreenergy and water per square foot than the typical office building. Explore the resources and links on this site to learnmore about sustainable laboratories.

OPPORTUNITIES AND BENEFITS

Chemical Management

- Chemical Use and Disposal: Reduce and/or eliminate the use of hazardous products, ensure proper disposal of chemical hazards and toxic materials within the health care facility to safeguard the health of building occupants. Use targeted strategies for: hand hygiene products, sterilization and high level disinfection; laboratories; and radiology.

- Pharmaceutical Management and Disposal: Safeguard human and ecological health through compliant management and disposal of pharmaceuticals and associated wastes.

- Polychlorinated Biphenyl (PCB) Removal and Asbestos-Containing Materials (ACM) Management: Implement a PCB and ACM removal and abatement program. Prevention of associated harmful effects of these hazardous materials in new and existing buildings.

- Leaks and Spills: Mitigate of leaks, spills and waterborne effluents to prevent releasing waterborne environmental, health and safety burdens to the site neighbors and surrounding community.

- Sanitary Sewer: Eliminate wherever possible, the drain disposal of chemicals, pharmaceuticals and toxic materials within the health care facility to safeguard the health of the local waterways and communities.

Indoor air Quality (IAQ) Management

- IAQ Performance: Develop an IAQ management plan that addresses ongoing operations and maintenance, as well as planned future upgrades related to smoking, IAQ performance, systems maintenance, and systems monitoring.

- Systems Maintenance: Support for appropriate training, monitoring, operations and maintenance for buildings and building systems to ensure they deliver target building performance goals over the life of the building.
- Systems Monitoring: Ensure ventilation system monitoring to help sustain long-term occupant comfort and well-being.\

Environmental Services

- Products and Materials: Minimize exposure of building occupants and cleaning personnel to potentially hazardous chemical, biological and particulate contaminants from cleaning chemicals, equipment and procedures, while ensuring effective infection control processes.
- Entryway Systems: Use and maintenance of entryway grills, grates, mats, etc. to reduce the amount of dirt, dust, pollen, and other particles entering the building.
- Indoor Integrated Pest Management: Emphasize non-chemical strategies for pest prevention that protect people from unnecessary exposure to pests and pesticides.

Chemicals and Waste Management

A worker at the Lagluja dumpe site, the biggest collector of persistent organic pollutants (POPs) in Georgia. UNDP and GEF are working with the government and local authorities to reduce the release of POPs from the stockpiles and, wherever possible, evacuate the toxic substances from the country. Photo: Vladimer Valishvili/UNDP.

Chemicals are critical to the manufacture of many products and protection of human health, and an important contributor to the GDP and employment. However, without goodmanagement practices, chemicals and their hazardous wastes can pose significant risks to human health and the environment especially the poorest members of the global community. In urban areas, low-income or minority populations are often exposed to hazardous chemicals and associated wastes in their jobs or because they reside in polluted areas. In rural areas, most chemical exposure and environmental pollution is linked to the misuse of agricultural chemicals and pollution brought by waterways, impacting the natural resources upon which these communities depend.

The sound management of chemicals and wastes is an important component of UNDP's efforts to achieve sustainable, inclusive and resilient human development and the Sustainable Development Goals (SDGs). UNDP advocates for integrating chemicalsmanagement priorities into national environmental and poverty reduction planning frameworks, helps countries access financial and technical resources, and provides assistance and implementation support to improve the holistic management of chemicals and waste at national, regional and global levels. We tackle unsustainable consumption and production patterns, including poor design and material choices, which lead to resource depletion, waste generation and pollution.

UNDP's expertise covers management of chemicals harmful to human and environmental health, including Persistent Organic Pollutants (POPs), Ozone Depleting Substances (ODS), Mercury, Lead, and other heavy metals. UNDP helps countries strengthen their wastemanagement systems, including waste prevention, reuse/recycling, treatment and disposal. Safe and effective treatment of hazardous medical waste (*e.g.* from the Ebola crisis in West Africa) through innovative technologies is also underway.

As an implementing agency for the Global Environment Facility (GEF), UNDP supports implementation of projects on POPs, mercury, lead and waste in 84 countries, addressingcountry priorities under relevant chemicals-related global environmental agreements; under these projects, 220,000 people have been safeguarded from high-risk POPs exposure, 65 POPs policies and regulations adopted, and 335,000 tonnes of POPs contaminated waste safeguarded. Under the UNDP-UNEP Partnership Initiative, 17 additional countries have assessed their capacity for the sound management of chemicals, identified needs, and integrated management priorities into national development policies and plans, using the UNDP SMC mainstreaming methodology.

UNDP activities on chemicals and waste management are carried out in cooperation with the GEF, other IOMC members, Secretariat of the Stockholm Convention on POPs, Interim Secretariat of the Minamata Convention on Mercury, and a broad range of bilateral, private sector and NGO partners.

SUSTAINABLE CHEMISTRY

The concept of sustainable chemistry exists to link preventative protection of the environment and health with an innovative economic strategy that will also result in more jobs. Sustainable chemistry is a broad-ranging area that concerns stakeholders in the scientific community, the economy, public authorities, and environmental and consumer advocate associations.

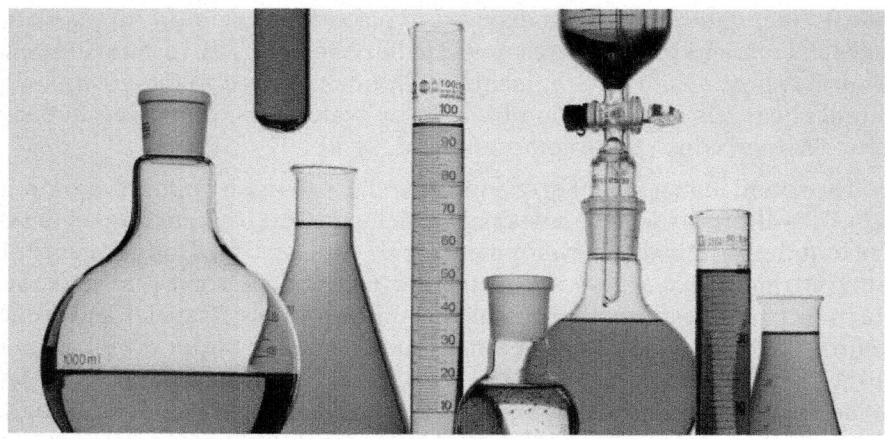

Fig. : Chemicals influence our life.

There are various different approaches to the implementation of sustainable (or green) chemistry. One well-known example are the twelve principles of "green chemistry" according to Anastas and Warner, dated 1998. At the European level 12 criteria for Best Available Techniques listed in Annex IV of the Council Directive concerning integrated pollution prevention and control (IPPC Directive) define the standards of sustainable production also affecting the chemicals industry.

The 12 main criteria in the IPPC Directive (96/61/EG)

1. the use of low-waste technology,
2. the use of less hazardous substances,
3. the furthering of recovery and recycling of substances generated and used in the process and of waste, where appropriate,
4. comparable processes, facilities or methods of operation which have been tried with success on an industrial scale,
5. technological advances and changes in scientific knowledge and understanding,
6. the nature, effects and volume of the emissions concerned,
7. the commissioning dates for new or existing installations,
8. the length of time needed to introduce the best available technique,
9. the consumption and nature of raw materials (including water) used in the process and energy efficiency,
10. the need to prevent or reduce to a minimum the overall impact of the emissions on the environment and the risks to it,
11. the need to prevent accidents and to minimise the consequences for the environment,

At a workshop on sustainable chemistry held in 2004, jointly with the Organization for Economic Cooperation and Development (OECD), the German

Federal Environment Agency (Umweltbundesamt) has developed criteria for a sustainable chemistry:

- Qualitative development: Use of harmless substances, or where this is impossible, substances involving a low risk for humans and the environment, and manufacturing of long-life products in a resource-saving manner;

- Quantitative development: Reduction of the consumption of natural resources, which should be renewable wherever possible, avoidance or minimization of emission or introduction of chemicals or pollutants into the environment. Such measures will help to save costs;

- Comprehensive life cycle assessment: Analysis of raw material production, manufacture, processing, use and disposal of chemicals and discarded products in order to reduce the consumption of resources and energy and to avoid the use of dangerous substances;

- Action instead of reaction: Avoidance, already at the stage of development and prior to marketing, of chemicals that endanger the environment and human health during their life cycle and make excessive use of the environment as a source or sink; reduction of damage costs and the associated economic risks for enterprises and remediation costs to be covered by the state;

- Economic innovation: Sustainable chemicals, products and production methods produce confidence in industrial users, private consumers and customers from the public sector and thus, result in competitive advantages.

Avoiding Hazardous Chemicals – Developing Safe Chemicals

One step towards creating more safety lies in chemicals with less hazardous properties that can be handled at lower risk. Chemicals no longer pose a risk once there is compliance with principles of safe handling.

The use of very hazardous substances must have limited application or must be banned entirely. These substances include carcinogenic, mutagenic or substances toxic to reproduction (CMR substances), or substances that are highly critical for the environment because they are persistent and bioaccumulative (PBT substances).

Concerning environmental protection policy sustainable chemicals may not cause any short or long-term problems upon release into the environment. Additionally, sustainable chemicals are not persistent, do not spread across large areas (short range chemicals) and have no irreversible effects.

Chemicals are considered sustainable when no hazardous properties are detected: they neither cause known damages nor persist in the environment long enough such that they might have a negative impact heretofore unknown.

A sustainable chemical is not characterised by the properties of its contents only. The conditions under which substances are produced, processed and applied must be evaluated throughout their entire life, which includes their specific demand for resources (in terms of energy, raw materials and additives), their yield upon production, emissions to the air, water and soil, and volumes of wastewater and solid waste.

Why is the German Federal Environment Agency Concerned with Sustainable Chemistry?

It is the Federal Environment Agency's objective to help prevent any negative impact on mankind and the environment issuing from the chemicals industry or the downstream processing and application of chemicals. When products and processes consume fewer resources there is less strain on the environment as well as cost savings for businesses.

The Federal Environment Agency sees sustainable chemistry as an important element of environmentally friendly innovation policy that protects the environment and health at the same time. The Federal Environment Agency acts as a forum for stakeholders in which there can be an exchange of ideas and approaches to sustainable chemistry, development of design options, and an opportunity to come to agreement on common goals.

OUR SUSTAINABILITY STRATEGY

Innovation has never been more important than it is today. It's not just about improving the products we make. It's about how we invent better ones, work with other companies and organizations to develop markets that value and encourage the creation of new sustainable processes and products, and improve lives by leveraging sustainability as the world's greatest innovation opportunity.

NIKE, Inc. continues to deliver strong growth. What fuels our success and challenges us? Innovation. It's our growth engine. It's what our athletes, consumers and investors expect of us – and always have. It's what we expect of ourselves. It enables new heights in performance and raises the bar for our industry and beyond.

Our world faces unprecedented change: constrained resources, population growth, heightened connectivity, increased demand. In a world of finite natural resources, our growth is enabled by infinite human resources: innovation and inspiration.

We understand that innovation through the lens of sustainability is fundamental to achieving our vision of growth that is not dependent upon constrained resources.

Indeed, creating and building business models that not only recognize and accommodate but thrive on the constraints of the natural world is the only way we can achieve growth in the present that won't compromise our ability to grow and succeed for decades to come.

Of course, this transition is challenging for a company that sells physical products and has bold ambitions to provide those products to more people in more places worldwide. Transforming NIKE in this way will take time and is a long-term commitment. It touches every aspect of our business, from how we design and make our products to how we engage our employees and other businesses in our value chain. It begins with placing sustainability at the beginning of the innovation process.

We focus sustainable innovation on our biggest areas of opportunity and risk. We ask difficult questions of ourselves, our business model and our industry. We seek to understand the meta-trends and the signals – strong and weak – that add complexity and uncertainty to the future. Rather than working to avoid only known risk, we work to understand where risks and opportunities may emerge as well as what's needed to address them and enable new forms of innovation to take hold.

Based on extensive analysis of the impacts of our business across the value chain, we know that materials and manufacturing represent the greatest areas of impact on workers, communities and the environment, and the greatest potential for sustainable innovation.

While NIKE has focused on materials innovation for many years, most improvements across the industry have been incremental, not fundamental. The time is right for innovation that leads to a more sustainable palette of materials and chemistries . We have focused for many years on improving the working conditions of contract factory workers, increasing productivity, and reducing the environmental footprint of manufacturing, eventually bringing these threads together through implementation of lean manufacturing.

The future of lean for NIKE is to deliver profitable growth through sustainable manufacturing and sourcing. To do this, we are making lean NIKE, Inc.'s manufacturing standard. We require a commitment to lean as part of being accepted into our source base and a minimum commitment and progression for positive ratings by including it in our Sourcing and Manufacturing Sustainability Index, a component of our Manufacturing Index which assesses factories based on sustainability, cost, quality and on-time delivery. We are working with our supply chain to demonstrate the value of lean as a driver of sustained, improved business performance where workers are engaged and enabled to drive business success through continuous improvement. Some of the standard metrics we use to assess factory adoption include productivity, human resources management assessments, turnover, absenteeism, and factory implementation of and results from worker engagement and well-being surveys.

We are now driving a portfolio of advanced manufacturing initiatives that encompass changes to the traditional manufacturing business model that includes the role of workers in the production of better products and services. As described throughout this report, these two areas – materials and manufacturing – are central to our strategy.

Making Today Better while Designing The Future

Sustainable innovation takes many forms at NIKE. Incremental improvements play an important role in our efforts, since even small changes have large impacts across our portfolio. However, incremental changes on their own fall short of the progress needed.

The scale of today's challenges requires breakthrough innovations such as entirely new materials and ways to make products. As an example, in 2012 we introduced the first product made using the innovative new NIKE Flyknit technology and manufacturing process. The Flyknit Lunar 1+ running shoe, launched in 2013, reduces footwear waste in the upper by 80% on average compared with traditional production methods. To put a picture to that amount of waste, from its introduction in 2012 through FY13 the total combined material savings from just the uppers of the Flyknit Lunar 1+ compared to a traditional running shoe was 66,000kg – about the same as 12 adult male African elephants. Flyknit is a game changer and we believe there is significant potential ahead as Flyknit technology is implemented more broadly across our footwear offering.

NIKE's ability to deploy disruptive technologies like Flyknit at scale multiplies sustainability benefits. And as the game changes, we remain on the offense, always.

In some cases, where we have direct influence, we use our management tools, expertise and other resources to take a leadership role. One example is the creation and proliferation of the NIKE Materials Sustainability Index (NIKE MSI).

Other opportunities, such as improving working conditions within our global supply chain, are beyond the reach of any single organization and require collaboration with other supply chain participants. One way we are working on these changes is through our collaboration with the Fair Labor Association to develop the Sustainable Compliance Initiative. This initiative, still in development, provides tools to improve the quality, consistency and efficiency of efforts in our industry to comply with country laws/regulations, and company standards.

For challenges such as building a market for sustainable materials, we need input from a broad range of stakeholders and experts.

After all, risk is interconnected so solutions must be collaborative. Companies that win in the future will be those capable of accessing new sources of knowledge, creativity and capital to accelerate sustainable innovation and bring it to scale.

To further advance our efforts, at the start of FY14 we embedded our Sustainable Business & Innovation function into the company's Innovation group . This streamlines the sharing of sustainable innovation throughout the company and continues a nearly 20-year evolution to move corporate responsibility and sustainable innovation from the periphery of the organization to the core of the company. This move ensures that sustainability will be central to the questions asked and the solutions created in the innovation process.

Our Value Chain Footprint (FY13)

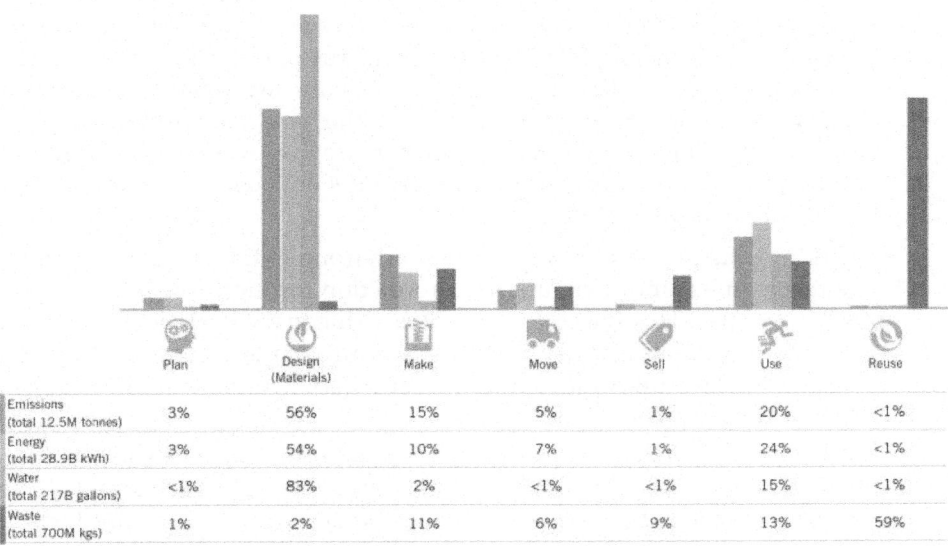

	Plan	Design (Materials)	Make	Move	Sell	Use	Reuse
Emissions (total 12.5M tonnes)	3%	56%	15%	5%	1%	20%	<1%
Energy (total 28.9B kWh)	3%	54%	10%	7%	1%	24%	<1%
Water (total 217B gallons)	<1%	83%	2%	<1%	<1%	15%	<1%
Waste (total 700M kgs)	1%	2%	11%	6%	9%	13%	59%

Our "big picture" data collection gives us a view of the overall impact of our business in key impact areas which helps us to focus our efforts on those areas where we have greater impact and understand the interaction between impacts from decisions at different stages of our value chain. This is a point-in-time reference. We continue to update information on the big picture and assess and improve on the more direct and concrete elements of a reportable footprint for year-on-year comparisons.

The purpose of this study was to determine the environmental footprint of NIKE, Inc. – including its products – across four impact areas: CO_2e, energy, water and waste. We undertook this work by identifying and quantifying water and energy usage, CO_2e emissions and waste created at each stage of the value chain. The impact of each individual product differs considerably, based on its profile, materials used, size and weight, method of manufacture, and location of production, use and disposal.

Understanding Our Impacts, Risks and Opportunities

The journey to integrate sustainability deep into NIKE has required us to show how it supports our company strategy, creates value for stakeholders, mitigates risk and makes our business more resilient.

To do this we must understand our business and its impacts, so we can make informed decisions that will bring about the greatest change. We also must consider the connectivity across all areas, as a decision in one area may have unintended consequences or enhancements in another. We take a broad view, as we recognize the impacts from our directly controlled operations are a small part of the total system.

An important factor is our outsourced manufacturing model, which is common across the apparel and footwear industries, as well as many others. The model brings complexities related to lack of direct control over factories that may serve multiple brands and incomplete information about performance and impacts. Nonetheless, our vast and interconnected value chain also offers great opportunities to raise expectations and standards across the sector, in areas such as labor practices and environmental protection, to increase our overall positive impact. This approach has been central to our relationships with contract manufacturers and other brands for many years.

In 2013, we furthered our study of our environmental footprint across our entire value chain – from raw materials production through consumer disposal of products after use. This is a complex analysis, due to the variation in the availability and quality of data, the need to make assumptions and extrapolations, and the dynamic nature of this system and its many participants. Our knowledge in this area continues to grow.

The analysis – which covers energy use, water consumption, waste generation and greenhouse gas (GHG) emissions – confirmed what we have long believed about the importance of the design process and materials, and how products are made. These issues are thus the main focus of our programs and this report.

The footprint graphic only tells part of the story. We also made discoveries within these categories, such as the significant effect of cotton and leather on our overall impacts. In the raw materials stage of the value chain, growing cotton represents 87% of water use, and it accounts for more than 63% of water use across our entire value chain. Leather production represents 56% of GHG emissions during the raw materials stage – mostly from methane as a result of cattle's digestive processes – and more than 18% of our total greenhouse gas footprint. These types of insights are valuable as we develop and evolve our programs moving forward. They compel us to look well beyond the surface levels of the systems we touch and deep into our value chain. This perspective reinforces that our span is much bigger and more complex than one might imagine, stretching even into the agricultural fields where the raw materials for our products are grown and raised, and where some of our biggest system impacts occur.

By nature, footprinting is backward-looking. Setting our sustainability strategy also requires looking forward, to navigate in the direction we believe the world is heading.

For many years we've focused on meta-trends – "strong signals" – pointing to the sustainability-related issues that pose the greatest business risks or opportunities. These meta-trends include issues such as water scarcity, materials cost inflation, climate change, rising labor costs as well as increased transparency and heightened levels of collaboration.

We also listen closely to "emerging signals." Some of the emerging signals we've identified – including micro-plastics in the environment and increasing and changing environmental policies – are broader than our company or industry.

Issues such as these have the potential to become even more relevant to NIKE in the future.

Within every sustainability challenge also lies a business opportunity. For example, to better manage constrained resources, we develop and use more recycled and more sustainable materials, and leaner manufacturing processes. To decrease our exposure to labor cost inflation and capacity constraints, we increase the efficiency of our supply chain. We have developed superior materials such as water-based adhesives through green chemistry, which also delivers health and safety benefits to workers. These types of innovations have already delivered substantial benefits to our business, and we expect them to increase in the coming years.

Sustainability Challenges and Opportunities

How NIKE, Inc. views, anticipates and responds to emerging signals alongside global trends, risks and opportunities, and business challenges to inform its plans

Emerging Signals

We also use scenario planning to sharpen our understanding of the potential impact of sustainability issues on our business and to inform decision making. Through scenario planning, we can assess the potential impacts that external issues such as climate change or resource scarcity might have on NIKE. We can model the rippling effect that a percentage change in our use of a more sustainable material might have across the value chain, or the impact of changes to our sourcing base as we fully implement our sustainability indices. We can also analyze how initiatives, such as those that improve energy or water efficiency, or decrease waste, could impact the company's competitiveness.

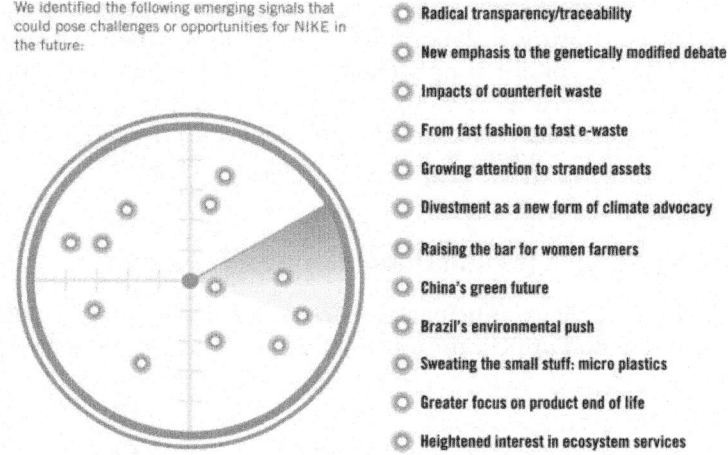

We identified the following emerging signals that could pose challenges or opportunities for NIKE in the future:

- Radical transparency/traceability
- New emphasis to the genetically modified debate
- Impacts of counterfeit waste
- From fast fashion to fast e-waste
- Growing attention to stranded assets
- Divestment as a new form of climate advocacy
- Raising the bar for women farmers
- China's green future
- Brazil's environmental push
- Sweating the small stuff: micro plastics
- Greater focus on product end of life
- Heightened interest in ecosystem services

Understanding these interconnections helps us prepare for a range of possible futures and improves our ability to develop innovative solutions to address emerging risks before they become more challenging to manage.

For example, in 2013 we developed a scenario tool to explore the impacts that climate change, and related water scarcity, could have on cotton, one of our main product inputs. We developed a Business and Environmental Scenario Tool which gives us the ability to assess overall and intersecting impacts from changes to different scenarios.

We work with stakeholders and experts from outside the company to validate this work, to develop a collective understanding of the systemic issues we face and to identify shared solutions.

Case Study: Best (Business and Environmental Scenario Tool)

BEST provides a 10-year quantified view of environmental and financial impacts from changes to scenarios such as materials used or changes in sourcing.

Before developing BEST, we:

- Calculated separately the impact that changes to the business had on each criterion (financial, water, energy, CO_2, waste)
- Took extended time to conduct the environmental analyses
- Had a limited view due to the focus on and analysis of one criterion at a time, therefore we made decisions based only on one criterion or continued with potentially conflicting analyses across different criteria
- Often excluded interactions between criteria

- Usually focused only on Tier 1 suppliers – those with whom we directly contract for goods or services – though a considerable portion of the impact occurs earlier in the value chain
- Saw each group within NIKE focus on the criteria that were important to them

With BEST we can:

- Input one scenario and receive data on simultaneous impacts to five criteria (financial, water, energy, CO_2, waste)
- Complete the analysis quickly, with a turnaround in minutes, rather than weeks
- Take a holistic view of all criteria and compare the return on investment for all criteria at the same time
- Capture how changes to one criterion impacts others
- Take into account a more complete view of the supply chain
- Provide users with a view of the entire business so they can see how their decisions impact other business areas

We now use BEST to assess various impacts and decisions. Some examples include:

- Impacts to costs, water usage and energy usage if we use less cotton and more polyester
- Changes to the amount of materials purchased, energy used and waste generated if we increase pattern efficiency by certain percentages

Nike's Sustainability Strategic Framework

NIKE's Sustainable Business and Innovation team focuses on enabling the company to thrive in a sustainable future, and provides the insights, tools and expertise to hardwire sustainable decision making into global business operations.

Given the trends and analysis described earlier, NIKE drives sustainable business innovation in three strategic ways:

1. Deliver a portfolio of sustainable products and services that enhance athlete performance
2. Prototype and scale sustainable sourcing and manufacturing models
3. Explore new sources of revenue not based on constrained resources

These pillars are underpinned by skills and capabilities that enable their success. The sections below describe these pillars and how they work together to drive the disruptive innovations that will shape our future.

It's important to note that NIKE also continues to focus on other important aspects of corporate responsibility including compliance, stakeholder management, employment practices and community investment. These play an important role in our overall performance, and our programs and progress in these areas are described later in this report.

Sustainability Strategic Framework

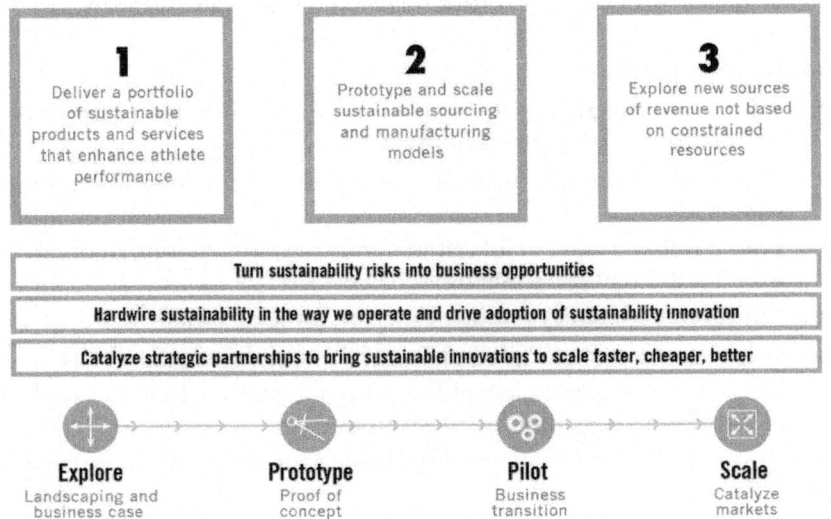

Materials Matter

NIKE's main environmental impacts, by far, are realized in the products we sell and the materials used to make them.

The materials we choose have impacts that ripple across every stage of the value chain. The approximately 900 million units produced annually through our supply chain are made from more than 16,000 materials selected from more than 1,500 different vendors, chosen from a staggering 80,000 material options.

From our analysis, the production of these materials – from growing cotton and harvesting rubber, to raising livestock for leather and extracting oil for polyester – represents 21% of the total energy use throughout our value chain, 73% of the water consumption and 33% of the greenhouse gas emissions. When you include materials processing activities, such as the dyeing and finishing of fabrics, those percentages increase to 54% of total energy use, 83% of water consumption and 56% of GHG emissions.

The Evolution of Our Product Sustainability Indexes

Reducing materials-related impacts is among the strongest levers we have for improving our overall environmental performance. However, it's challenging to reach far upstream into our value chain and influence the behavior of companies and individuals over which we have no direct impact. Therefore, we focus on product design, an area we do control. Decisions we make in the design phase determine the majority of a product's environmental impacts, and can have exponential effects up and down our value chain. These impacts are embedded in the creation of materials, from how much they weigh to transport, to how much water and energy is used in washing them, to what is left when a product's useful life is over. However, our materials decisions are limited by what materials are available and their expected performance attributes. So, we also seek to understand and engage others who play critical roles in the process of developing, creating and bringing materials to market.

We launched new and updated evaluation tools for rating the sustainability of footwear and apparel designs in FY13. Described below, these indexes represent the next generation of our Considered Indexes.

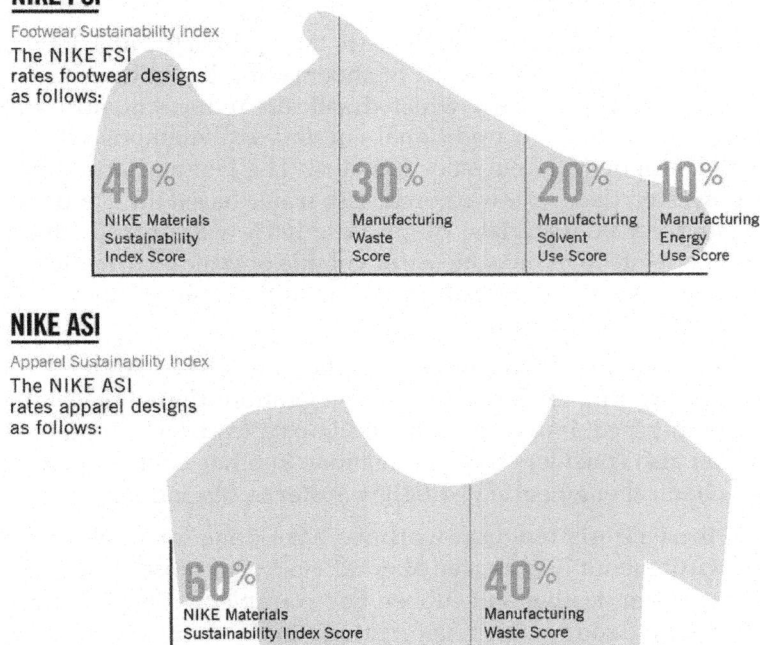

NIKE FSI

Footwear Sustainability Index
The NIKE FSI rates footwear designs as follows:

40%	30%	20%	10%
NIKE Materials Sustainability Index Score	Manufacturing Waste Score	Manufacturing Solvent Use Score	Manufacturing Energy Use Score

NIKE ASI

Apparel Sustainability Index
The NIKE ASI rates apparel designs as follows:

60%	40%
NIKE Materials Sustainability Index Score	Manufacturing Waste Score

Over the past several years, we have created scoring tools and indices that give our product creation teams the information they need to make better decisions about materials based on sustainability as well as performance characteristics. Two key improvements help a design's score: reducing waste by improving pattern efficiency, and the choice of more sustainable options (*e.g.*, recycled polyester rather than virgin polyester). We tested and rolled out our updated footwear and apparel indices in FY12; both include the NIKE Materials Sustainability Index (NIKE MSI). The NIKE MSI measures energy and water use, waste generation and chemical use in materials. We have communicated with and trained our material vendors to help them understand the NIKE MSI and how they can improve their materials' scores by providing more sustainable options, which makes them more attractive to work with as a source for materials.

Collaborating with others is key to fueling disruptive innovation. Our strategy includes strategic partnerships and making selective investments in breakthrough technologies in the materials space. In 2012, for instance, we made a minority investment in DyeCoo Textile Systems B.V., a Netherlands-based company that has developed the first commercially available waterless textile dyeing machines. By using recycled carbon dioxide, DyeCoo's technology eliminates the use of water in the dyeing process. This holds great promise for NIKE, since dyeing and finishing represents about 5% of water use across the value chain. In 2013, we worked with Far Eastern New Century Corp. and DyeCoo to launch the "dye house of the future" in Taiwan to optimize the dyeing process. This new process, which we call ColorDry, provides the most consistent color results to date and eliminates water and process chemicals from the dyeing process.

Decreasing materials use is also core to our approach. NIKE Flyknit technology revolutionizes the age-old craft of shoemaking by knitting the shoe upper with individual strands of yarn, which drastically reduces manufacturing waste and materials compared to traditional cut-and-sew methods while providing strength and support where it's needed most. The Flyknit Racer's laces are 30% recycled polyester, the upper is colored with water-based inks, and the sock liner is made from recycled materials. We also use 100% water-based adhesives in the midsole and outsole to reduce the use of volatile organic compounds. We're only just beginning to see the potential for Flyknit and we plan to increase use of this technology in the coming years.

To continue to drive innovations like these, we have a Materials Science Innovation function that explores the next generation of materials – for instance, non-petroleum-based, low environmental impact feedstocks for apparel fabrics. This function also considers how innovations in other fields – such as medicine and biomechanical engineering – might transfer to our industry.

All of these efforts build on work we have done for years to identify and develop environmentally preferred alternatives to our most-used materials, such as cotton, polyester, leather and rubber. For example, we helped to establish the Leather Working Group, which has created standards for better environmental management of leather processing, and are Pioneer members of the Better Cotton Initiative.

New Methods of Manufacturing

Manufacturing is another major part of the equation. We estimate that more than 2.5 million people work at various stages throughout our supply chain, including more than 1 million in the factories we contract with directly. This makes manufacturing our biggest area of impact on people. That's not a new insight, and we have been working for years to help raise the bar for working conditions, not only in our own supply base but across our industry. We have done this by developing and communicating our Code of Conduct and Code Leadership Standards, and by assessing suppliers' compliance with our requirements and legal standards. We have also worked with our suppliers to help them develop their human resources management capacity so they can proactively manage and engage their workforce.

These efforts led to an insight that has been critical in our work with factories: that lean manufacturing benefits factory owners and workers, increases productivity, reduces environmental impacts, enhances our brand, and that workers are key to the successful implementation of lean.

This integrated approach is reflected in how we assess performance. Factories are rated using our Sourcing & Manufacturing Sustainability Index (SMSI), a component of our Manufacturing Index, which puts sustainability considerations on equal footing with quality, cost, delivery, and is one tool we use to select factories with which we do business. This approach serves as a way to identify factories to engage with more collaboratively and to which we direct more attention, resources and business. The SMSI incorporates results of other scoring tools that measure progress in worker health and safety; labor compliance; human resources management; lean implementation; energy and carbon management and other environmental sustainability issues.

We are now working to implement an enhanced vision of lean manufacturing as part of a portfolio of initiatives we're calling the "manufacturing revolution." This transformation is a response to significant trends in our supply chain, including advancements in technology and engineering, and ongoing volatility in labor and materials pricing. It aims to redefine both how our products are made and what they are made of. It includes innovations in advanced technology, as well as manufacturing excellence and modernization.

As we look ahead to a new era of manufacturing, we also see opportunities to create a more sustainable, stable supplier base. To implement the changes we're anticipating, the workers employed by our suppliers will need to learn and use multiple skills. During 2013, we conducted two pilot studies in Indonesia that tested both the technology and human aspects of lean manufacturing implementation. The pilots measured changes to productivity, cost and worker engagement, and found significant improvements in a number of areas when lean manufacturing lines were compared to control lines. We are collaborating with other organizations and contract factories to encourage them to look at additional opportunities to improve the lives of workers outside the factories.

Beyond Products

In 2009, we founded our Sustainable Business and Innovation Lab to focus on materials and manufacturing innovation partnerships as well as revenue sources that are decoupled from constrained resources, including through digital services. Among other initiatives, the Lab helped to fuel our vision to extend NIKE's leadership in athletic footwear, apparel and equipment into the digital realm of fitness, coaching and training services.

One example of NIKE's efforts that focuses on services is the NIKE+ digital platform that enables athletes to track and analyze movement – for example, to share their workouts and calories burned and personal targets and performance. NIKE+ has grown into a full ecosystem, delivered with different apps and services, with more than 18 million members spanning the globe at the end of FY13. Collectively, members have run more than 1 billion miles, the equivalent of 40,000 times around the earth at the equator. This expanding business supports athletes' health and wellness, opening additional opportunities to NIKE.

In early 2013, we launched the first NIKE+ Accelerator, which hosted 10 companies for a three-month, immersive, mentor-driven startup program. Leveraging the success of NIKE+, participants worked to create products and services to inspire athletes across a broad range of activity and health goals, including training, coaching, gaming, data visualization and "quantified self." Participating companies pitched their business concepts to more than 1,000 investors, potential strategic partners and prospective collaborators. Through the program, we received more than 1,000 requests for access to the NIKE+ development platform.

Strengthening Our Core

For more than 15 years, we have been on a journey of sustainability integration. This task is essential. To achieve ongoing, profitable growth, sustainability considerations must be deeply embedded throughout the company. At NIKE, sustainability is not just about vision and values. It depends equally on having the systems, structures, people, responsibilities and accountabilities in place to ensure our commitments are reflected in our day-to-day business activities.

During the last two years, we have advanced business integration on several fronts. Some changes relate to refining our internal organizational structures to more effectively drive sustainable innovation. These include the following:

- The Sustainable Business & Innovation (SB&I) team focuses on a broad range of areas including finding and deploying new chemistries, training supplier factories on energy and water efficiency programs, partnering with product teams to promote the use of more sustainable materials and designs, developing scenario planning tools, gathering information about social and environmental performance within NIKE and its supply chain, and evaluating and reporting that performance. The team also drives several of our external engagement activities, collective action efforts and open innovation agendas to accelerate and scale game-changing solutions. Importantly, SB&I became a part of NIKE's Innovation organization in 2013, streamlining the process of scaling sustainability innovation throughout the company.

- The Materials Science Innovation function, which grew out of the sustainable product research and design function within SB&I and focuses on sustainable materials, now reports into NIKE's product innovation team. It includes a dedicated sustainable product R&D team, making it even better aligned with broader apparel and footwear materials research.

- The Sustainable Manufacturing Excellence team is formally aligned to both SB&I and NIKE's sourcing functions. This enables a strategic outlook for the future of manufacturing and a holistic view of the impacts and capacity-building work done with factories as well as joint planning for future growth.
- A new product sustainability team works directly with NIKE's product groups to drive sustainable innovation deeper into the company's product creation processes and pipeline.

Other recent changes reflect the deeper integration of several important processes and activities throughout the company. For example:

o Our NIKE Apparel Sustainability Index and NIKE Footwear Sustainability Index have become part of the standard tools used by our global product creation teams. This ensures that design decisions take environmental factors into account.

o We integrated our Manufacturing Index which includes the Sourcing & Manufacturing Sustainability Index into our sourcing selection and evaluation criteria. The Index assesses suppliers on relevant dimensions related to lean manufacturing, such as on-time delivery and defect rate. As a result, sustainability is a factor in all supplier ratings.

o We created and launched a business simulation experience – NIKE 2021 – a half-day strategy immersion in which participants adopt roles as chief executive, chief financial officer, and vice presidents in supply chain, product, brand, innovation and sustainability. Together they compete against other teams to plot strategic investments and decisions that consider impacts to cost of materials and manufacturing, product availability, revenue and net income. Nearly 2,000 employees from senior executives to teams in product development, finance, supply chain and geographies have participated. NIKE continues to evolve the experience and deliver it to more teams.

o We have publicly discussed and published our perspective on the risks of climate change and sustainability-related issues in broader forums. This illustrates the increasing importance of sustainability issues to the company and our stakeholders.

During 2013, NIKE also aligned community engagement activities more closely with our businesses and geographies. This embeds planning and performance tracking into the appropriate business cycles and increases accountability.

As we integrate sustainability even more deeply throughout the company, the role of the SB&I team continues to evolve. Using a unique set of capabilities and expertise, the team focuses on activities that enable more sustainable decision making companywide that often involves a longer time horizon than the typical business planning cycle. This includes:

- Providing a holistic perspective on sustainability opportunities and risks

- Working in partnership with business functions to set and deliver against companywide sustainable innovation priorities
- Delivering analysis and guidance on sustainability-related investments
- Building a center of excellence in sustainability data capture, analysis and disclosure
- Developing tools and capabilities such as environmental footprinting, scenario planning and systems thinking
- Facilitating companywide sustainability target setting, performance measurement and reporting (both internal and external)

In addition to having the right structures in place, integration is also about engagement. Everyone at NIKE has an important role to play in making the company sustainable and we encourage leaders and employees to grasp the issues and opportunities ahead, consider the sustainability implications of their decisions, and are accountable for their actions.

To help educate internal audiences, and to improve our ability to measure sustainability integration across key business levers, in FY13 we further developed a two-part Sustainable Business Roadmap, consisting of a framework and an assessment. The framework describes what sustainability integration looks like across the areas of strategy, structure, operations and employees. It identifies levers that enable integration within those areas, such as leadership commitment, resourcing and education. The framework also describes tactics – such as establishing goals, budgeting for sustainability efforts as a part of standard business processes, and embedding sustainability learning opportunities in employee development – that provide context and a means to measure integration. Business teams will use the assessment to identify current levels of integration and opportunities to drive business performance within their functional areas.

This Roadmap not only provides us a better understanding of where we are today in terms of integration, but it will also guide us in building the organizational capabilities we will need to realize our sustainability objectives. The tool will be piloted in 2014 and is expected to inform further iterations of our sustainability integration strategy.

Integrating sustainability enterprise-wide is challenging, and remains a work in progress. We have learned a great deal in the more than 15 years we have been working toward a more sustainable supply chain and business. To share our experience in this area with others, in 2013 we engaged with professors at Harvard Business School who produced Governance and Sustainability at NIKE – a case study about our journey. This publication, which is being taught at business schools, describes how we transformed from a company that reacted to external pressures while working to minimize reputation risks to one that views sustainability as among our leading innovation opportunities. Another case study, published by Stanford Graduate School of Business – NIKE: Sustainability and Labor Practices 1998-2013 – highlights our long-term journey of sustainability and labor.

Table: Identifying and Developing System-Change Initiatives

SYSTEM BARRIER	DESCRIPTION	STRATEGY	NIKE ACTIONS
Lack of common standards	Without a shared understanding of materials-related impacts, designers cannot make efficient decisions.	Establish a universal standard through industry collaboration, leveraging our extensive experience in this area.	In 2012, to enable alignment on a common language and set of definitions related to materials footprinting, we opened the NIKE Materials Sustainability Index (NIKE MSI) for public use. In 2013, the NIKE MSI became a core part of the cross-industry Sustainable Apparel Coalition's Higg Index.
Continually changing information about materials	Information about the impacts of more than 16,000 materials used in our industry is continually evolving.	Create an open technology platform that enables experts worldwide to contribute and share information related to materials.	In 2011, we engaged external data experts to help us envision the future of information sharing and tools innovation in this space. We created a mechanism to allow experts and brands worldwide to contribute to and benefit from a shared, current dataset of materials information.
Lack of access to information	Designers and other decision makers cannot make informed, sustainable choices about materials unless related information is easily accessible	Develop tools for non-technical users to integrate sound information about materials into their decisions.	In 2013, NIKE collaborated with our design networks including, the London College of Fashion through a sponsored studio project to test the prototype app which became MAKING in 2013. The app helps designers and product creators make informed decisions about the environmental impacts of the materials they choose. Leveraging information in the NIKE MSI, MAKING ranks materials used in apparel based on four environmental impact areas: water, chemistry, energy and waste. Through direct comparisons, designers can quickly see how material choices stack up. We envision that this innovation will help to catalyze other advances in this area.
Insufficient focus on materials innovation	While we have focused significantly on materials design and use in past years, our journey to find, create and innovate new materials has really just begun.	Engage system participants in a process of collective transformation.	In April 2013, NIKE hosted the LAUNCH 2020 Summit. Along with NASA, the US Agency for International Development, and the US Department of State, we convened 150 materials specialists, designers, academics, manufacturers, entrepreneurs and NGOs to catalyze action around the sustainability of materials and how they are made. This included introducing the LAUNCH 2020 Challenge Statement, an open call for innovation to transform the system of producing fabrics.

Sustainability is a Team Sport

At NIKE, we increasingly see sustainability as a source of competitive advantage and business value. However, we also see a bigger picture. Ultimately, our collective future depends on solving problems that are much larger than any individual company or organization can tackle on its own – issues such as climate change, resource scarcity and substandard working conditions.

We recognize we must do our part to help transform entire systems rather than just addressing our role within the parts of the systems we touch. There is no substitute for the collective action this requires. Significant collaboration is necessary to organize the diverse capabilities, resources, insights and political will needed to develop solutions and bring them to scale. In sustainability, we win or lose together.

For systems to change, we must first understand them and our role in them, and other participants must improve their understanding as well. Two years ago we launched a more formal approach to systems innovation within SB&I. Drawing on systems theory and capitalizing on the insights and involvement of experts within and outside the company, we have developed an approach to mapping complex systems, building cross-sector networks, mobilizing resources and fueling open innovation. This helps us to understand how a system works – the key players, relationships, transactions, points of leverage, barriers and opportunities – and to create a shared vision for change. It also supports our efforts to engage with the broader network of stakeholders.

Our understanding in this area continues to increase, as we gain insights into how we can most effectively create change depending on our role in a specific system. Our ability depends on our role, the type of business relationship, regulations and many other factors.

We are working to apply systems thinking to materials and manufacturing. We have collaborated for many years with other companies in our industry through organizations such as the Sustainable Apparel Coalition, with multistakeholder groups including the Fair Labor Association, and with organizations such as the International Labor Organization and the International Finance Corporation. Our efforts have been aimed at raising the bar for performance across our industry, which requires actors across the value chain to collaborate, share information and innovate more effective, efficient and equitable ways of doing business. Based on this analysis, we developed strategies to address the main barriers to progress in this complex system, and launched initiatives to tackle them.

We have also worked to positively change systems in other areas of need:

- Created by the Nike Foundation in collaboration with partners, and informed by insights from thousands of hours of research about the impacts of girls living in extreme poverty around the world, the Girl Effect leverages the unique potential of adolescent girls and provides them with resources to end poverty for themselves, their families, their communities, their countries and the world.

- In 2012, with more than 70 organizations from government, civil society and companies, we launched Designed to Move, a study on the effects of physical inactivity in childhood and the need to build physical activity into the daily life of children. The report articulated the decline in physical activity and economic costs of that decline. With a coalition approach we provided research and defined a physical activity agenda and framework for action, including creating early positive experiences for children and integrating physical activity into everyday life.

Helping transform entire systems is complex, difficult and requires long-term investment and effort. We're committed to the process – it's essential to achieving our vision of a sustainable business. We look forward to learning more in the coming years.

A VISION AND ROADMAP FOR SUSTAINABILITY THROUGH CHEMISTRY

Charting the Path Forward: An Education Roadmap for Green Chemistry

With ever-increasing student enthusiasm, publications, and industry investment in green chemistry and engineering, now is an opportune time to chart a community-driven vision for the future of chemistry education. The roadmap project for green chemistry will outline both the long and short-term needs of a motivated community. Using past accomplishments as a springboard, a roadmap will help focus and coordinate individual and regional efforts through comprehensive planning to move the field of green chemistry forward. Collaboration within this dedicated and diverse community is a key step to consolidating knowledge, defining gaps in existing resources, creating strategies to overcome challenges, and ultimately achieving goals to build a more sustainable future.

CHALLENGES FOR CHEMICAL EDUCATION: ENGAGING WITH GREEN CHEMISTRY AND ENVIRONMENTAL SUSTAINABILITY

Although sustainability is a contemporary concept of immense significance, it is an intangible abstract concept, a concise definition of which remains elusive. It essentially represents a paradigm shift in understanding the relationship between humanity and the environment and recognition that the present excessive demands of humanity on natural resources, coupled with excessive waste and pollution generation by human beings, constitute an unsustainable, stressed and threatened global environment. This has been brought into sharp focus over the past decade by increasing awareness of the consequences of global warming, believed to be due to increasing levels of greenhouse gases in the atmosphere derived from anthropogenic sources. However, although there are a number of global initiatives to reduce greenhouse gas emissions, there are by comparison limited initiatives underway to seriously address environmental sustainability. This may be due to a lack of a comprehensive theoretical framework for understanding 'sustainable development' and its multi-disciplinary complexities, since the range of published definitions is vague and it remains a contested topic, fraught with contradictions.

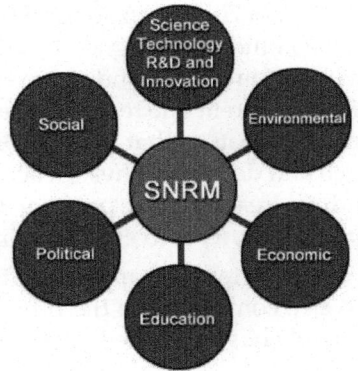

Fig. 1 : Concept framework of SNRM.

The many definitions of sustainable development are largely benchmarked to the World Commission on Environment and Development (WCED) definition: 'maintaining the needs of the present generation without compromising the ability of future generations to meet their needs'. There is much debate on 'what are the needs (and scale) of present generations' in the context of the 'haves' and 'have-nots' disparity and 'what are the needs (demands) of future generations', as the global population continues to escalate and the 'haves and have-nots' divide widens. Furthermore, there is much concurrent debate on how simultaneous harmonization of the needs of humanity with the needs of the environment can be achieved. It is intuitively obvious that 'sustainable development', and hence 'environmental sustainability', are progressive objectives and that metrics of these concepts need to be established in conjunction with tangible and workable definitions.

In this context, Hill and Mustafa have argued that sustainable development is essentially consistent with sustainable natural resource management (SNRM) that embraces several dimensions including 'social', 'political', 'scientific', 'technological', 'economic', 'research, innovation and development' and, most importantly, 'education', which collectively form a concept framework, as shown in Fig. 1.

Fig. 2 : The IYC 'themes'.

It emerges that since environmental degradation is largely due to the excessive needs of humanity, environmental sustainability is the responsibility of humanity, and thus sustainability education is of paramount importance. Science education and chemical education in particular are crucial components of sustainability education since the former informs and enables the climate change debate and the latter emphasizes how green chemistry and green chemical industries enhance the transition to environmental sustainability. Also, the UN declaration of 2011 as the International Year of Chemistry (IYC) established a roadmap for the future of chemistry (and hence chemical education) by defining a series of 'themes', all of which relate to addressing the most critical problems facing humanity in the 21st century, of which 'environmental sustainability' is the primary theme.

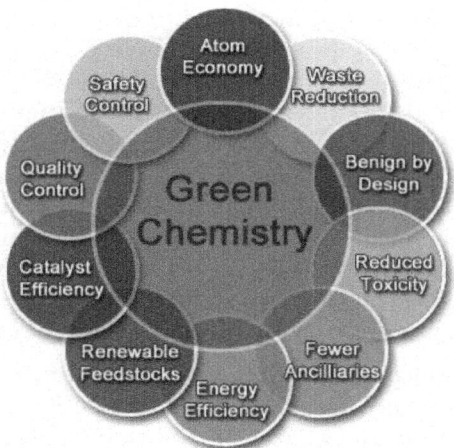

Fig. 3 : Principles of Green Chemistry.

The IYC themes offer significant challenges for chemistry and hence for chemical education. It is evident that the rapid emergence and development of 'green chemistry' over the last decade has dramatically enhanced a positive (social) image of chemistry and the IYC themes not only continue this trend but also enable a transition towards sustainable chemistry. However, although it appears that green chemistry and sustainable chemistry are synonymous terms, Tundo has suggested that there is a subtle difference in that sustainable chemistry has chemical processing connotations involving more energy efficient, less polluting chemical manufacturing processes which may generate greater profit margins, whereas green chemistry is more focused on 'greening' chemical reactions to produce products which are environmentally benign, but are not necessarily of industrial interest or significance. This differentiation of these terms widens the scope and the challenges for chemical education since the sustainability of the chemical enterprise has to be addressed using sustainable chemistry as the major driver.

The challenges for chemical education in enabling the transition of the chemical enterprise towards sustainability. Such a paper is timely since although environmental sustainability is a contemporary concept of immense significance and there are widespread calls to embed sustainability into tertiary curricula, courses emphasizing sustainability are currently limited and tend to be in non-science areas, with emphasis on economic, social and political sustainability. Furthermore, the challenges for science education and chemical education in particular are formidable since embedding sustainability into chemical education involves identification of the major chemical concepts that relate to sustainability of the chemical enterprise and hence to environmental sustainability; consequently, a paradigm shift in chemical education pedagogy is envisaged.

Influence of Green Chemistry

As for 'sustainable development', there have been many definitions of 'green chemistry' which relate to the synthesis of environmentally benign molecules and materials, new chemical (energy efficient) processes and new quality control technologies which reduce effluent and waste. These definitions imply that 'green chemistry' and 'sustainable chemistry' are closely inter-related since the vision of green chemistry is holistically aligned with environmental sustainability. With its direct linkages to other major science disciplines, such as the life sciences, materials science, chemical engineering and environmental science, and its indirect linkages to economics and ethics, together with its principal aim to provide benefits to society, green chemistry is rapidly changing the negative public image of chemistry which has prevailed for decades.

The guiding principles of green chemistry, offer significant challenges for chemical education, since the mindset of students and researchers has to be changed to think and learn in terms of environmental sustainability rather than in terms of 'pure chemical sustainability', and the latter has to be addressed with reference to terms and concepts such as 'atom economy', 'waste reduction', 'toxic versus benign', 'energy efficiency', 'renewable feed-stocks', 'quality control' and 'safety management'. Traditionally these terms and concepts have not been included in chemistry education. If the 'benign by design' philosophy is introduced progressively into chemical education pedagogy at all levels, this will greatly assist in the production of trained personnel for sustaining the chemical enterprise.

Sustaining the Chemical Enterprise

The chemical enterprise is an amorphous concept composed of a myriad of interconnected parts, each of which has to be considered from a sustainability viewpoint.

Green chemistry strategies target each stage of the lifecycle of a chemical product to continuously enhance its biological and ecological safety, reduce energy consumption associated with its production and eliminate the co-production of waste. Hence, green chemistry offers substantial reductions to the environmental footprint of chemical processes, improvements in the health and safety of those exposed to the manufacture of chemicals and safe handling of hazardous materials leading to greater public recognition of the beneficial significance of the chemical enterprise.

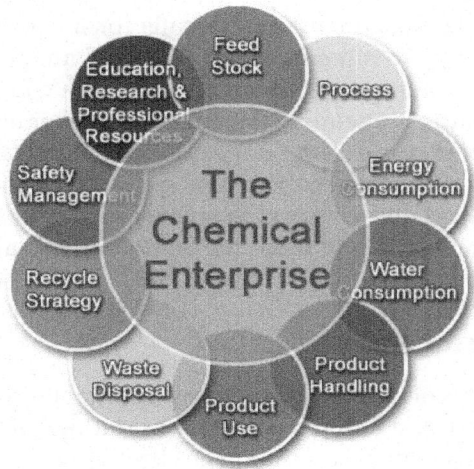

Fig. 4 : Structure of the Chemical Enterprise.

Poliakoff and Licence have argued that although the chemical industry has over recent decades made significant efficiency improvements to the manufacture of 'essential chemicals', finite natural resources will inevitably limit sufficient quantities of essential chemicals to meet the demand of a rapidly increasing global population. Thus, the chemical industry is currently not sustainable, as shown by 'E-factors' greater than zero. (The E-factor for a chemical process is the ratio of the amount of waste generated compared to the amount of product produced, and thus it is a measure of the 'green-ness' of the process.) Since natural resources feed-stocks limit chemical industry E-factors, a sustainable chemical industry depends on achieving E-factors approaching zero by maximizing product yield and minimizing feed-stock input.

The E-factor metric presents further challenges for chemical education, since most chemical manufacturing processes produce by-products which are typically regarded as waste materials. E-factors can be reduced by recycling and/or reusing such by-products, but this has to be economically viable. Thus, chemical education has to address not only the chemical reactions involved in an industrial manufacturing process, but also reveal the complexities and intricacies of 'stoichiometric economics', which is the basis of the so-called 'measurement science'.

Greening the chemical industry is obviously a progressive exercise. Ananda *et al.* have proposed a roadmap to a green (sustainable) chemical industry, based on a set of interactive principles broadly termed 'economic', 'social', 'technological', 'cultural', 'political' and 'environmental', which collectively form the 'drivers of change' in chemical industry policy planning and incorporate green chemical technology. This roadmap has considerable potential for moving the chemical enterprise towards sustainability, and by inference, reducing E-factors.

A definitive and comprehensive study by Grassian and Meyer has proposed that education and basic research in renewable energy resources, green chemistry and the environment play pivotal roles in the quest for sustainability, and have argued that a sustainable future calls for a carbon-neutral economy based on renewable (non-fossil) energy supplies and an enhanced understanding of the environmental impacts of increasing human consumption of natural resources. Thus, to address chemistry sustainability, chemical education has to be multi-disciplinary, encompassing not only new chemical concepts, but also the myriad of connections between these concepts and the chemical enterprise in order to reveal how the latter is linked to environmental sustainability. In view of the complexity and diversity of these connections, the challenges for chemical education are formidable.

Sustainable Chemistry Education

The accepted status of chemistry as the central (enabling) science has progressively been augmented by a rich research history dating back over at least a half century. Traditionally, chemical education has been considered as a combination of three major dimensions.

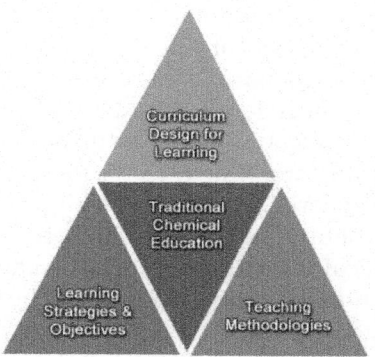

Fig. 5 : Structure of Traditional Chemical Education.

Over the last two decades, there have been some notable developments of these dimensions. With respect to curriculum design, Atkins has proposed that 'chemistry is based on just a few simple ideas', which has not only led to finalizing the constitution of the long-debated 'core of chemistry knowledge', but also to the re-evaluation of the content and context of secondary and tertiary chemistry courses in conjunction with more effective learning strategies. Also, Fensham has proposed that 'science has social responsibilities', which has been interpreted by Hill from a chemistry perspective. Furthermore, with respect to addressing a new vision for tertiary basic chemistry courses, Hill has designed a curriculum framework for these courses which embraces the proposals of Atkins and Fensham and which leads to defined learning outcomes. However, it should be noted that "chemistry knowledge" alone is not sufficient to resolve issues such as toxic waste disposal, climate change consequences and nuclear energy concerns.

Chemical education is critical as "education is an essential element of all aspects of a transition to sustainability".

With respect to teaching methodologies, Bedgood has asked, 'Why are we still teaching (Chemistry) the way we were in the 1980's?' He has been involved in a pioneering project aimed at enhancing science learning and teaching in Australian universities with a focus on first year science programs, which are characterised by large numbers of student participants, didactic teaching methods and multi-cultural learning environments. This program is piloting student-centred teaching methodologies in conjunction with student group learning strategies, fostering knowledge accumulation and enquiring minds. Further, Reid has asked, 'What do we know about how students learn in the sciences and how can we make our teaching match this to maximize performance?' He has proposed new strategies for correlating teaching methods more directly with learning rates and learning outcomes and how well core concepts can be linked to form a meaningful outcome. But such well-intentioned ideas for learning chemistry are often met with challenges in the wake of more computer-oriented online approaches to teaching chemistry.

With respect to learning strategies and objectives, Mahaffy has shown that there is an integral connection between chemical reactivity and human activity, and proposed that the traditional three levels of learning chemistry – 'macroscopic', 'symbolic' and 'molecular' – be extended to a fourth dimension 'human element'. This proposal is consistent with the earlier proposal of Fensham and also, most significantly, links chemical education to the IYC themes and their promulgation.

However, it is timely to examine whether education in chemistry as it is currently practiced is sufficiently equipped to address the IYC themes and hence the sustainability paradigm. In this context, Hill and Mustafa have proposed that the most significant manifestation of 'sustainability' is 'sustainable natural resource management' (SNRM), which correlates a sustainable environment with human endeavour. Furthermore, SNRM has many dimensions, a major one of which is 'scientific'. Chemistry, and in particular, environmental chemistry is obviously a central component of the scientific dimension of SNRM, which, when combined with the 'social', 'economic', 'technological' and 'research, innovation and development', dimensions becomes a major driver of SNRM. Further, Hill and Warren have shown that the sustainability theme can be embedded into the curriculum framework which Hill has proposed for the tertiary foundation chemistry course.

Mahaffy has campaigned for the integration of environmental sustainability into chemical education consistent with chemistry being an integral influence on the global future of humanity in terms of secure energy supplies, and the consequences of climate change and of diminishing food and fresh water resources. Similarly, Nocera has proposed that 'carbon-neutral energy resources' and 'efficient energy use' are the most pressing issues facing planet Earth in the 21st century, and that chemistry and chemical education have pivotal roles to play in addressing these challenges, since 'the chemical bond is the currency of energy'. Further, Fanzo, Remans and Sanchez have argued that 'chemistry' is the backbone of food and nutrition security and that hunger is one of the greatest threats to the sustainability of humanity; thus, chemical education has a pivotal role to play in revealing how food security is directly related to sustainable agriculture and aquaculture. Also, the cultural dimension of chemical education must not be overlooked. Kumar makes a compelling case in Wilderer *et al.*'s collection of essays 'Global sustainability: The importance of local cultures' – that local cultures markedly influence transitions towards sustainability, and in this context, sustainable chemical education plays a most significant role.

Thus, it appears that a new vision for chemical education is required, encompassing many new dimensions, if it is to address the challenges inherent in engaging environmental sustainability.

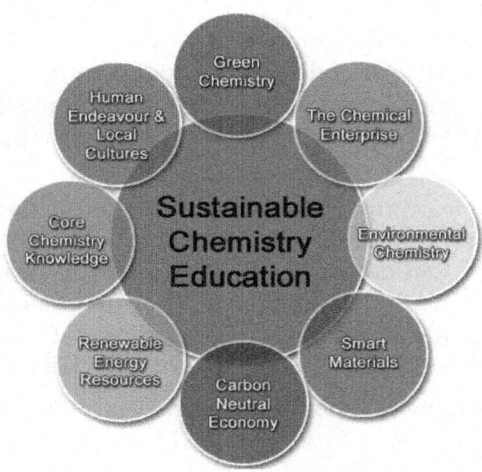

Fig. 6 : Dimensions of Sustainable Chemistry Education.

It is evident that sustainable chemistry education involves different methodologies in teaching fundamental chemistry concepts, whereby new terms and new philosophies are introduced. For example, chemical systems involving multiple chemical reactions need to be discussed in addition to examples of different types of single chemical reactions. Furthermore, chemical reactions need to be discussed in terms of 'atom economy' to illustrate the principle of chemical efficiency.

The core topic of thermodynamics needs to be discussed in terms of energy efficiency of chemical processing and manufacture in addition to energetics and spontaneity of chemical reactions. The core topic of kinetics needs to be discussed in terms of selective catalysts, which maximise product yield by decreasing by-product formation. Such discussions interlink core chemistry knowledge with green chemistry principles and form the foundation on which sustainability of the chemical enterprise is progressed. As a consequence of such inclusions in chemistry curricula, a suite of new terms emerges such as 'feedstock' replacing 'reactant' and 'E-factor', which is the ratio of the mass of 'waste' compared to that of 'product'. The latter is a simple empirical measure of the 'green-ness' of a chemical process and hence its sustainability.

Similarly, a discussion of 'renewable energy resources' must be prefaced by a discussion of present primary energy resources, namely fossil fuels, in order to address climate change; arguing that these have to be replaced progressively by clean, green, renewable energy resources, such as solar energy. However, the dilemma of this challenge must also be discussed, namely that all known commercially viable renewable energy resources combined are unlikely to meet global energy demands over the next two decades, and hence fossil fuels are most likely to dominate the global energy market for the foreseeable future. This leads to a discussion of clean coal technology and the contemporary concept of a 'carbon-neutral economy'; both are outside the scope of contemporary chemical education, but are essential inclusions in the quest to extend chemical education to engage with sustainability.

Sustainable chemistry also embraces environmental chemistry, whereby fundamental chemical concepts such as the p-block elements – C, N, O, P and S – are termed 'nutrients' and 'salts' are responsible for 'salinity' of soils and surface waters. Pollutants disturb the natural nutrient cycles and salinity reduces soil and freshwater quality with overall degradation of the natural environment. Similarly, increasing acidity of rivers and oceans disturbs aquatic ecosystems and is a direct consequence of increased levels of carbon dioxide in the atmosphere. Furthermore, increasing toxicity of the environment due to chemical waste in soils, air and surface waters is of greatest concern in terms of addressing environmental sustainability.

Sustainable chemistry intuitively involves engagement with the generation of new smart materials and hence with nanotechnology and its envisaged linkages to global clean energy requirements. The rapidly advancing nano-chemistry is perhaps the most significant exemplar of leading edge sustainable chemistry with its focus on the development of new smart materials for energy storage, production and conversion, for advancing agricultural productivity, water purification and desalination food processing, building construction, health monitoring and for pest control. Of these applications, rapid advancement in the production of photo-voltaic devices and carbon nano-tube solar cells is accelerating the solar energy industry. Similarly, the development of nano-catalysts for hydrogen production, coupled with carbon nano-tube hydrogen storage systems are promoting hydrogen as a viable, alternative clean energy resource. Thus, sustainable chemistry *via* nano-chemistry directly engages with environmental sustainability by providing processes and products which directly benefit humanity without harming the environment.

However, all of these dimensions of sustainable chemistry present formidable challenges for chemical education, both in terms of future direction and scope. It is clear that 'sustainable chemistry' cannot be considered as a single academic course, but requires the concept and philosophy of sustainability to be progressively introduced into all chemistry courses, both at the secondary and post-secondary/tertiary levels. Furthermore, the complexity of sustainable chemistry and the diversity attached to its implementation demand flexible teaching methodologies, such as Problem Based Learning supported with multimedia anchors, leading to carefully designed learning outcomes (research into which is, at best, embryonic).

In conclusion, since 'sustainability' and 'sustainable development' are complex, multi-dimensional' concepts, sustainable chemistry is also multi-dimensional in character, embracing disciplines not normally aligned with it such as economics, accounting, humanities, sociology, cultural studies, health sciences, food science and agricultural science. Hence, successful engagement of chemical education with sustainability involves developing partnerships with these disciplines to form a united educational platform for moving towards environmental sustainability. Fundamentally, sustainable chemistry education is a powerful philosophy integrating 'chemistry' into the 'sustainable future' syndrome and offers challenging educational opportunities to achieve identifiable sustainable outcomes.

SUSTAINABILITY THROUGH CHEMISTRY

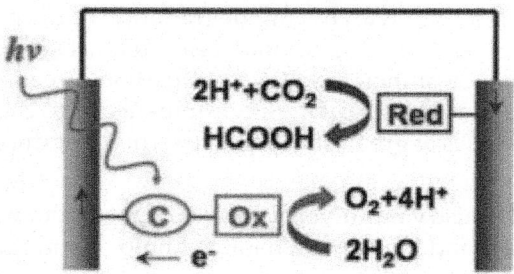

An urgent challenge faced by the world is to find viable solutions to meet our energy needs while maintaining the quality of our environment. The discovery of new materials with improved properties will play a pivotal role in achieving a sustainable future. In Professor Gonghu Li's group at the University of New Hampshire, students are interested in utilizing the principles of catalysis and nanoscience to develop sustainable, functional nanocomposite materials for energy and environmental applications.

Surface molecular catalysis deals with the post-synthetic derivatization of solid materials with well-defined molecular catalysts. Such "supramolecular chemistry" represents a unique approach that brings together the robustness of solid surfaces and the molecular understanding of catalysis. Crystalline aluminosilicates such as zeolites are excellent supports for molecular catalysts. Professor Li's current research involves the synthesis of nanozeolites with particle sizes less than 100 nm, followed by surface functionalization of nanozeolites with molecular catalysts. The functionalized materials will be characterized with a variety of techniques including X -ray diffraction, microscopy (SEM and TEM), UV-visible, FT-IR, and EPR spectroscopy.

One potential application for Professor Li's functional nanocomposite materials is in the field of artificial photosynthesis. Solar energy remains the largest unexploited renewable energy resource. In his research, transition metal (Ru, Re, Ni, Co, *etc.*) complexes will be synthesized and supported on nanozeolites. The surface molecular catalysts will be applied to solar fuel production by reducing CO2 into CH4 and CH3OH as energy-rich fuels. Other applications of functional materials include dye-sensitized solar cells and photochromism.

Surface molecular catalysis deals with the post-synthetic derivatization of solid materials with well-defined molecular catalysts. Such "supramolecular chemistry" represents a unique approach that brings together the robustness of solid surfaces and the molecular understanding of catalysis. Crystalline alumino-silicates such as zeolites are excellent supports for molecular catalysts. Professor Li's current research involves the synthesis of nanozeolites with particle sizes less than 100 nm, followed by surface functionalization of nanozeolites with molecular catalysts. The functionalized materials will be characterized with a variety of techniques including X-ray diffraction, microscopy (SEM and TEM), UV-visible, FT-IR, and EPR spectroscopy.

One potential application for Professor Li's functional nanocomposite materials is in the field of artificial photosynthesis. Solar energy remains the largest unexploited renewable energy resource. In his research, transition metal (Ru, Re, Ni, Co, *etc.*) complexes will be synthesized and supported on nanozeolites. The surface molecular catalysts will be applied to solar fuel production by reducing CO_2 into CH_4 and CH_3OH as energy-rich fuels. Other applications of functional materials include dye-sensitized solar cells and photochromism.

THE APPLICATION OF TECHNOLOGY ROADMAPPING METHOD RELATED TO SUSTAINABLE ENERGY

In recent years, the growing worldwide concern with environmental issues, coupled with rising oil prices and uncertainties associated with its market, gave rise to the need to find sources of energy cheaper and less harmful to the environment. There is a stimulus for the use of new technologies, in particular, the so-called biofuels to replace fossil fuels.

The world economies have great dependence on oil, the main input of the global energy mix, accounting for about 36%. However, such non-renewable fuel has caused serious damage to the environment, endangering the survival of life on Earth. As an example of damage to nature, there is global warming, this issue on the agenda of all nations, which is caused by the greenhouse effect .

But this threat is not restricted to environmental issues, also affecting the economic plan. Considering that oil is not renewable, each barrel is a barrel processed unless the world reserves. By being concentrated in a few countries, and find themselves increasingly scarce, are checked successive increases in its international price.

The continued use of sources of non-renewable energy such as oil and coal, runs counter to the sustainable development, according to the Brundtland Report, "is one that meets the needs of the present without compromising the ability of future generations to meet their own needs".

In this context, renewable energies are gaining importance in the discussion based on a new paradigm of sustainable development. The use of renewable energy, such as biomass, wind, solar and geothermal for instance, began to be permanently on the agenda of the governments of all countries and policies to promote research to develop new technologies for their production. To guide these policies it can use of technological forecasting tools in order to explore the dynamics of emerging technologies in the industrial sectors. Among examples of such tools it could cite the Technology Roadmap, which is a widely used technique in the industry for the development of long term planning strategies, making it possible to align market, product and technology over time.

Sustainable Energy

Renewable energy, derived from natural cycles of conversion of solar radiation, the primary source of almost all the energy available on Earth, are practically inexhaustible and do not alter the thermal balance of the planet. They are configured as a set of energy sources that can be called non-conventional, ie, those not based on the fossil fuels .

These fossil fuels are classified as non-renewable energy are those sources that will eventually cease, making it very costly or fatal to the environment with irreversible consequences. In contrast, renewable energy sources are offered by the environment with an abundance .

Thus, it can say that renewable sources of energy are clever ways of using the resources of the planet. The main characteristics of renewable energy sources and the major described in detail below :

Table : Main characteritics of renewable energy source.

	Solar	Wind	Geothermal	biomass	Marine
Scale	extremely large	large	too large	too large	too large
Distribuition	worldwide	coastal, mountains, plains	tectonic boundaries	worldwide	coastal, tropics
variation	depends on the time, day and season	highly variable	constant	depends on the time, day and season	depends on tide and season
Improvenments	materials, cost, efficiency, and data source	materials, design, location, source data	exploration, extraction, use	technology, management of agriculture and forestry	technology, materials and cost

Biomass

It is the chemical energy produced by plants in the form of carbohydrates through photosynthesis. Plants, animals and their derivatives are classified as biomass. Wood products and agricultural residues, forest residues, animal dung, charcoal, alcohol, animal oils, vegetable oils, poor gas and biogas are forms of biomass used as fuel .

Renewal in biomass occurs through the carbon cycle. The decomposition or burning of organic matter or its derivatives causes the release of CO_2 into the atmosphere. In photosynthesis, plants convert CO_2 and water into carbohydrates, which make up your living mass, releasing oxygen. Thus, the use of biomass does not alter the average composition of the atmosphere over time provided that not occur in a predatory way.

Solar

It is the energy from the sun. It can be used directly for heating the environment, heating water and for the production of electricity, with the possibility to reduce by 70% the conventional power consumption.

Furthermore, the radiation can be used directly as a source of thermal energy for heating fluid, heating environment and to generate mechanical or electrical power. It can also be converted directly into electrical energy through effects on certain materials, especially for thermoelectric and photovoltaic . It should be noted that other energy sources cited (biomass, wind, ocean energy, *etc.*) are indirect forms of solar energy.

Fuel Cells - Hydrogen

The fuel cell consists of an electrochemical device that directly converts the chemical energy of a fuel into electricity and heat. Typically, the hydrogen is used as fuel. In this case, hydrogen and oxygen are recombined at the surface of a catalyst and producing only water and heat as by-product .

There are numerous advantages of using fuel cells in aviation. One is to reduce emissions by almost one hundred percent. Another benefit is caused by decrease in the intensity of noise in the aircraft itself and around airports.

Moreover, the electricity generated can also be used to power electric pumps in hydraulic systems, and even replace the APU - Auxiliary Power Unit (usually installed in the tail device, capable of providing power to an aircraft on the ground or in flight). The pure water derived from this process can be used in various ways, for example, in sanitary systems or even to use by passengers, resulting in a considerable reduction in weight of the aircraft since it is not necessary supply it with this input .

However, there are some limitations when employing these cells with hydrogen. How is chemically very active, it is not found in gaseous form in nature (H_2). Thus, the procedures for production and purification become also expensive. Besides, large amounts would be required for storage, requiring thus redesign of the aircraft.

Geothermal

It is obtained from the heat of the Earth which the magma, located below the Earth's crust, heats the deposits or streams of water at temperatures above 140°C. The water vapor, to find a crack in the crust, forming the emerging geysers, fumaroles and hot springs .

This energy source has almost no CO_2 emissions, a low operating costs and the area for the installation of the plant is small and can supply isolated communities. However, it is an expensive energy that needs to be put into use in the field and emites hydrogen sulphide, which has an unpleasant odor.

Wind

It is the kinetic energy of air masses (wind) caused by uneven heating in the Earth's surface. This type of energy has proven itself as a great alternative in the composition of the energy matrix of many countries because it is abundant, renewable, clean and available everywhere .

Its use for electricity generation on a commercial scale began in 1992 and, through knowledge of the aviation industry, equipment for wind generation have evolved rapidly in terms of ideas and preliminary concepts for high-technology products.

Marine (Wave/Tide)

Tidal power also has a potential energy to be harvested by humans, mainly in Brazil due to its vast coastline. The tides vary from place to place because they are influenced by the shape of the coastline and the seabed and the existence of bays and estuaries .

The largest power plant built in the world is located in France, La Rance, and operates more than 40 years with an installed capacity of 240 MW divided into 24 turbines. There is also a sizable one in Canada.

Despite such initiatives, these countries have not been investing in new ventures. Technology leads to stagnant tidal energy is not economically and environmentally feasible yet. The disadvantages are that energy is not available all the time, only during periods of the cycle. Salt also has a high power corrosive, requiring the use of special materials, raising the price of equipment. Although not generate greenhouse gases, their impact is on marine ecosystems, fauna and flora .

In general, the main barriers to greater use of renewables are economical, as the technologies employed are new, many are still under development, and therefore have very high implementation cost. This barrier can be overcome by government support being necessary technology investments in order to achieve economies of scale and become more competitive . Thus, in order to organize and direct studies of this nature it is used technological forecasting tools.

Method TRM

The technology roadmaps are part of the tools that aim to explore the dynamics of emerging technologies in industry, in a long-term horizon and, especially, develop, implement and execute strategy maps to align the company's strategy to its technological capabilities .

The technology roadmap is a tool that provides support to an organization charged with developing a product or process, providing the method to turn your strategy to future actions and explicitly include a plan for the infrastructure, skills and technologies needed available at the right time.

Usually, the term roadmap refers to a layout of paths or routes exist or may exist in a particular geographic area to help travelers in planning the trip in order to reach a particular destination . This definition helps to understand the Technology Roadmapping Method, which consists of plotting the roadmap for the evolution of technologies, products and existing markets (today) and will be built (the future), helping leaders (travelers) in an organization alignment of planning and development activities with business goals (target) . The definition of TRM adopted is a flexible method whose main objective is to assist in strategic planning for market development, product and technology in an integrated manner over time, while the term technology roadmap is the document generated by the process.

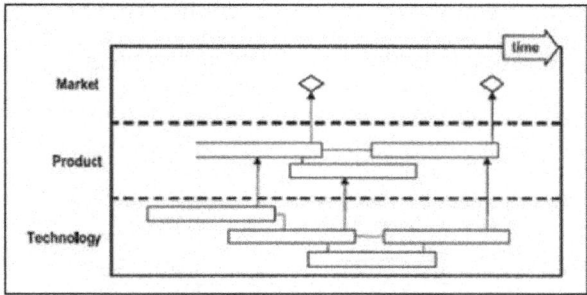

Fig. : Schematic technology roadmap showing how technology can be aligned to the development of products and services, business strategies and market opportunities

Studies of Kappel and Garcia & Bray suggest that the roadmap can be represented in two levels: industrial or corporate. Some organizations do technology roadmapping internally as one aspect of your technology planning (corporate technology roadmapping). However, the industrial level, the technology roadmapping involves multiple organizations, whether individually or in consortium (industrial technology roadmapping).

Organizations that use the technology roadmapping as a strategic planning process to benefit from the following possibilities :

(i)   to produce greater alignment between research and development (R & D) initiatives in product development,

(ii) clarify the views strategic, resulting in better informed decision making,

(iii) Manage data, product plans and objectives at a high level,

(iv) To interact markets, products, technologies, customers and suppliers,

(v)   To enable the discovery of re-use technology and opportunities for synergy,

(vi) Reveal gaps, challenges and uncertainties related to product, technology and training plans,

(vii) Reveal weaknesses of long-term strategy before they become critical,

(viii) Communicate and provide visibility toward strategic program around the organization,

(ix)   Enable growth of product portfolio in line with business demands and market, and

(x)   Provide guidance to project groups of people and enable them to quickly see changes in events or strategic directions.

ROADMAPS ON SUSTAINABLE ENERGY

World

Based on studies Phaal , referring to 1,300 roadmaps published online in English and covering the most varied fields of science, technology and industry, Loureiro conducted a count of roadmaps in each of these large areas.

Table : Roadmaps public domain selected by Phaal : number and percentage of large area.

Area	Number of Roadmaps	%
Software, computing, information and communications technology	385	21.9
Energy	242	13.8
Sustainable energy systems	94	
Hydrogen & fuel cells	36	
Electricity	27	
Fossil fuels	25	
Nuclear	23	
Others	37	
Science	242	13.8
Policy, government and community	233	13.2
Industrial, business and other organisational	196	11.1
Transport	103	5.9
Eletronics	94	5.3
Materials	62	3.5
Defense	61	3.5
Manufacturing	51	2.9
Construction	45	2.6
Nanotechnology	23	1.3
Chemistry	22	1.3
TOTAL	1759	100%

Phaal updated its research roadmaps identifying more than 2,000 public domain. From this list it was made new counts, as shown in Table. This table reveals that the area has the largest number of roadmaps isSoftware, Computer and Information Technology and Communication with about 21.9%, in the second place come the areas of Energy and Science with 13.8%. Comparing with the results of Phaal , consolidated by Loureiro , it can note that the area of Information Technology and Communication remained in first place with the largest number of roadmaps, but there was an increased use of TRM in Energy and Science, areas considered strategic for nations and companies around the world.

From the pointed count was a survey of the characteristics of 94 roadmaps classified by Phaal and sustainable energy systems, which were extracted the following information:

The most of the roadmaps were performed by one or more organizations of the United States being the country that applies the tools of TRM in studies of technological forecasting. Second, there are the countries of the European Union (EU), noting that in some cases for the joint participation of the Organization for Economic Cooperation and Development (OECD).

Various types of organizations perform or commission studies roadmap, which features businesses, universities, research centers, institutes, foundations, industry associations, government departments, ministries. Among the organizations emphasize the participation of the United State Department of Energy (USDOE) and the International Energy Agency (IEA).

It appears that, despite the roadmaps exist from the 70's, have the documents reviewed publication date from 1999, therefore, the TRM can be considered a tool that has relatively recent use.

The vast majority of these roadmaps adopts the format of text, but some contain, in addition to text, representations in the form of bars, single and multiple layers, tables, flowcharts and/ or graphics.

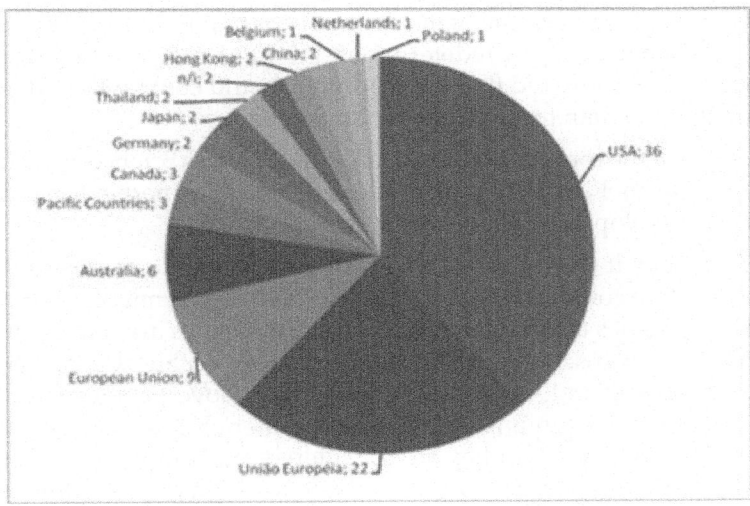

Fig. : Number of roadmaps related to sustainable energy.

About 32% of the roadmaps identified make predictions for the year 2020, using a long-term period range from 15 to 20 years. However, documents show 15 roadmaps by 2050, which highlights the importance given to global sustainable energy in the energy analysis of countries, institutions, companies and universities.

Table : Roadmaps for the types of sustainable energy systems.

Type of Energy	Number of roadmaps
solar photovoltaic	30
mix of renewable energies	27
biomass (biogas,biofuels)	19
wind	5
marine (wave, tidal)	4
fuel cells	1
geothermal energy	1

It is observed that the roadmaps addressing various types of sustainable energy, as shown in Table, with emphasis on solar energy, with emphasis on photovoltaics; mix of renewable energy (including solar, tidal, wave, hydro, geothermal, biomass, wind, fuel cells, hydrogen, *etc.*) and biomass with 30, 27 and 19 roadmaps respectively.

Some roadmaps found not only make predictions, but also define the technological barriers that need to be overcome. There are, in some cases, proposals for actions to encourage and stimulate the sector study, identifying those responsible for such actions and establishing a time frame for making them. It should be noted also that certain roadmaps have also proposed government policies for the sector object of study in order to support the actions of industry, serving as a roadmap to guide the sustainable development mechanism.

Some documents take the name does not contain forecasts roadmaps structured desired goals over a predetermined time horizon, contrary Kappel , the researcher who emphasized that the roadmap should contain the explicit revelation of the time domain for each element presented.

The most of the roadmaps found can be classified as Science and Technology, according to Kappel , as they seek to identify trends, generate forecasts and set goals for the development of the sector studied.

Additionally, it was observed that in other major areas classified Phaal there are roadmaps that could also be related to the field of sustainable energy systems, for addressing issues that have interface. Some examples are found in other classifications of roadmaps: Energy (Electricity, Hydrogen & Fuel Cells, Other Energy, Science (Life Sciences and Agriculture, Land) Policy, Government and Community, Industry, Business and Organizational others; materials, Construction, Nanotechnology, and Chemistry. So if they are considered, the sample space of roadmaps to be analyzed would increase considerably, requiring a greater effort in research.

Brazil

In studies Phaal , shown in Table, there are no roadmaps developed in Brazil on the issue of sustainable energy systems in English. However, there are some Brazilian institutions such as Centro de Gestão de Estudos Estratégicos (CGEE) and Serviço Nacional de Aprendizagem Industrial do Paraná (SENAI/ PR) among others, are conducting studies in various areas of national performance in order to subsidize the decision-makers by identifying the dynamics of innovation and competitiveness of the domestic industry. To achieve this goal, some make use of the TRM to provide an overview of the sectors of short, medium and long term, yet they often have not received such a designation.

Serviço Nacional de Aprendizagem Industrial do Paraná SENAI/PR

An example of a successful initiative in Brazil is the series of roadmaps designed by SENAI/ PR called "Strategic routes for the future of the industry of the state." This project was created by the System of the Federation of Industries and Companies of Paraná (FIEP) in 2006 to draw maps of the routes to be traveled toward a sustainable industrial future for each of the sectors and areas most promising for the industry of Paraná on the horizon 10 years. The specific project objectives are:

(i) to sketch future visions for each of the selected sectors and areas,

(ii) prepare a schedule of actions converged to focus efforts and investments,

(iii) identify key technologies for the industry of Paraná and

(iv) to prepare maps of the paths and desirable for each of the sectors/ strategic areas.

The Strategic Routes project in its first phase addressed the sectors/ areas: food processing industry, Consumer Products, Forestry and Agricultural Biotechnology, Animal Biotechnology and Microtechnology. It was later given to continuity in a second phase, designed the sectors/ areas: Energy, Pulp and Paper, Metal Mechanic, Plastic, Health, Tourism, and Environment.

All roadmaps of the project were executed according to the same work methodology that consists of four steps:

- Stage 1- Preparatory studies: surveys were prepared of the current situation of each of the sectors/ areas worked and studies on the technology trends that could impact the subjects/ objects of roadmappings sectors.

- Step 2- Organization: the works were designed using the technique "Expert Panel" for each selected theme.

- Step 3- Conduct: We performed the following activities: brainstorming on the current status, visions of the future challenges, identify critical success factors, solutions and actions and agents involved.

- Step 4- Consolidation of Results: this step aims to systematize the end of all materials generated during the process. The roadmaps outlined during the meetings were completed and validated by the participants and consolidated information gave rise to technical reports.

For each view a roadmap was generated containing the proposed actions. Each roadmap is presented in the form of multiple layers, each layer refers to a critical factor identified, and the time horizon was divided into three periods (short, medium and long term). For each of these views have been identified to overcome challenges, critical success factors and actions to be implemented in the short, medium and long term, in order to achieve the desired future state in the industrial sector analyzed.

The following are the roadmaps regarding Forestry and Agricultural Biotechnology, Energy, Pulp and Paper and Environment to show respect to sources of renewable energy that is the subject of this work.

Roadmapping Biotechnology Applied to Agriculture and Forestry Industries

The study on Biotechnology Applied to agriculture and forestry industries Paraná and experience of participants of the Expert Panel supported the initial debate that culminated in the shared perception of the group on the current context in the state sector, key item to enter the stage of elaboration of visions the future.

It was developed and validated a set of four complementary views, shown below, comprising a desired scenario in which the industry of the state of agricultural and forestry sectors is entrepreneurial in Biotechnology and becomes reference in research, development, technology and innovation in the area.

Vision 1: Solution Provider in bioenergy

Vision 2: Reference in genetics and plant breeding

Vision 3: Innovation in plants with nutraceutical properties

Vision 4: Reference for plant biotechnology

Roadmapping Energy 2015

Similarly to the previous roadmap, the expert panel developed and validated a set of five complementary views, set out below, which make up the desired scenario of a strong energy sector to support innovative and sustainable growth of the industry of the state.

Vision 1: Reference Planning in Systemic Energy Affairs

Vision 2: Reference Generation Distributed Renewable Energy

Vision 3: Model for Energy Efficiency Competitiveness

Vision 4: Solution Provider in Energy from Biomass

Vision 5: Energy and Logistics for Sustainable Transport

Environmental Roadmapping - until 2018

Following the same dynamics of previous roadmaps, the expert panel developed and validated a set of four complementary views presented below, which set the scene for a desired industry of the state that respects the environment and seek sustainable development.

Vision 1: Environmental management in the industrial chain internalized

Vision 2: Environmental management in the industrial chain internalized

Vision 3: Excellence in public policies for sustainable development of the industrial chain and society

Vision 4: Reference in education in sustainability

Vision 5: Model of interaction academy-industry-government on behalf of the Environment

Roadmapping Pulp and Paper - Until 2018

From the same methodology as the previous examples, the expert panel developed and validated a set of four complementary views, displayed below, which make up a scenario where the desired industry of the state's pulp and paper industry becomes reference in research, innovation and technology, with the north, sustainable development.

Vision 1: Sustainable Industry Pulp and Paper

Vision 2: Excellence in R & D & I Fiber

Vision 3: Pole of Competitiveness in Packaging

Vision 4: Biorefinery for a Global Market

Centro de Gestão e Estudos Estratégcios (CGEE)

In the publication entitled "Green Chemistry in Brazil: 2010-2030" from CGEE a study was conducted in order to establish a dynamic of innovation and competitiveness for the Brazilian industry based on chemical processes that use renewable raw materials within the context of Green Chemistry.

It was analyzed the key technology platforms related to Green Chemistry, namely:

    (i) biorefineries - Route biochemistry,

   (ii) biorefineries - Thermochemical Route,

  (iii) Alcoholchemistry,

  (iv) Oleochemistry,

   (v) Sucrochemistry,

  (vi) Conversion CO_2,

 (vii) Phytochemistry,

(viii) Bioproducts and biofuels and bio-processes and

  (ix) renewable energy.

To conduct the studies were prospective information about the world scene and a national survey conducted by international databases Web of Science and Derwent Innovations Index on scientific production and patents related to this issue in the period 1998-2009 on various topics associated with. In addition to these surveys were used bibliographic information about the state of the art in order to identify the technological gaps to be overcome.

As a result of the work, maps were prepared technology (technology road-maps) on the above key technology platforms in the world and in Brazil, the range of 2010-2030, divided into the periods 2010-2015, 2016-2025 and 2026-2030. For the subjects of biochemical and thermochemical biorefineries, different themes were allocated over the time interval in the different stages that encompass research on bench, pilot stage, demonstration stage, scale-up, innovation/ deployment, production/ processing and marketing.

Others-Technology Roadmapping for Ethanol

The work done on the premise that if Brazil is to maintain leadership in the production of fuel ethanol will have to do planning in the actions of R & D throughout the production chain sugarcane ethanol, which meets the goal of any process of technology roadmapping.

Thus, the roadmap proposed in the study aims to offer innovative propos-als covering the whole cycle of the production chain - farming, manufacturing, products (sugar, alcohol, energy and others) and external environment through forward analysis. In the study the four components that were considered are the genetic improvement, handling, hydrolysis and thermochemical processes. The approach of these components are explained below:

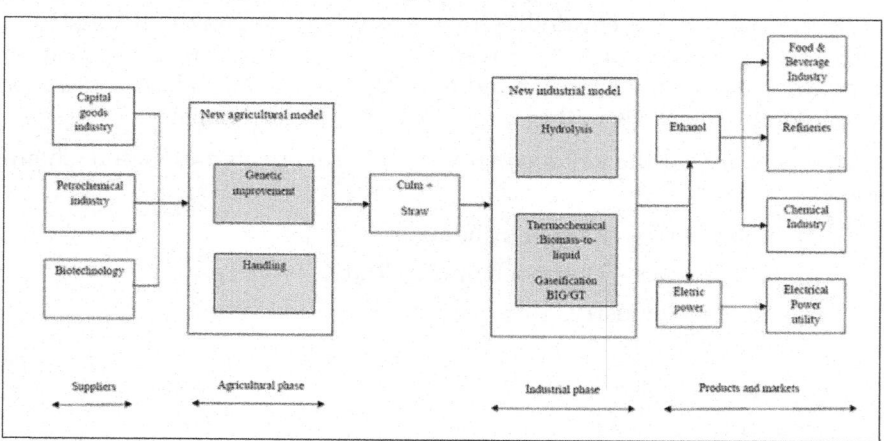

Fig. : The four components of the technology roadmap.

- Breeding: incorporates new techniques of molecular biology and genetic engineering, knowledge areas that are rapidly developing for the genera-tion of cultivars of sugar cane improved technologically specific purpose of ethanol production (sugar-energy) and electricity also incorporating requirements such as high agricultural productivity and resistance and tolerance to pests and diseases and adverse factors (drought, flooding, soil acidity, etc.).

- Management: addresses the till, the recovery of straw, agricultural mechanization and management alternative. It should be noted that the first three issues are closely interlinked through the straw, which represents about a third of the biomass contained in the cane sugar that has historically been discarded, posture, no more accepted in the current scenario.
- Hydrolysis: The process of transformation are of lignocellulosic materials for their conversion on Ethanol.
- Termoconversion: involves issues of gasification and BTL.

In addition, the paper also considers the external environment, analyzing the economic, social, cultural, international, demographic, environmental and political-legal influence the development of the productive chain.

Finally, it should be noted that all developed roadmaps presented as short, medium and long term period of five, ten and twenty years respectively, covering the time interval from 2010 to 2030.

Technology Roadmap on Renewable Raw Materials

Braskem held during the year 2009, a reflection process involving not only its internal departments but the university and research institutions. One result was the development of a technology roadmap on renewable raw materials (rMPR) as proprietary methodology created by the authors. The roadmap does not represent Braskem's strategy but above all intended to be a document of departure for discussion, to be submitted to the various interest groups involved in the future of technology and innovation in Brazil.

The objective was to identify the products and technologies related to MPR that the horizons of 5, 10 and 15 years, could be developed. The identification of products and technologies from the document prepared by the university and was supplemented in internal discussions and interaction with the group itself. The information in this consolidated case was passed to the area of Corporate Innovation Braskem, which then built the first version of rMPR.

Chapter 2

METHODS AND TOOLS OF SUSTAINABLE INDUSTRIAL CHEMISTRY: CATALYSIS

The Earth has finite resources and they're not evenly distributed. As we use up the easy-to-access sources of ores and fossil fuels, some of the key foundations of modern society risk becoming rare and prohibitively expensive. Until we actually perfect fusion and asteroid mining, these are realities driving our push to develop sustainable practices.

At the meeting of the American Association for the Advancement of Science, researchers talked about the progress they're making when attempting to put industrial chemistry on a sustainable path. The overall belief of the panel is that it's not simply enough to make any one part of the process sustainable. Using a cheap and easily available catalyst to drive reactions that require fossil fuels will only buy us so much. It's only when we make every step of the process sustainable — including what happens to the chemicals when we're done with them — that we can really make progress.

The session's organizer, UCSB's Susannah Scott, set out the scale of the problem. Industrial chemistry needs account for something like a quarter of US energy use. The metals Ru, Te, Pd, Rh, Au, Pt, Re, Os, and Ir are all fantastic catalysts, but they are the least common elements in the crust relative to silicon. Princeton's Paul Chirik added a few more details: the current US lifestyle requires something like 80 different elements (GE, for example, uses 72 of the first 82 in the periodic chart). Right now, 19 of them come exclusively from different countries.

While supply is a challenge, disposal is a problem as well. Stephen Miller from the University of Florida said that, globally, we produce more than 200 billion kilograms of plastics each year. Very little of that gets recycled or will degrade naturally and this leaves us with a major disposal problem. The primary source of the raw materials used in these plastics is fossil fuels, and those are subject to the same sorts of price chaos that other energy uses have faced.

Cheap Catalysts

Most of these reactions, whether sustainable or not, require catalysts that push the reaction forward, allowing them to proceed under milder conditions or ensuring that they produce the desired product. And, right now, catalysts typically mean rare earth metals, like the ones Scott listed in the introduction. Chirik's working on getting rid of them.

The advantage of these catalysts, he said, is two electrons that they can donate to the reactants, which are key for the interesting organic chemistry. The downside? Metals like platinum cost morethan $10,000 per mole, and some inevitably gets lost in each reaction. But living systems catalyze interesting reactions using metals like iron, which only has one electron to donate — but only costs about $4 a mole. The trick, Chirik said, is that organisms complex the metals in an organic molecule that donates a second electron to iron (typically from a nitrogen atom), making it behave like it's amore expensive metal.

Shifting to iron-based catalysts (or those made with other cheap and abundant metals) has the potential to make industrial processes far more sustainable.

Chirik gave some specific examples of the reactions he'd like to fix. One of them involves breaking a carbon-carbon double bond to form a link to silicon, a reaction catalyzed by platinum. Each year, $350 million worth of platinum gets lost in the resulting polymer, which is used for things like the no-stick surface that you peel stamps and adhesive labels off. Chirik estimates that 30 to 40 percent of the cost of these labels comes from the platinum left behind in the backing. But his group has developed an iron-organic catalyst that is more active than platinum and doesn't produce unwanted side-products; it can also catalyze a wider range of reactions than platinum.

INTRODUCTION: SUSTAINABILITY AND CHEMICAL ENGINEERING

As the Third Millennium draws near, sustainability is increasingly becoming a key social, political, scientific and engineering issue. Indeed, there are increasing signs that sustainability will become a major new paradigm influencing the society of tomorrow and the engineering it requires. With their knowledge of chemistry and physics, mass and energy flows, and process technology, chemical engineers are in a pre-eminent position to play a major role in implementing sustainable development. The particular attention is given to the role of chemical engineering and industrial chemistry. This role is wide. Traditionally it concerns the design and operation of chemical process plants. Nowadays it also concerns ethical and rational public policy involving science and technology.

The sustainable development, that can very simply be defined as a process in which one tries not to take more from nature than nature can replenish, can be obtained without sacrificing the many benefits that modern technology has brought.

To reach the objective of sustainable development reconciling dynamic industrial growth with low environmental impact, highly advanced and sophisticated technologies must be made accessible to developing countries. This not only implies the transfer of advanced and new technologies, but also indicates the necessity to stimulate innovation and sound research applicable to industrial development. Sustainable development depends on harmonization of economic growth and environmental conservation and protection.

The only problem is that technology respects the imposed constraints. Engineers are asked to do this by designing new processes and/or by modifying existing processes aiming at using renewable resources and producing by products that can be returned to the earth and biologically degraded.

Chemical Engineers have been dealing with areas such as safety, health and environment for years. These areas are actually comprised in the definition of sustainability. By extending the concept of scale, sustainability can easily be related to safety, health and environmental questions, which already receive a great deal of attention in (chemical) engineering and industry. Sustainable development is logically related to traditional well-developed areas in chemical engineering, like those relating to environment, safety and Loss Prevention. Existing techniques, such as used for hazard and risk analysis (*e.g.* QRA) and Environmental Impact Assessment, can be combined with newer techniques, like Life Cycle Analysis, to create powerful new tools for the design of chemical processes and products.

Molecular modeling on the other hand can provide useful and unique information to process engineers developing sustainable processes when thermophysical data are missing or are scanty. In fact, one of the most important parts of any process simulation is the data bank of physical properties or the predictive methods (such as group contribution methods) included in the package. Recently it has been shown how it is possible to combine the molecular simulation techniques with process simulation for an integrated view of modeling in chemistry and engineering. In this way a very powerful set of tools will be available to process engineers for developing safe, clean and sustainable processes.

But sustainability can also be viewed quantitatively. Interesting is to observe quantitative trends in very long historical perspective - sometimes going back thousands of years. Examining the numbers in this way leads to startling results. Interesting is to compare these trends with more recent data, like numerical indicator data concerning social sustainability. There is no doubt that sustainable industrial development requires a balance between the economic and the environmental aspect. Thus the necessity of cleaner technologies is becoming an imperative simply because the efficiency of the processes (especially those using new efficient catalytic systems) leads to the reduction of environmental impact and, at the same time, they should imply economic advantages. This is particularly important for developing countries and countries in transition, where "turn-key" solutions are not feasible due to prohibitive economic requirements. In this context, modern tools like modeling and process simulation are of increasing importance, because they are able to evaluate possible technological solutions with much lower costs and time involved. On the other hand, one must be aware of the limits of this approach because of appropriate nature and models imposed.

Process Simulation Goals and Definitions

A process simulator solves material and energy balances by means of a computer code. A description of a typical process simulation code in which all the most important elements of a simulator and their connections are evidenced. The clearly that a process simulator include cost estimation routines as well as economic evaluation. The importance of the database is shown in the figure as a necessary source of information for different objects in the structure.

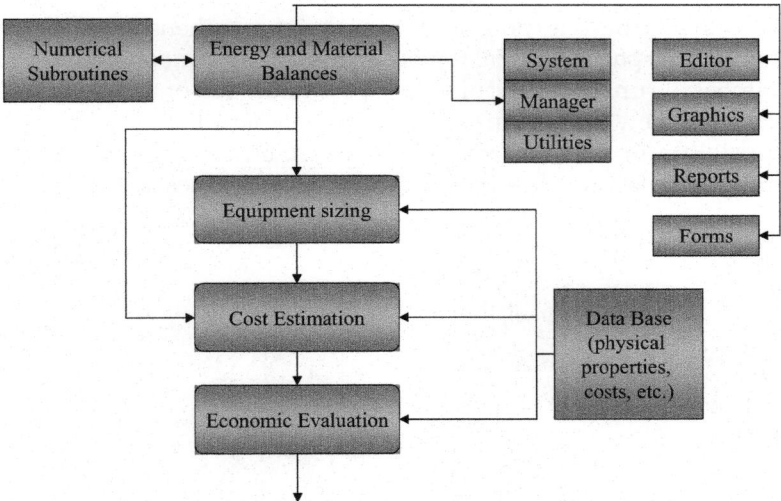

Fig. : General scheme representing a steady state process simulator.

The following different approaches are available in the process simulation:

- Steady state simulation which considers a snapshot in time of the process
- Dynamic simulation which considers the evolution in the time domain of the equations describing the process
- Integrated steady state - dynamic simulation, which combines the two previously, described approaches.

The approaches listed above may be used in three different philosophies when dealing with process simulation. One possibility is to perform a process analysis, in which an existing process is studied and alternative conditions as well as dynamic behavior is investigated for the appraisal of effectiveness of design. The second approach is the process synthesis, in which different process configurations are compared aiming to the identification of the optimal choice of units and the connections between them. The third possible stage in design a process is the process design and simulation, aiming at the establishment of the optimal operating conditions of a given process.

In all the possible philosophy of application process simulation impact on industry is pervasive rather than restricted to a single moment in the development of the process. Process simulation has effected strongly the way engineering knowledge is used in processes. The traditional way of using process simulation was mainly focused on flow sheet design and on equipment critical parameters definition, such as distillation column stages, column diameter, and so on.

Today engineers are oriented to a more comprehensive use of process simulation in the entire 'life' of the plant such as the control strategies design, the process parameters optimization, the time evolution of the process for understanding start up and shut down processes and performing risk analysis, the operator training and the definition of procedure to reduce the unsteady state operations.

The main benefits one can gain from such a comprehensive use of the process simulation are the partial or total replacement of Pilot Plant operations (reduction of the number of runs and runs planning), the reduction of time to market for the development of new processes, the fast screening of process alternatives to select the best solution in terms of economic aspects, environmental aspects, energy consumption aspects and flexibility of the proposed process.

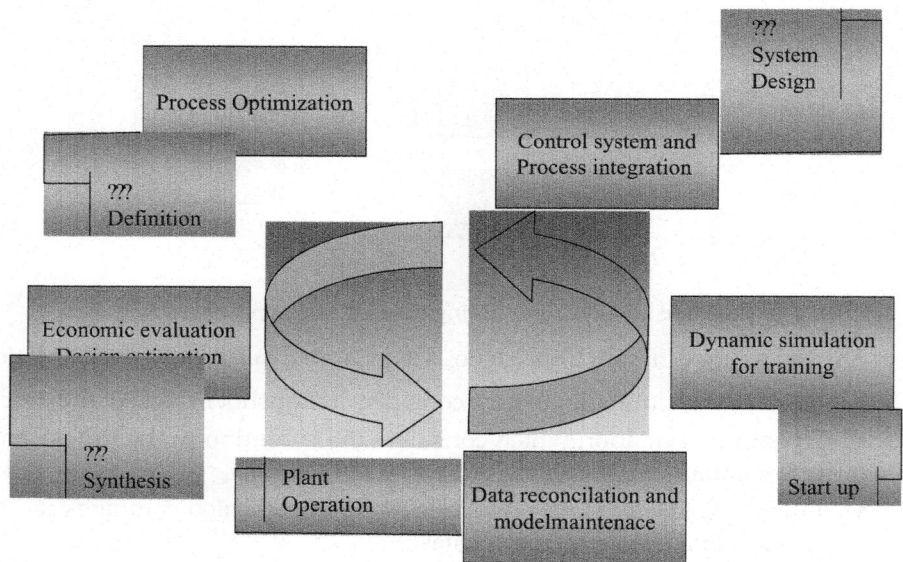

Fig. : Process simulation and the life cycle of a process.

Due to the high complexity of chemical process, to get those benefits one must critically simplify the process and apply process simulation techniques in the entire life cycle of a process. How process simulation methods can help engineers in different periods of the life cycle of a process, from the process synthesis to the control strategies design.

The solutions of the material and energy balance equations can be performed by an equation-oriented strategy, namely the simultaneous solution of all the model equations or the sequential modular approach. In the first case, one must write down the entire set of equation, identify the constraints and solve the non-liner system. In the Sequential Modular approach each subsystem is solved independently, starting from the first one: the output streams for the solved subsystems are input streams for the next subsystem. In the sequential modular approach the main problem is to deal with recycle streams (recycle of material, energy and information).

There is another possible approach, which is a mixture of the two fundamental approaches in which equations can be lumped into modules that can be represented by polynomials that fit input-output information. The main advantages of the sequential modular approach are the following:

- The flow-sheet architecture is easily understood because it closely follow the process
- Individual modules can easily be added and removed
- Modules of different levels of accuracy can be substituted

The drawbacks of this approach are:

- The input of a module is the output of a module: you cannot arbitrarily introduce an output or input
- The modules need extra time to generate derivatives (perturbation of the input)
- The modules may require a fixed procedure for the order of solution: slow convergence
- Parameter specification is done with control loops: possibility of introducing nested loops
- Phase equilibrium instability during the convergence of the process

Steady state simulators are considered the core products of process simulation and it is used for process design, evaluating process changes and analyzing what-if scenarios. All the other kind of simulation are normally performed after a steady state simulation: dynamic simulation, process synthesis with Pinch technology, detailed equipment design, off-line and on-line equation-based optimization, application technologies for vertical markets, *e.g.* polymers.

The problems involved in a process simulation run are the definition of an accurate thermodynamic model (Equations of state or Excess Gibbs Energy model), the necessity of defining dummy operations, non always easy to identify and the tear streams identification to achieve rapid convergence.

As far as the Dynamic Simulation is concerned, applications can be found in continuous processes, concurrent process and control design, evaluation of alternative control strategies, troubleshooting process operability, verification of process safety.

The most important benefits of Dynamic Modeling is the capital avoidance and lower operating costs through better engineering decisions, the throughput, product quality, safety and environmental improvements through improved process understanding, the increased productivity through enhanced integration of engineering work processes.

An important feature of modern process simulators is related to its usability and interoperability with other applications. A user familiar graphical interface and easy and standard inter-process communication techniques are key elements to obtain the Workflow integration in the process design.

To obtain the engineering Workflow Integration it is necessary that the software can exchange information with other application. This is done through the basic Windows interoperability by using the Microsoft COM/ OLE Automation technology.

A two-way data transfer between the software and other Windows applications *via* copy, paste, paste link should be performed, Windows should give access to all inputs & results and give access to plots and flow sheet graphics. It should be possible to copy data tables and spreadsheets into the simulator for Data Regression, Data-Fit, *etc.*

Another important point in the workflow Integration is the support to interfaces to specific 3rd-party engineering applications such as equipment design (B-JAC, HTRI, HTFS), engineering databases (Aspen Zyqad, PASCE), costing packages (ICARUS) and in-house technologies. In summary, the benefits of the workflow integration are the following:

- Support for engineering infrastructures that integrate engineering work processes
- Error-free data transfer into 3rd party Windows engineering programs
- Quick and consistent use of simulation results throughout the engineering lifecycle
- Improved engineering quality
- Work Flow Integration

Recent Trends in Modeling: Bridge between Molecular Modeling and Process Simulation

More than seventy five per cent of the code in process simulators is dedicated to physical properties estimation, calculation and predictions. Data banks storing pure component parameters and binary interaction parameters for phase equilibrium calculation are extensively used and continuously implemented in modern process simulators. This gives an idea of the important role that physical property availability plays in process simulation. This role become of paramount importance when dealing with new products, with new processes and/or with the revamping, in terms of more sustainability, of existing processes. In most cases substances are used or produced whose physical properties are not known and in some cases are not even measurable.

Molecular modeling can be seen as a very useful tool to process engineers, by providing the results of virtual experiments, pure components physical properties in a wide range of process conditions, phase equilibrium and kinetic data as well as an interesting tool for screening new and interesting products and reaction schemes.

There are several possibilities for coupling molecular modeling and process simulation. Among them a very interesting one has been recently proposed for combining molecular modeling and chemical engineering oriented semi-empirical models such as equations of state and excess Gibbs energy models. The general idea here is to use molecular dynamics and Monte Carlo methods for generating the parameters of an equation of state that will subsequently be used in process simulation. In this way one can still perform process simulation with a reasonable hardware (a PC, as well as a workstation) and with a reasonable CPU consumption. In fact, the present computer technology do not allow us to perform a process simulation with integrated molecular modeling tools. The general scheme in which two possibilities are shown:

- Generation of thermo-physical data by means of virtual experiments
- Generation of parameters to be used in semi - empirical model built in process simulators.

Traditionally experimental work has been and is still carried out to provide information to process engineers. The cost of collecting one vapor-liquid equilibria (VLE) data point (*i.e.*, one temperature and composition for just one binary mixture) has been estimated to be around $ 2,600 and to take 2 days. Thus, we can reasonable hope to perform an experimental characterization of VLE for a minute fraction of the total possible mixtures, temperatures and compositions.

With the computing power and technology available nowadays, computer molecular simulations can be considered cheaper and faster than true experiments, especially for simple molecular fluids. Therefore, provided we properly account for the two major problems encountered in any virtual experiment (*i.e.*, the size of the configurational space that is accessible to the molecular system and the accuracy of the molecular model or atomic interaction function or force field that is used to model the molecular system), we can think of computer simulation at least as a first way of screening among the plethora of possible system candidates.

Indeed, during the last decades virtual experiments based on quantum/ molecular (QM/MM) mechanics calculations and molecular dynamics (MD) simulation techniques have opened avenues in the estimation and prevision of thermophysical properties (both under equilibrium and non-equilibrium conditions) of simple molecular fluids. The treatment of molecular systems in the vapor/ gas phase by quantum mechanics is quite simple, due to the possibility of reducing the many-particle problem to a few-particle one based on the low density of a system in the gas phase. If the classical statistical mechanics approximation is permitted, the problem becomes even simpler.

Nevertheless, for both amorphous solid states and liquid systems such as solutions and polymers we remain faced with an essentially many-particle systems, for which no simple reduction to a few degrees of freedom is possible, and a full treatment of many degrees of freedom is necessary to adequately describe the properties of molecular systems in the fluid-like state. In such cases, to obtain reliable estimations of dynamic and non-equilibrium properties dynamic simulation methods that produce trajectories in the phase space are to be used. The method of MD solves Newton's equation of motion for a given molecular system, which results in space trajectories for all atoms in the system. From these atomic trajectories, a plethora of thermophysical properties can then be calculated as time averages from the relevant microscopic relationships expressed in terms of molecular positions and momenta.

The aim of computer simulations of molecular systems is to compute macroscopic behavior from microscopic interactions. The major contributions a microscopic consideration can afford are:

(a) the understanding and

(b) interpretation of experimental results;

(c) semiquantitative estimates of experimental results and, last but not least,

(d) the possibility to interpolate or extrapolate experimental data into regions that are only difficultly realizable, if at all, in the laboratory.

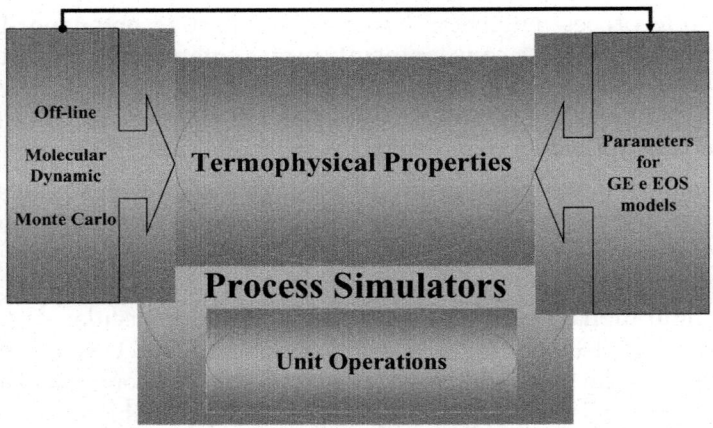

Fig. : Process simulation and molecular modeling.

The general connection scheme and the possible interactions between molecular modeling and process simulation. The fundamental starting point of the scheme is the Quantum chemistry and the quantum mechanics ab initio calculation. Fundamental studies in this field are of paramount importance for defining a theoretical framework within which to develop expressions and parameters for the interaction potential and force fields. Structure and properties of simple materials are used both for developing methods and for the validation of the obtained force fields. Structure and properties may directly be used for defining reaction mechanism and kinetics, which are one of the key input to any process simulation run involving reactions and chemical transformation.

Force fields, on the other hand, has a direct application to molecular dynamics and Monte Carlo calculations which are the key techniques for determining pure component and phase equilibrium properties for simple and complex substances. These calculations may be directly applied to the design of new materials and new drugs, to the calculation of parameters of semi empirical models such as equations of state and excess Gibbs energy models or the model development and model validation. Parameters obtained by this procedure are then stored in data base and used by process simulation coupled with semi empirical models, which are by far computationally simpler than any molecular dynamic or Monte Carlo method and therefore affordable in a process simulation.

The availability of data that are necessary in different parts of the scheme: force fields and model validation, reaction modeling, process simulation *etc*. A service database is provided for the entire process containing data on thermodynamic properties, phase equilibrium, kinetics, costs and profits, structure and geometry. Furthermore another general service to scheme should contain subroutines implementing all the numerical methods needed for parameter estimation and for solving differential and algebraic equation.

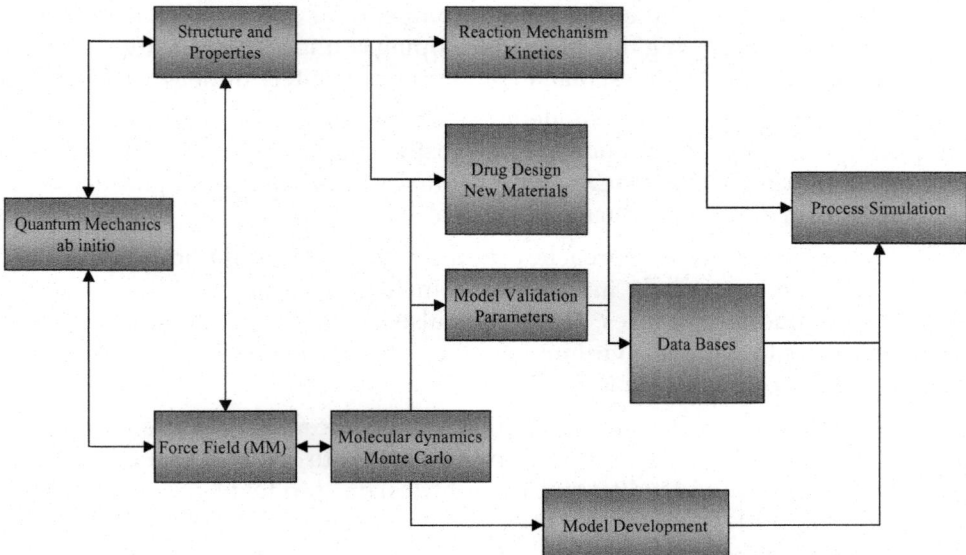

Fig. : General scheme of bridging molecular modeling and process simulation.

Another traditional method to obtain physico-chemical properties used for screening different possibilities in process synthesis is based on the approach belonging to the group of methods called Computer aided molecular design. It provides a means for determining molecules (CAMD) or mixtures of molecules (CAMMD) having a desirable set of physico-chemical properties. As the physico-chemical properties are directly or indirectly related to the structure of the molecule(s), methodologies for CAMD and CAMMD are typically based on "exploiting" these relationships. Currently, CAMD/CAMMD has found useful industrial applications. However, the application range is restricted because of the limitations on the complexity of the generated molecular structures and on the availability of suitable models for property prediction. Recently, molecular-level information has been combined with the current group contribution based methods for opening new horizons of applicability and accuracy of CAMD. At the same time, CAMD approaches, including molecular mechanics, are extensively used for interpretation of molecular structure and properties and design of new molecules with the desired properties (drugs, agrochemicals, new materials *etc.*)

Importance for Developing Countries, ICS-UNIDO Activity

Developing and in transition countries are fully aware of their urgent need to become acquainted with modern technologies, so that local enterprises can remain competitive and economically viable in the coming decades and gain expertise in application practice. Both molecular modeling and process simulation are unanimously considered to be powerful tools for the implementation of a country's capabilities in drug design, agrochemistry, new materials and upgrading of industrial processes. These are essential if a developing country wants to have the possibility to enter the market competition. The above considerations are reinforces considering that many developing countries have a large potential in terms of natural resources, that are at present definitely under-exploited.

Furthermore, molecular modeling and process simulation are approaches requiring low investment, and can thus easily be applied in R&D as well as in industries (even SMEs) in developing countries and working in pharma, agrochemistry and natural product exploitation.

Evaluation of these approaches should indicate their applicability in R&D institutions and especially in industries (including SMEs) in developing and transition-economy countries. This should also facilitate optimization of production processes to reduce environmental impact, which is particularly important for obsolete processes often in use in these countries.

The role and applicability of modeling have recently been evaluated in its overall sense during an Expert Group Meeting (with the focus on Central and Eastern European and Mediterranean countries) on "Modeling in Chemistry and Chemical Industries", which was held at and organized by ICS-UNIDO (International Centre for Science and High Technology of the United Nations Industrial Development Organization), in Trieste, Italy, on 14-16 October 1998. The subjects dealt with ranged from modeling of molecular structures and properties, modeling of molecular processes and complex molecular systems to process simulation at a macroscopic level complemented by preventive risk analysis of industrial pollution. It was concluded that although several highly qualified groups exist in these regions, the industrial applications are rare, probably because of the transforming economies characterizing these countries. The plan of networking Central and Eastern European and Mediterranean countries with the involvement of the Italian Institutions has been agreed upon. The networking activity will be coordinated by ICS-UNIDO in Trieste, Italy.

CATALYSIS AS ENABLING FACTOR OF SUSTAINABLE CHEMICAL PRODUCTION

The aim is to increase efficiency with a reduced number of unit operations *via* process integration. The concept goes beyond state-of-the-art microreactors that employ microchannels to confine chemical reactions and enhance speed, yield and safety. It exploits nano-size channels and an ordered sequence of catalytic sites along the axial direction of those channels in a membrane providing a vectorial pathway for multi-site catalytic reactions. The concept applies to reactions, where cascade processes are not possible or not effective.

The use of nano-designed catalytic membrane for transient generation of risky intermediates will go beyond the on-site/on-demand production concept for safer operations. Toxic reactants produced as a result of transformations are immediately converted into harmless entities to completely eliminate storage, which is minimized but not eliminated in on-site/on-demand microreactor production concepts.

The project Consortium to achieve these challenging objectives is formed from 11 beneficiaries, 6 of which are companies, 4 research & education Institutions and one a non-profit research association.

The project final expected impacts are to develop:

i.   new approaches in process intensification through a novel concept of multiphase nanoreactor design,

ii. new approaches in multifunctional catalyst design by integrating catalyst and membrane functionalities in an approach aimed at process intensification,

iii. new approaches for intrinsically safer design for reactions involving risky reagents.

The final result of the project was to verify the applicability and scalability of new concepts in catalysis related to the development of novel nanoreactors and related catalysts for the two listed target reactions and how they can improve (in these industrially-relevant multistage reactions) process intensification, sustainability (in terms of resource and energy efficiency) and safety of operations. Functional to this general objective was the development of catalytic nanomembranes, and of the associated novel reactor concepts and engineering. Due to challenging objective of developing novel nanoreactor concepts, the demonstration activities in the project were limited to the proof-of-the-concepts.

In the first two years the focus was on the lab-scale experimentation to develop and tests the reactor concepts, while in the 2nd part of the project the activities were centered on one side on the specific investigation of the performances of the nanoreactors in the two lines of activity and from the other side on the investigation of the issue of scaling the catalyst/membrane and nanoreactors.

Project Context and Objectives

Promoting a sustainable chemistry and more eco-efficient chemical syntheses requires new efficient tools and working examples, which catalyze the transformation of the energy- and capital-intensive industrial chemical production. A key factor to foster this change is to develop integrated eco- and energy-efficient syntheses, which imply the capacity to innovate in the field of process intensification, a key for industrial competitiveness, improved use of resources, reduction of the amount of waste and emissions, as well as better safety.

The project context is to address these challenging objectives by integrating nanoreactor, membrane and advanced catalytic concepts. The key concept is the use of nano-designed catalytic membrane for transient generation of risky intermediates, going beyond the on-site/on-demand production concept for safer operations.

The other key concept is to have a step-forward in the integration of microreactor technology with catalysis, two of the crucial pillars to foster a sustainable industrial chemistry. They are key components of process intensification to transform from energy- and capital-intensives to safer and eco-friendly processes, with positive impact also on the competitiveness of chemical industry.

A solution is a modular process design, based on the concept of parallel reaction units, as opposite to scale-economy approach used currently in the chemical industry. Between the advantages:

   (i) *Safety:* less storage and/or transportation of hazardous materials, reducing the likelihood of accidents, and any accidents that do occur will be smaller and less catastrophic. Possibility of realize a delocalized and on-time production.

  (ii) *Cost and time in introducing new processes:* capital costs, transportation costs, and inventory costs are all lower. Easier introduction of new processes on smaller scale; productivity can be increased later by adding new units. Faster time from discovery to industrial production. Easier customization.

 (iii) *Cycle avoidance:* existing large plants consistently suffer from over- and under-capacity scenarios, leading to typical boom-bust cycles.

This new approach requires a next step in an integrated microreactor and catalyst design. In fact, a key aspect for success, and a reduction in energy and capital intensity, is to realize in a single unit complex multistage processes. This is a key for safer and environmental-friendly operations, because there is a reduction of the stages with consequent decrease in energy consumption and waste generation, and the elimination of the storage of possible dangerous intermediates. The approach typically used in microreactor technology is the miniaturization of the reactor and other unit operations, to make compact devices where the reduction of mass and heat transfer leads to process intensification. However, the real potential of using this technology to realize in a single unit complex multistage reactions has been scarcely investigated, because requires a further step in an integrated microreactor and catalyst design, the core aspect of this proposal.

Usually, microreactors have channels in the micrometer diameter range (around 100-150 micron), while the more effective integration between micro-reactor and catalysis require developing reactors with channels of nanometric size. They are interesting also in general terms, because would further increase of one-two order of magnitude the possible process intensification. In terms of safety, the passage from micro- to nano-reactor would bring several benefits. It is known that the higher wall-to-volume ratio in microreactors with respect to conventional reactors, allows to effectively quenching radical-type reactions and runaway effects. In nanoreactors, the further increase of one-two order of magni-tude in the wall-to-volume ratio will further increase safety of operations, allowing, for example, operating without risks inside the explosion region, or with highly exothermic reactions. The integration between nanoreactor and catalysis is thus a key factor towards scientific and technological breakthrough in synthesis for a sustainable chemical production.

The use of catalytic membrane as nano-reactor offers a further possibility to implement novel concepts for safer industrial syntheses. Phosgene is central to the chemistry of pharmaceuticals, polyurethanes, and polycarbonates; a huge market sector generating 8 million tons of products. Phosgene, however, is a highly reac-tive chemical, and very toxic (TLV = 0.1 ppm), and its manufacture poses serious problems. For this reason, from about two decade there is a large effort in find-ing safer substitutes such as DMC. Despite these concerns, 5–6 million tons y-1 of phosgene are still produced and used worldwide, while a greener substitute such as DMC has a market about 20-25 times lower, due to techno-economic rea-sons. From the industrial perspective, the effort has been thus mainly directed towards a safer use of phosgene (on-site or on-demand production) more than to its substitution. The use of catalytic membrane allows to further progress in this direction, implementing the concept of "dynamic nanoreactors" where phosgene is produced and immediately converted in the phosgenation reaction. We indicate this concept as transient phosgene generation. The storage of phosgene is elimi-nated, and the amount of toxic chemicals becomes negligible. Reduced dimensions allow also confinement of the reactor and supply system, further minimizing the risks and possible leakages.

The project vision for safer and sustainable industrial chemistry is thus cen-tered on these aspects:

(i) to realize efficient modular-design of industrial chemical synthesis, also for large-scale processes, as a decisive factor to accelerate the introduc-tion of novel processes, and enhance competiveness;

(ii) to progress with respect to actual microreactors as necessary step to move in this direction and realized by integrating nanoreactor, mem-brane and advanced catalytic concepts;

(iii) to utilize this novel design as a key element to improve safety, as more effective (time, cost) to accelerate the transition to a safer and sustainable industrial chemistry with respect to redesign the process using alternative reactants requiring large investments costs.

Project Main Objectives

The project aims is to explore, with reference to the two keys industrial processes listed below :

- line 1: the direct synthesis of H_2O_2 and its in-situ use in propene oxide synthesis;
- line 2: the safer synthesis of dyphenilcarbonate DPC *via* in situ synthesis of $COCl_2$ the possibility to achieve more efficient processes with a reduced number of unit operations by realizing process integration through the use of novel nanoreactors concepts.

The aim is to increase efficiency with a reduced number of unit operations *via* process integration. The concept goes beyond state-of-the-art microreactors that employ microchannels to confine chemical reactions and enhance speed, yield and safety. It exploits nano-size channels and an ordered sequence of catalytic sites along the axial direction of those channels in a membrane providing a vectorial pathway for multi-site catalytic reactions. The concept applies to reactions, where cascade processes are not possible or not effective.

The general objective of INCAS project is to provide new tools for process intensification and enhanced selectivity by catalysis. In addition, the methodologies facilitate safer and more energy-efficient production routes. Higher-yield, cleaner and more resource-efficient synthesis of large volumes of chemicals will be a benefit not only to the process routes highlighted but also to applications including fine chemical production and environmental clean-up or remediation. Finally, the project will provide a much-needed framework to analyze the sustainability of various manufacturing processes.

The use of nano-designed catalytic membrane for transient generation of risky intermediates will go beyond the on-site/on-demand production concept for safer operations. Toxic reactants produced as a result of transformations are immediately converted into harmless entities to completely eliminate storage, which is minimized but not eliminated in on-site/on-demand microreactor production concepts.

The project has highly challenging objectives, because no proof of the concept of the possibility to implement these concepts was available at project start. It also has many materials challenges, for example in realizing (nano)membranes, their scale-up, in development (nano)carbon materials for phosgene synthesis, *etc.* Finally, many technical challenges were also present: how scale-up reactors, design of (nano)-reactors, catalysis in membrane interface. As a consequence of this high risk/high gain project, the project objectives evolved during the project, although maintaining the focus on the general project aim.

Project Results

The project structure is organized in four main scientific WPs, plus one WP dedicated to dissemination and exploitation of the results and one WP project management and coordination. The main objectives for the core scientific WPs were the following:

- WP1: prepare reliable catalytic nanomembranes to be used in the nano-reactors for the H_2O_2 direct synthesis, PO synthesis with integrated H_2O_2 generation and DPC synthesis via in situ synthesis of $COCl_2$.

- WP2: assemble, test and optimize the nanoreactors for the H_2O_2 direct synthesis, PO synthesis with integrated H_2O_2 generation and DPC synthesis with integrated $COCl_2$ generation.

- WP3: use the catalyst/membrane and nanoreactors developed in WP1 and WP2, and tests in the reactions of H_2O_2 direct synthesis, and PO synthesis with integrated H_2O_2 generation.

- WP4: use the catalyst/membrane and nanoreactors developed in WP1 and WP2, and tests in the DPC synthesis with in-situ transient generation of phosgene.

Some of the key results obtained in these WPs are the following:

In WP1:

i)   produced, characterised, modified and assessed AAO (anodic alumina oxide) membranes,

ii) synthesised and deposited Pd nanoparticles for catalyst substrates,

iii) synthesised carbon-based membranes and zeolite coated systems,

iv) engineered testing equipment facilities,

v)   modelled substrate and catalyst system configurations,

vi) reviewed, identified and progressed alternatives *e.g.* tubular ceramic

In WP2:

i)   developed flat-type nanoreactor for line 1 as well as tubular-type membrane (nano)reactor,

ii) fabricated mock-up prototype,

iii) developed hollow-fiber modules for scaling-up reactor in line 1,

iv) developed membrane reactors to work in gas-liquid and gas-gas phases,

v)   simulations of the membrane model for the phosgene synthesis + phosgenation of phenol to diphenyl carbonate (DPC),

vi) design of the prototype reactor for line 2

vii) fabrication of the reactor-prototype for line 2

In WP3:

i)   tested different reactor configurations and combinations of multilayer membranes,

ii)   analysis of the different feeding possibilities to operate in safe conditions,

iii) scale-up and testing of hollow-fiber configuration,

iv) patentability analysis.

In WP4:

i)   understanding the mechanism of COCl2 formation over carbon catalyst, particularly what are the active sites on the carbon catalyst surface and what is the nature of the transient intermediates,

ii)   understand the mechanism of deactivation, expecially what triggers deactivation of carbon catalysts and identification of the measures to avoid deactivation/ regeneration,

iii) identification of stable catalysts and of alternative, non carbon catalysts,

iv) design and performing a lab-scale demonstrator, identification of the concept for scaled reactor and application of insight gained for current $COCl_2$ manufacture.

In terms of integrated reactor for in situ $COCl_2$ synthesis and immediate conversion (to DPC), different options were analyzed to identify the preferable configuration.Patents for novel recator configuration were filed.

WP5 was related to project dissemination and exploitation. Several publications and presentations at conferences resulted from the project, 3 patents, and a list of exploitable items. Between the activities made:

i)   maintenance of the website for public and inside consortium activities,

ii)   presentation of the project's results and activities during the project's meetings,

iii) drafting and maintaining the dissemination and exploitation plan following the EC's requirements,

iv) establishing the actions for dissemination and exploitation,

v)   organizing and managing the activities of the Exploitation Committee for protecting the project results,

vi) continuous monitoring of the progress of the research within the project to verify which results:

(a) should be protected by patents and

(b) are relevant to exploitation in fields outside the specific scope of the project, take all the actions necessary for successful exploitation of the results of the project.

WP6 was related to project management and coordination. Between the activities made:

i)   realization and implementation of a coherent and efficient project management and co-ordination of the project in accordance with the budget and the schedule of milestones and deliverables;

ii)   design and implement the management tools necessary for an efficient and timely project management.

In addition to preparing the various contract report and meeting, as well as maintenance of the communication and exchange with the European Commission, it was implemented a continuous monitoring the status of the project's deliverables and milestones, the handling of legal issues, IPR issues and maintenance of the consortium agreement, the management of knowledge generated by the project and the handling of the project correspondence and the day-to-day requests from partners and external bodies.

Selected Main S&T Results

Synthesis, characterization, modification and assessment of membranes

The first two years of the project were focused on catalyst/membrane preparation for lab-scale tests while later the effort was directed towards scaling of the catalyst/membrane. Ultimately such membranes were evaluated for their performance as nanoreactors in H_2O_2 direct synthesis, PO synthesis with integrated H_2O_2 generation and DPC synthesis with integrated $COCl_2$ generation.

During the first 24 months of the project, more than three basic membrane types have been produced: several with straight-through vertical porous channels and several with mesoporous structures. Two types of membrane with vertically-aligned straight channels have been generated by anodizing either titanium or aluminum: these are generally referred to as anodic titanium oxide (ATO) or anodic aluminum oxide (AAO) membranes. The mesoporous membranes (*i.e.* no straight-through porous channels) have been prepared by either polymerising siloxanes, TS-1 or with carbon powder.

Initially 60 x 40mm AAO membranes with a capacity to produce 2-4 membranes of this size daily, due largely to the anodizing protocol employed requiring a 2-hour dwell time and constant current and voltage conditions, were prepared. The electrolyte used in the production of AAO membrane types is a mixture of phosphoric and oxalic acids and a carbon counter-electrode. Samples of these types of AAO membranes have been supplied to the relevant partners for their further characterization, modification and assessment.

In parallel, planar ATO membranes (by anodising titanium foil) were also prepared (size about 3 cm diameter). However, this type of membranes lacked a sufficient degree of mechanical or dimensional stability to be employed effectively with the project as a catalyst support material in a reactor assembly. In addition, these ATO membranes adversely affected hydrogen peroxide decomposition.

Novel synthetic routes for both carbon-coating membranes and producing carbon monolithic membranes were also investigated, because the tested commercially available C membrane materials lack of the necessary characteristics. One of the preferred route to prepare these C membranes involves starting from powder type nanocarbons, process in solvents and then cast onto glass plates at room temperature prior to vacuum oven treatment.

POSS (Polyhedral Oligo SilSesquioxane) membrane materials, that are both completely and incompletely condensed variants through different synthesis routes, were also synthetized as well as TS-1 membranes were synthetized and assessed.

Synthesis and Deposition Pd Nanoparticles for Catalyst

Nanoparticle synthesis. In order to identify suitable synthetic methods for the production of palladium nanoparticles started from a review of the open literature. By using H_2PdCl_4 and PVP, palladium nanoparticles with a narrow particle size distribution and, in the size range desired <5nm diameter, were obtained. A consistent a reproducible synthesis protocol for palladium nanoparticles has been achieved with initial problems identified and resolved. This procedure has been used in all subsequent experiments with palladium nanoparticles

The deposition of palladium nanoparticles on membrane was initially attempted by dropping the palladium nanoparticles solution on the membrane and allowing the solvent to evaporate, but this method proved inconclusive in terms of the deposition of the particles as the amount was too low. After evaluation of different variations on the method, and the investigation of alternative methods such as by homogeneous deposition precipitation (HDP; a solvated metal precursor is deposited on the support surface by the slow and homogeneous introduction of a precipitating agent, generally hydroxyl ions), a method by ligand exchange was adopted. A modification of this method regarded the ligand functionalization of AAO membrane. A specific novel apparatus was introduced to avoid the generation of gradients in nanoparticles. The methodology allowed to produce a consistent and reproducible protocol for palladium nanoparticle deposition on the AAO membranes to prepare samples with uniform and concentrated deposition of the nanoparticles throughout the membrane. The novel catalytic membranes showed good stability.

Synthesis of Carbon-based Membranes and Zeolite Coated Systems

Carbon coated AAO alumina membranes were investigated. However, a major problem derived from the need to avoid from one side to completely block, or significantly alter the AAO nanomembrane porosity, and from the other side to avoid rupture of the membrane particularly during the annealing procedure. Two types of preparations (differing in the type of carbon-precursor) were selected by using two different precursors for the carbon: glucose and triton X-100. After pyrolysis at 250°C, the best results were obtained with the Triton X-100, where during decomposition the Al_2O_3 membrane doesn't stress and the overall appearance was smooth and homogeneous. Permeability tests, as well as characterization of the surface roughness, confirmed that the membranes are suited for testing.

Self-supported carbon membranes were also prepared and evaluated. The method developed consisted in the physical mixture of a carbon powder with an epoxidic resin, co-adding high surface area SiO_2 or Al_2O_3 as "hierarchical agent". In fact, further dissolution of them by HF, confers the introduction of higher porosity to the composite. This method is particularly advantageous for several reasons:

i) The carbon powder source can be selected: we chose to test a particular carbon onion-like structured, because it seems to be active for the phosgene synthesis. Alternatively, any kind of carbon powder can be added.

ii) The membrane does not need the use of a support: the resin acts as self-structuring agent.

iii) The membrane is mechanically and thermally resistant: the HF attack and the treatment at 220°C generate a stable structure.

iv) The size of the membrane can be easily tuned: since the moisture Carbon/resin/HA is fluid, it can be placed in a specific mold in a way to obtain the desired shape.

The choice of the carbons to incorporate in the membrane was given by the experience of carbon materials used in the $COCl_2$ synthesis. So, as carbon precursor a "onion like" carbon prepared within the project and a carbon supplied from one of the partners were used for membrane preparation.

Characterization techniques and gas permeability tests showed the good accessibility to inner core of membranes. These membranes were utilized in the catalytic tests.

TS-1 membranes. Different deposition procedures to obtain improved alumina membranes supported with thin TS-1 layers were investigated. Thick TS-1 membranes (125μm) through direct in situ synthesis of TS-1 nanocrystals onto AAO membranes were initially prepared. The method was later optimized to produce TS-1 layers with much reduced thickness (35μm). Further studies have carried out high temperature treatments with the TS-1 supported membranes to study their hydrothermal stability and the results showed that no cracks are observed after calcination to 550°C but the membrane was curved. Following this initial membrane synthesis work, further studies focused on two different approaches: using standard and polymer assisted dip-coating methods. These latter studies have been performed with AAO membranes and the results obtained have shown that the TS-1 seeding was more effective when a polymer electrolyte was employed.

Modelled Substrate and Catalyst System Configurations and Engineered Testing Equipment Facilities

For an optimized concept of the membrane nanoreactor, the steps in the reaction need to be de-coupled. In line 1, first H_2O_2 has to be in situ generated over Pd catalyst and then diffuses over titanium silicalite-1 for the further reaction with propene to produce propylene oxide. The limiting step of the overall reaction (H_2O_2 production) is taking place in the membrane itself where H_2 and O_2 are fed over the channels of the membrane. Since hydrogen peroxide is a chemically instable compound due to fast decomposition over different types of materials, this limits the material choice for the membrane. The first requirement has then to ensure that no (or limit) decomposition of H_2O_2 over the material occurs. A series of initial studies thus focused on this aspect.

An issue in using the flat-type nanomembranes described above is fragility. A possibility investigated regarded the utilization of a non ordered macroporous oxide support on the top of which deposit the ordered membrane layer that will act as a catalytic one. The second suggestion presents the advantage of overcoming the problem of increasing the thickness that can lead to a mass transfer limitation during reaction operation. In fact, by increasing the thickness one of the most important characteristics of the membrane itself is limited and reduced: the gas permeability. In order to understand the performance of the membrane, permeability tests were conducted. Also with suitable modeling it was evaluated the minimum value of the membrane thickness needed for catalytic application. The membrane thickness is an important parameter to be integrated with the catalytic requirements from the reaction. A further modeling was utilized to improve the design of the membrane nanoreactor concept and analyze the characteristics for industrial scaling up.

An important element to optimize and scale-up the integrated reactor in line 1 was to develop a modelling approach to determine the optimal characteristics necessary and the potential results. The reactor model deals with a high pressure gas/liquid system. The purpose of the model is the optimization of the reactor concept by knowing the concentration-diffusion profiles of the different reactants and products along the membrane layers. A working model for the direct synthesis of propylene oxide has been developed. The model shows the dependence of H_2O_2 conversion and selectivity on the membrane parameters such as the thickness. Also the PO synthesis is strictly connected to the first catalytic layer. In order to reach the scaling up requirements, a Pd/SiO_2 membrane thickness between 100-250 µm has to be ensured. The model is also available to understand the influence of other parameters such as the gas partial pressures, liquid and gas flow rates and membrane areas in order to further understand the membrane reactor behavior.

In parallel, density functional theory was used to investigate the direct synthesis of H_2O_2 over Pd based nanoparticles. The aim of the study was to elucidate the reaction mechanisms for H_2O_2 obtaining on two representative surfaces of an active Pd nanoparticle in reaction conditions:

(i) $O/(\sqrt{5}x\sqrt{5})R27o$ Pd(100) and

(ii) PdO(101).

Development of Tubular Ceramic Nanomembranes

Due to issues presented by flat-type membranes, the focus on the last part of the project was on the development of alternative tubular-type ceramic nanomembranes, to utilize for the scale-up of catalysts/membrane, in relation to reactor design identified in other WPs. In line 1, limited robustness of flat-type membranes caused the need to develop alternative tubular-type membranes, but limited performances of the various configurations tested do not allowed to pass to larger scale prototypes. However, it was explored the use of a reactor based on hollow-fibers as scaled-type prototype. In parallel, investigation was related to deposit the catalyst (Pd and TS-1) on tubular membranes as next step in scaling this type of membranes and to prepare and functionalize with catalysts anodising aluminum foams as scaled flat-type catalyst/membrane systems having higher mechanical robustness. In line 2, the activity was in relation to scaling the catalyst/membrane in relation to prototype and model of reactor defined in simulation studies, and to develop the zeolitic membrane for the two-step phosgene-free DPC synthesis.

Procedures to improve Pd dispersion on alumina tubular membrane were developed and the samples characterized in terms of properties and performance. The procedure developed for Pd incorporation was based on impregnation-decomposition method. The membranes produced gave the best performance in H2O2 direct synthesis, a narrower particle size distribution; and no deactivation as had occurred previously with samples prepared by sol-immobilization procedure. Additionally, the effect of different cycles of thermal treatments on the same membrane was evaluated. A second heat treatment led to a narrower particle size distribution, retained the same degree of metal dispersion on the support and provided an improved product selectivity.

The high selectivity observed on the membrane prepared by impregnation-decomposition (D-I) after heat treatments has been explained on the basis of FTIR characterization where it has been observed that more exposure of homogeneous facets (*i.e.* less defect sites) is a consequence of preferential exposure of the facets. Adsorption of reactants on planar sites seems to be the optimal condition to get direct formation of H_2O_2. This agrees with theoretical modelling results by DFT method.

Membranes by using two preparation technique (sol immobilization and reduction by hydrazine) were prepared to test alternative configurations for palladium arrangements. The scaling of the deposition of TS-1 on tubular membranes, in order to improve the quality and coverage of TS-1 membranes on alumina tubular supports, was also investigated with the objective being to prepare suitable bi-functional catalytic systems Al_2O_3@Pd@TS-1 for direct one-step PO production from H_2, O_2 and C_3. A novel type of composite membranes based on palladium thin layers externally coated with TS-1 nanocrystals, using α-alumina tubes as support substrates was prepared.

Important advances were achieved by identification of the need to seed coated Pd films with TS-1 nanocrystals *via* a polymer dip-coating process followed by a secondary growth step. In this latter case, micellar gel for secondary growth of the seeded supports was investigated and the resulting membrane consisted of a well-intergrowth layer of TS-1 zeolite with a thickness of 0.7 μm. The TS-1 membrane is strongly attached to the palladium film and no peeling was observed. Electron microscopy characterization confirms the effective presence of outer TS-1 layer onto internal palladium film which is coating the tubular alumina supports.

The possibility to obtain a dense zeolitic membrane on the polymer-coated metallic surface by direct growth of the zeolite crystals and without the previous TS-1 seeding deposited by polymer assisted dip-coating was also analyzed. A densely packed TS-1 membrane was grown on the palladium surface, although visible discontinuities were detected.

Scaling the Tubular-Membrane Approach to Develop Hollow-Fiber Reactor

Scaled-up nanoreactor (line 1) was based on the use of hollow-fiber modules. A new reactor prototype for testing them was assembled. A new scenario for the production of PO could be the use of hollow fiber membranes with a lower wall thickness (0.5 mm) then the previous membranes (wall thickness 1.5 mm). Although as known selectivity in H_2O_2 is limited for operations at low pressure as in the following tests, the results show that hollow-fiber membranes under the same conditions allow about three-four times higher productivity and selectivity with respect to tubular-type membrane R70 benchmark. The approach is thus well suited for scaling-up performances and intensifies the process.

Scaling the Preparation to Produce Anodising Aluminum Foams

A thinner, more open-pored structure on aluminum foam using alternative anodizing conditions (with phosphoric electrolytes) was successfully produced for subsequent catalyst deposition and evaluation: initially film thicknesses of <1µm were achieved but it was considered that these also were likely to be too thick and, consequently, oxalic electrolyte options were produced with anodic film thicknesses of 1-2µm but with very fine pore widths of ~30-40nm.

Further studies were then subsequently undertaken on pore widening options with anodized aluminum foam but, despite the "imperfect" anodic film structures initially grown on the anodized aluminum foam, Pd nanoparticles were deposited successfully within the pore structure. Alternative anodising processes were also explored with the aim of identifying open-pored structures of the "required" structure; one of these is hot AC phosphoric acid processing. This procedure provides an anodic film geometry on anodised aluminum which might be considered closer to product application requirements.

In addition, polymer replication from an anodised aluminium substrate – resulting in the creation of a textured polymeric substrate which may have applicability as a catalyst substrate, was developed.

Scaling the Preparation of Carbon Catalytic Membranes for Line 2

The scaling of the preparation of carbon membrane (line 2), with the aim to improve the synthesis procedures for C-containing membranes was investigated with a focus on:

- thermal stability up to 300°C
- control of membrane thickness in the range 0.3-0.5 mm
- high permeance
- high porosity
- higher carbon loading

Different types of nanocarbons starting materials were utilized. The results demonstrated that a synthesised variant can show comparable and even better properties to commercial systems.

Scaling the Preparation of Zeolitic Membranes for Two-Step Phosgene-Free DPC Synthesis in Line 2

The scaling of the preparation of zeolitic membranes for two-step phosgene-free DPC synthesis (line 2), one of the most promising alternative pathways for DPC synthesis, was also investigated. In the first step the formation of dimethyl carbonate (DMC) from methanol and CO_2 takes place, whilst in the second reaction DMC is converted *via* transesterification with phenol to form DPC as the target molecule. Both of the reaction steps are highly equilibrium constrained. Therefore, besides rational design of efficient catalysts, the selective product separation (*i.e.* water removal in the 1st step and methanol removal in the 2nd step) is demanded. In a first attempt, LTA zeolitic wafers were synthesised and thoroughly synthesised prior to membrane growth. Similar the case of DMC, in order to enhance the DPC formation, efficient membranes for the methanol removal from phenol abundant solution are required. Therefore, a similar approach as in the case of water/alcohol separation membranes was applied. For a selective MeOH removal for improvement of transesterification yield was achieved by using newly prepared LTA zeolitic membrane. However, the pore-opening ought to be different as in the case of water removal from alcohol. This modification of pore-opening was fulfilled by different cation exchange.

A continuous-flow reactor with on-demanding product separation has been proposed. At the bottom part of the reactor, the reaction would be initialized by DMC formation from methanol and CO_2 (catalyzed by ZrO_2-based catalyst), co-produced water would be eliminated simultaneously by membrane made by zeolite. Till DMC yield reach certain high level, transesterification of phenol and DMC would started (catalyzed by MoO_3- or TiO_2-based catalyst), another zeolitic membrane would be applied for methanol removal. At last, DPC would be obtained with high yield..

Development of flat-type nanoreactor for line 1 as well as tubular-type membrane (nano)reactor, fabricated mock-up prototype, developed membrane reactors to work in gas-liquid and gas-gas phases, design of the prototype reactor

The following types of flat-type nanomembranes have been investigated:

- for line 1,

i)   alumina (AAO) nanomembranes of about 25 mm in size produced by Innoval and

ii) titania nanomembranes of similar size prepared by INSTM.

Both were produced by anodic oxidation, but the investigation of the titania type membranes was stopped due to both problems of material compatibility and of preparation (severe curling).

- for line 2,

i) carbon-coated nanomembranes (on alumina substrate) and then coated with carbon.

ii) carbon-only membranes prepared using nanocarbon materials and then converted to the flat membranes by casting procedure after addition of suitable additives.

For testing the characteristics of these flat-type nanomembranes (robustness, sealing, permeability), a simplified mock-up reactor was constructed and used.

A new mock-up reactor prototype was also fabricated to investigate the fluid-dynamics of the gas-liquid contact and mixing using the flat alumina AAO membranes. The reactor is designed to carry out reactions in gas/liquid phase:

i) one side is prearranged to feed gas and the other to feed a liquid flux in continuous mode;

ii) the 2 chambers are separated by the membrane. The reactor was designed to have a minimal thickness for the liquid side (about 1 mm), in order to simulate the original design.

Some of the main conclusions from these activities are the following

Line 1:

-    a SS prototype was already fabricated, and two reactors were already sent to partners for testing.

-    the decision concerning the scaled-up nanoreactor (line 1) was based on the use of hollow-fiber modules. A reactor prototype for this experimentation was designed, assembled and tested.

-    the tubular reactor prototype has also been modified to work in gas-liquid phase conditions. This modification was carried out taking account that the formation of H_2O_2 intermediate is favored in liquid phase at softer reaction conditions. The apparatus were updated for testing membrane in another configuration; feeding a non-explosive mixture $3\%H_2$-$97\%O_2$ through the membrane while in the other side the methanolic solution phase is feed as in the previous tests.

In Line 2:

-    the design of the prototype reactor was assessed and the reactor-prototype together with o-ring in kalrez, thermocouple and some carbon based membranes send to partners for testing.

-    engineering modeling and safety aspects have been considered for the nanoreactor design. They were also used to consider the carbon thickness necessary and to estimate the heat release, also in consideration of the heat control necessary to minimize deactivation.

-    based on the elaborative analysis of the integrated system, the final reactor concept of a multi-tubular reactor was put forward and assessed for its ability to reach a realistic commercial capacity of the targeted product DPC.

Development of hollow-fiber modules for scaling-up reactor, fabrication of the reactor-prototype. A nanoreactor (and related apparatus) for using hollow-fiber modules was designed and assembled as new scale-up reactor prototype.

Increase in PO performances were studied through the use of further membrane catalytic systems studied considering the influence of the following parameters: use Pd films or nanoparticles (seeds), combination of active layers in different order such as Pd@TS-1 or TS-1@Pd membranes, passivation of acidic external groups, increase in the content of active phase and reactions with high oxygen regimes, *i.e.*, using non-explosive H_2-O_2 mixtures. All these parameters were considered in membrane reactor suitable to work in gas phase. Furthermore, novel tubular reactor prototypes were assembled to better analyze different reaction variables and increase the PO productivity. For this, liquid-gas reactor prototype was assembled and optimized to compare with the most favorable reaction conditions to generate intermediate in situ H_2O_2. Moreover, tubular prototype adapted to work with alumina hollow fibers was also designed and optimized trying to increase the productivity and efficiency of the two-step one-pot catalytic process. Both approaches (i and ii) are complementary, being the advances achieved in batch catalytic studies useful to be applied in the membrane reactor proto-type tasks.

Tested Different Reactor Configurations and Combinations of Multilayer Membranes

Different configuration/arrangement of Pd/membrane were tested. The best results in the direct synthesis of H_2O_2 were obtained with the benchmark configuration giving a better productivity, activity (hydrogen conversion) and selectivity respect to arrangements of Pd on the external side. All the membranes tested shown activity toward hydrogen conversion (H_2O productivity), which means that the membranes are completely wetted by the methanol solution. The lower selectivities obtained by the catalytic membranes prepared by external Pd deposition are attributed at the higher hydrogen peroxide retention time in the membrane pores (pore length: 1.5 mm, without taking into account the tortuosity factor) in contact with Pd particles. Similar results are obtained with the membranes prepared by N_2H_4 reduction technique, where we have also a higher H_2O production rate in the case of the external palladium arrangement compared with our benchmark.

We might conclude that the higher concentration locally achieved inside the pores of the membranes prepared by external palladium deposition are detrimental in our experimental condition, but might be a key factor for the PO production. A new scenario for the production of PO could be the use of hollow fiber membranes with a lower wall thickness (0.5 mm) than the previous membranes (wall thickness 1.5 mm).

Higher productivity in the PO synthesis were shown by using a specific type of membrane with the reactor configuration as follows: TS-1 layer deposited on the internal side). Based on this result some membranes by using two preparation techniques (sol-immobilization and reduction by Hydrazine) were prepared.

Four different reaction configurations were investigated for tubular membranes:

1. Reaction configuration IA: Pd seeds or film are located in the internal side of the alumina supports, and TS-1 membrane is grown over it. Hydrogen is fed from the outer side of the membrane in gaseous phase at constant pressure (2 bar). Propylene and oxygen saturated methanolic solution is continuously circulated on the inner side by means of a peristaltic pump (25 ml/min).

2. Reaction configuration IB: Pd seeds or film are located in the external side of the alumina supports, meanwhile TS-1 membrane is in the inner side. Hydrogen is fed from the outer side of the membrane in gaseous phase at constant pressure (2 bar). Propylene and oxygen saturated methanolic solution is continuously circulated on the inner side by means of a peristaltic pump (25 ml/min).

3. Reaction configuration IIA: Pd seeds or film are located in the internal side of the alumina supports, and TS-1 membrane is grown over it. Oxygen and hydrogen are fed together in a non-explosive regime by the outer side of the membrane in gaseous phase. Propylene saturated methanolic solution is continuously circulated on the inner side by means of a peristaltic pump (25 ml/min).

4. Reaction configuration IIB: Pd seeds or film are located in the external side of the alumina supports, meanwhile TS-1 membrane is in the inner side. Oxygen and hydrogen are fed together in a non-explosive regime by the outer side of the membrane in gaseous phase. Propylene saturated methanolic solution is continuously circulated on the inner side by means of a peristaltic pump (25 ml/min).

Reaction configurations IIA and IIB forced to work with only 3% of H_2 in order to be in a non-explosive regime, resulting in lower H_2O_2 productions, much higher selectivities. This approach could be an interesting alternative, because its use could significantly reduce the amounts of propane formed, being, in any case, necessary to increase the H_2O_2 productivity to improve the final PO production. On the other hand, reaction conditions IA and IB lead up to now the best H_2O_2 production results using this kind of tubular membrane reactors. Configurations IIA and IIB, in which H_2 and O_2 are feed together, did not lead to any production of propylene oxide. These results agree with the results obtained, which showed a low H_2O_2 production in both configurations when H_2-O_2 non-explosive mixtures are used.

When the reactions were carried out with the typical configurations, IA and IB, feeding H_2 from the outer side and O_2 together with C_3H_6 in the inner side dissolved in methanolic solution, two different results were observed. On the one hand, membranes in which Pd is in seeds form located in the inner side (R73), the PO production does not exist or is too reduced, with propane being the main product. On the other hand, the membrane with Pd film located in the internal side, led to better catalytic results, methoxypropanols were also observed during the reaction. The results indicate that PO production is favoured when Pd seeds in the outer side or Pd film in the inner side are coating the alumina support, always feeding separately H_2 and O_2.

Conclusions of Tubular Membrane Reactor Tests

- TMR prototype in gas reaction conditions is not effective to produce PO in one-pot process (only PO traces are obtained) due to the complete oxidation of Pd species in the catalytic system.

- TMR prototype is effective to produce PO in one-pot process in liquid-gas conditions when H_2 and O_2 are separately fed in the reaction system. Specifically, in these conditions, the best performances have been obtained when Pd film or Pd nanoparticles are supported into the internal or external side, respectively, of alumina tubular membrane, in presence of internal TS-1 layer. In these conditions, the stabilization of in-situ generated H_2O_2 is enough to produce PO in the second reaction step.

- Reproducibility test show a strong dependence with the membrane used in all cases. This issue is still a matter of study.

- Feeding together O_2 and H_2 (using H_2-O_2 non-explosive mixtures at high O_2 regimes) did not led to PO production in any case, due the poor stability and low productivity of in-situ generated H_2O_2 under these reaction conditions.

- Process optimization is necessary to increase PO yields, analyzing different parameters such as catalyst amount, incorporation of promoters (Au, Pt, Ag), pressure gradients, reduction of propane formation as main reaction product and use of H_2O_2 stabilizers among the most relevant matters to be considered in the close future.

Tests in batch reactor were also made. Some of the conclusions are the following:

- Bi-functional Pd@TS-1 catalysts are effective to produce directly PO through one-pot two-step process from H_2, O_2 and C3=, using supercritical CO_2 conditions (high pressures ~ 75 bar).

- Best results are obtained using MeOH as co-solvent, ammonium acetate as acidity inhibitor and C3=/H_2/O_2 molar ratios close to 1/1/1. Higher selectivity to propane was observed when H_2/O_2 ratio was higher than 1, resulting in a decrease in the PO selectivity.

- In general, H_2O as only co-solvent does not favor the PO production, although its negative effect is strongly minimized in the presence of MeOH, overall when Pd@TS-1 catalysts are used.

- The presence of Au/Pd alloy nanoparticles supported onto TS-1 allows the preparation of effective bi-functional catalysts which are more reactive than only Pd counterparts, achieving C3= conversions of ~15%, and PO selectivities and yields close to 82% and 11% respectively, reducing significantly the propane formation (SC3~10%).

As general conclusions from these tests it may be indicated:

- Bi-functional catalytic systems (powder or membrane system), based on Pd species (nanoparticles or films) and TS-1 zeolite (powder or dense layer) are effective for direct PO production from H_2, O_2 and C3 = through in-situ generated H_2O_2.

- High PO performances are obtained in batch catalytic processes with supercritical CO_2 conditions at high pressures and with the additional presence of metallic promoters (Au, Pt).

- Tubular membrane reactor (TMR) prototype is valid to carry out the one-pot two-step process in liquid-gas conditions, using bi-functional Al_2O_3/Pd/TS-1 membranes, feeding separately H_2 and O_2 into the reaction system. The stabilization of in-situ generated H_2O_2 in methanolic solutions, containing saturated O_2 and C3 =, is necessary.

Engineering of the Scaled-up Nanoreactor and Assessment of the Nanoreactor Prototypes

Aspects analyzed regarded:

1. The detailed design of the reactor (dimension, shape, manner of heat transfer *etc.*).
2. The total mass balance of the reactor.
3. The process scheme including operating conditions.

These specifications were derived from and based on the earlier work of the other partners. However, the results of the process in line 1 are highly depending on some assumption made, which the engineering modelling developed later showed the necessity to be modified. In particular:

i) the use of very diluted conditions, about 1:500 mainly to remove the heat of reaction with the flow of methanol itself; by using a different approach for heat removal and working under more optimized conditions the ratio can be decreased to a value estimated around 1:20, determining a complete change in costs;

ii) the possible significant larger optimization in performances, by using a hollow-fiber type of reaction and optimized membrane configuration, as well as operative conditions, leading to an improvement by a factor about 20 in the productivities. Therefore, present assessment cannot be considered definitive, but as a case limit. It is possible a significant decrease in the costs and reactor sizes, but further experimentation is necessary.

Understanding the mechanism, particularly what are the active sites on the carbon catalyst surface and what is the nature of the transient intermediates. The activities are to:

i) Develop concepts for the mechanism of $COCl_2$ formation over the carbon catalyst surface;

ii) verify the proposed mechanistic concepts by molecular modeling applying DFT analysis,

iii) study experimentally the carbon-chlorine interaction;

iv) study CO interaction with the active sites and activated carbon – chlorine complex;

v) study experimentally $COCl_2$ synthesis in small scale batch reactors.

The scope of these activities was to elucidate the mechanism of phosgene formation over carbon catalyst, which goes along with the identification of the catalytically active sites. This knowledge will support other tasks dealing with catalyst deactivation, the development of stable catalysts and the identification of process conditions inhibiting deactivation. A concept for Cl_2 activation over carbon surfaces was developed and supported with corresponding DFT calculations. Experimental evidence for the postulated mechanism and catalytically active sites were also provided. Between the aspects investigated:

i) *C...Cl_2 interactions was investigated experimentally, using various advanced spectroscopic techniques and a pulse setup, to gain experimental evidence for the proposed mechanism/ active sites and to get insight into elementary processes occurring during interaction of Cl_2 with C.

ii) In situ temperature-dependent ESR measurements of Cl_2...C interaction using model carbon materials were performed, to evidence the nature of transiently formed chlorine intermediates.

iii) The interaction of CO with Cl_2...C complexes was investigated, to understand the reactivity of various activated chlorine molecules towards CO.

iv) Synthesis and thorough characterization of new model carbon materials for phosgene synthesis was made to identify potential active sites on carbon material surface to enable a better understanding on the contribution of various sites on the Cl_2 activation and $COCl_2$ synthesis.

v) The influence of various process parameters and material pre-treatments on the formation of chlorinated side products during phosgene synthesis was analyzed.

v) Synthesis and mechanistic investigations on N-doped carbon materials were performed.

Understand the mechanism of deactivation, especially what triggers deactivation of carbon catalysts and identification of the measures to avoid deactivation/ regeneration, identification of stable catalysts and of alternative, non carbon catalysts:

i) Thorough characterize state of the art carbon catalysts (unused/ fresh) for phosgene synthesis;

ii) derive structure-activity relationships by comparison of analytical data obtained from unused and used carbon catalysts;

iii) synthesize model carbon materials with an emphasis on given structural/ chemical features to prove mechanistic assumptions,

iv) experimentally investigate factors influencing carbon catalyst deactivation. The scope is to investigate and understand the deactivation of carbon catalyst during phosgene synthesis.

Between some of the activities for this scope:

i) DFT analysis on the thermodynamics of Cl_2 interaction with highly pre-chlorinated fullerene model carbon surfaces, to investigate if energetics of chlorination depend on degree of pre-chlorination of the model

ii) Investigations on catalyst regeneration. Carbon catalyst regeneration by treatment with CO, N_2 and H_2 was investigated at different temperatures

iii) Testing of further carbon materials in phosgene synthesis. Relevant carbon materials were investigated for their stability during $COCl_2$ synthesis at different temperatures

iv) Comparison of catalytic properties of powder-type carbon materials and carbon extrudates, to get closer to the real catalyst material in phosgene plants

v) Testing of Al_2O_3 as a potential non-carbon catalyst for phosgene synthesis, to identify a non carbon catalyst with potentially higher stability regarding deactivation

As result of these activities, Stable catalysts for $COCl_2$ synthesis were identified, based on the following studies:

i) process conditions for reduced/ inhibited deactivation of carbon catalysts;

ii) properties of tailored carbon materials that are active and stable catalysts for $COCl_2$ synthesis;

iii) alternative non-carbon catalysts that can be used within membrane concept for DPC synthesis.

The results suggested that N-doping of the carbon structure increases the stability of the catalysts during $COCl_2$ synthesis. However, at the same time catalyst activity is significantly reduced. A different mechanism for Cl_2 activation observed for pure and N-doped carbon materials may serve as an explanation for their different deactivation properties.

Activated carbon materials with strongly improved stability when compared with state of the art activated carbon catalysts were identified in the course of catalyst screening experiments at lab-scale. The identification of carbon catalysts within improved stability is a major advance not only in terms of a better understanding of the factors responsible for carbon catalyst deactivation in the future but also regarding possible implementation of the identified materials in current $COCl_2$ generators.

Alternative to carbon based catalysts, various metal based catalysts were selected and tested for catalytic Cl_2 activation for phosgene productions. In-situ MS measurements conducted during chlorine adsorption in the presence of 5 wt. % CO containing He atmosphere at 150 °C have shown that among others, some type of alternative materials are a potential non-carbon catalyst for the $COCl_2$ synthesis.

Design and performing a lab-scale demonstrator, identification of the concept for scaled reactor and application of insight gained for current $COCl_2$ manufacture.

The activities are to:

i) study the performances of lab-scale catalyst/membrane and nanoreactor developed in other WPs,

ii) analyze the effect of the reaction conditions and operative parameters (flux through the membrane, liquid flow, design of nanoreactor, *etc.*),

iii) compare the behavior (kinetics, stability) with that using conventional reactor/catalysts, and

iv) optimize the performances. The scope is to test lab-scale nanoreactor prototypes both in phosgene synthesis and coupled phosgene and DPC synthesis.

Between the activities for this objective:

i) membrane tester, with safety check and heater

ii) integrated fixed bed reactor, with design and assembly, cold commissioning

iii) integrated membrane tester, with tubular carbon membranes, membrane holder for integrated reactor, permeability tests and catalytic activity.

Various approaches for lab scale testing and demonstration were followed. Goal was to realize and test an integrated reactor combining in situ $COCl_2$ synthesis and phosgenation of phenol to DPC. Regarding $COCl_2$ synthesis two approaches were followed: $COCl_2$ generation in a membrane or in a fixed bed reactor. A membrane tester was designed to test carbon membrane stemming from partners, a separate inlet is intended to provide the possibility to feed phenol into the reactor.

An integrated reactor for DPC synthesis at elevated temperatures and pressures *via* in situ generated $COCl_2$ was designed and assembled. Both, the utilization of carbon membranes and a fix bed for $COCl_2$ generation were possible with the setup. Graphite cylinders were used as a (potential) membrane. The alternative reactor approach, based on an integrated carbon catalyst cartridge for phosgene generation, was driven up to a successful proof of principle, *i.e.* reproducible and selective DPC formation *via* in situ generated $COCl_2$. Technical challenges encountered during experimentation were caused by geometrical limitations in the setup used (responsible for low gas flows), PhOH penetration into the carbon cartridge (deactivating the carbon catalyst) and adsorption of significant fractions of the Cl_2 introduced during a typical experimental time on the activated carbon (reducing DPC yield).

In terms of integrated reactor for in situ $COCl_2$ synthesis and immediate conversion (to DPC), different options were analyzed to identify the preferable configuration. Patents for novel recator configuration were filed.

Basic Techno-economic Assessment of Integrated Reactor Concept

The production of DPC in a reactor fed with carbon monoxide and chlorine besides phenol was demonstrated. In order to assess the impact of the new reaction and reactor concept it is important to investigate the impact of reaction conditions on all process steps required to build a complete DPC process plant. One important aspect would be the cost for phenol separation from the product. Selectivity is another important issue. The reactor modelling was performed for a 20kt DPC system. The proof of concept was demonstrated successfully. Safety assessments took place required to run the lab system.

Phosgene-free DPC Synthesis

The activities are to:

(i) design efficient water and methanol removal units to minimize the thermodynamic limitation in the DMC and DPC synthesis;

(ii) develop stable and selective catalysts for the DMC and DPC synthesis;

(iii) study the complex reaction network and establish structure-activity correlations;

(iv) study experimentally the influence of critical process parameters.

Between the spects investigated:

1. Synthesis and characterization of various supported MoO_3 and TiO_2 mixed-oxide catalysts, to find an adequate selective catalyst for the DPC production from DMC and phenol, and to identify main reaction pathways on which DPC is formed.

2. Detailed kinetic and mechanistic studies, to understand complex reaction network

It should be finally mentioned that all expected Deliverables were produced in time.

Potential Impact

The expected final project impacts are to develop:

1. new approaches in process intensification through a novel concept of multiphase nanoreactor design,

2. new approaches in multifunctional catalyst design by integrating catalyst and membrane functionalities in an approach aimed at process intensification,

3. new approaches for intrinsically safer design for reactions involving risky reagents.

The expected final result of the project is to verify to applicability and scalability of new concepts in catalysis related to the development of novel nanoreactors and related catalysts for the two listed target reactions and how they can improve (in these industrially-relevant multistage reactions) process intensification, sustainability (in terms of resource and energy efficiency) and safety of operations. Functional to this general objective are the development of catalytic nanomembranes, and of the associated novel reactor concepts. Due to challenging objective of developing novel nanoreactor concepts, the demonstration activities in the project are limited to the proof-of-the-concepts.

In line 2 the expected impacts are achieved, with a new type for scaled reactor for intrinsically safer design, new more active and stable catalysts developed, relevant knowledge/catalysts for alternative processes. Further research, however, is needed to exploit these results.

In line 1 the integration between the two stages, even if many different configurations have been explored, has been not proven successfully. However, project results in this line showed new concept of ceramic hollow fiber reactor which are relevant and innovative for a new process of direct H_2O_2 synthesis, solving potentially issues of current approaches – autoclave, fixed bed, microreactors – that have inhibited commercial direct H_2O_2 synthesis. In the integrated process approach, progresses and new advances have been made in membrane reactor modelling, in preparing multicomponent membranes and in understanding role of gradients on performances.

These results indicate the need to extend current examinations and consider further alternatives in process design, in particular regarding alternative heat removal solutions, membrane optimization from modeling, alternative reactors as hollow fiber, *etc.* Some of the results development within the project may also be exploited outside the project specific field. In addition, the research was proven quite successful in having a decisive impact to SMEs involved in the project to overcome general crisis and open new markets. Therefore, research in this line has generated new knowledge and concepts which translate (later) to innovation.

In terms of dissemination, the following activities have been presented and uploaded on the web site and on the Participant Portal:

- 28 posters

- 16 oral communications

- 18 publications and about other 10 in preparation

3 patents have been prepared. All these 3 patents have not been published yet and for this reason they could not be uploaded on the Participant Portal

Analysis of the Impact

In terms of impact, different results have been obtained in lines 1 and 2, as indicated above. In line 1, there is still the need to further develop the materials and process. Actual techno-economic assessment was based on assumptions that need to be revised. Nevertheless, interesting materials in particular and nanoreactor configurations have been developed, which may be later exploited. In particular:

- Novel concept of ceramic hollow fiber reactor which are relevant and innovative for a new process of direct H2O2 synthesis, solving potentially issues of current approaches – autoclave, fixed bed, microreactors – that have inhibited commercial direct H2O2 synthesis

- Progresses and new advances have been made in membrane reactor modelling, in preparing multicomponent membranes and in understanding role of gradients on performances.
- Novel multilayer-type tubular membranes
- Novel flat-type AAO nanomembranes functionalized with nanoparticles, which may be utilized also for other type of applications such as in water purification and desalination
- Novel type of anodising aluminum foams as scaled flat-type catalyst/ membrane systems having higher mechanical robustness.

The list of results with exploitable foreground reports for line 1 seven results going from design of advanced reactor prototype to deposition method of nanoparticles within structured materials (foams, membranes) and new POSS catalysts. The possibility of PO production using the tubular membrane reactor is also under discussion, including regarding patentability. A patent has been issue by one of the partners (a SME) on the synthesis of POSS catalysts.

In line 2, both new catalysts, carbon membranes and new process configurations with membrane reactor have been successfully developed. The list of results with exploitable foreground reports for line 2 thirteen entries, going from new carbon-type and non-carbon-type catalysts to the novel reactor configuration (covered by two patents) and process.

The polycarbonate production capacity worldwide in 2012 was estimated to be about 4.7 million tons. The vast majority (about 80%) of capacity installed is phosgene based. Only a smaller portion (ca. 11%) of the phosgene based polycarbonate processes uses DPC as intermediate. In 2012 capacity installed exceeded demand. While demand for polycarbonate is growing worldwide several companies have announced projects for installation of additional capacity. At the current technical status of the project it is too early to be able to assess the likelihood of implementation of this particular reactor concept/system in future polycarbonate production plants. More detailed investigations of all aspects of DPC production are necessary, *i.e.* downstream processes. The potential will then depend on the general growth rate of the polycarbonate market and the economics of the competing processes at the locations of the future. Other important aspects are which of the companies operating worldwide will have access to the technology and at what cost/license fee.

Within the scope of the project, the production of DPC in a reactor fed with carbon monoxide and chlorine besides phenol was impressively demonstrated. However, an economic assessment of this new process step requires information beyond the results obtained so far. In particular, in order to assess the impact of the new reaction and reactor concept it is important to investigate the impact of reaction conditions on all process steps required to build a complete DPC process plant. One important aspect would be the cost for phenol separation from the product. Therefore a separation column should be installed. Its size is a direct result of phenol excess. Low phenol excess requires low energy for a later phenol separation step. The higher the phenol excess, the larger equipment size is required not only for the reactor but also for the following work-up.

Selectivity is another important issue. Could phosgene formation in the reactor lead to a different number or type of impurities? The first results of INCAS demonstrator are very promising, since minimal amount of byproducts have been observed.For an initial estimate of the economic potential the following cases could be compared regarding the invest cost:

A: Invest cost for a "stand alone" phosgene production unit plus a reactor of the size and running conditions with 4 fold phenol excess and

B: Invest cost for the newly developed reactor design with installed interior, in particular the required number of tubular membranes

The apparent delta to be considered regarding the reactor includes:

The number of required tubular membranes multiplied by the cost of each tube and the lifetime factor of the catalyst. While there is information available on the cost of graphite tubes, the additional cost for their installation in the reactor is still to be determined. The tube lifetime is assumed to be as long as that of the typical phosgene catalyst due to the same material and operating conditions. The exchange of a membrane tube is an issue since it requires the shutdown of the whole DPC plant and should be taken into account.

Considering invest cost estimates published by SRI consulting in their PEP yearbook 2010 for phosgene production units suggests that combining in situ phosgene generation and reaction with phenol appears economically favorable from an invest cost point of view. However, further detailed analysis of a complete DPC production process including *e.g.* separation of excess phenol, analysis/separation of side products is required to provide a meaningful economic assessment.

Regarding engineering (basic flowsheet, operation conditions, targets for scaling-up, safety, energy and material balance, process integration), available documentation was included in the technical part. The reactor modelling was performed for a 20kt DPC system. Experiments were conducted at lab scale size. Safety assessments took place required to run the lab system. The proof of concept was demonstrated successfully.

Regarding opportunities. About 16% of the worldwide phosgene consumption in 2013 (CEH report phosgene 2013) was used in polycarbonate production plants around the world. While the proof of principle and reactor design for in situ phosgenation was demonstrated and described for the reaction step in the production of DPC, the key principle and reactor could potentially be successful for other phosgene using production porcesses, too.

Currently, the new INCAS concept is being proposed and assessed as an alternative to the existing phase separation process used for DPC synthesis. The in-situ phosgene generation allowing for sparing the reactor block(s) for phosgene synthesis seems attractive at first view. Opportunities among other companies are also investigated for pursuing an INCAS device where in-situ phosgene is available to activate any secondary reaction.

Strategic Impacts

The general objective to provide a strategic advantage and added value to society, in terms of competitiveness, reduction of environmental footprint and industrial safety may be difficult to be reached by developing an alternative safer reagent or feed-stock (the classical example of substituting phosgene), or a synthesis methodology (the other classical examples of using supercritical CO_2 or ionic liquids instead of the common sol-vents). The quantitative potential impact of these approaches is limited and not in line with the expected impact, as well as limited their pushing role towards a major change in sustainability of the chemical production, *e.g.* they have a minor effect in providing a strategic advantage as requested in the initial call. The changes in macroeconomic scenario and in society demands require, on the other hand, to really progressing in the development of new enabling mechanisms towards sustainability. For a "strategic advantage" it is necessary to provide new mechanisms for designing the industrial chemical processes, for example which enable a parallelized multi-modular design.

Therefore, the project expected impacts can be summarized as follows:

- A new approach in process intensification through the novel concept of multiphase nanoreactor design; this novel design improves of one order of ma-nitude the process intensification with respect to actual microreactor and thus of about two-three order of magnitude with respect to conventional reactors. This novel concept enables an improvement in resource efficiency (energy) of over 25-30% with contemporaneous lowering of the costs, and waste production, and improves safety of operations.

- A new approach for intrinsically safer design for reactions involving risky reagents, with a nanoreactor design with integrated transient generation of the reagent (phosgene, in our case). This design also allow to operate the phosgenation in gas phase, and thus solvent-free operations, with great benefits in terms of quality of the product (avoided chlorinated byproducts) and reduction of over 20% in energy costs.

The results of INCAS project showed that for line 2 these strategic impacts can be considered to be met, as indicated before, although further experimentation would be necessary to develop the developed technology to implementation. In line 1, some promising results have been obtained, but there is the need to further develop the concept.

It should be remarked that all the studied reactions are of great industrial relevance. The reaction of clean phosgenation (*e.g.* with transient generation of phosgene) for DPC synthesis is another illustrative example of the approach used in this project. As commented before, it is better to reduce the risk of using phosgene, instead of finding novel phosgene-free processes or phosgene-substitutes, to improve the sustainability in several industrial processes of synthesis of large-volume chemicals (DPC - diphenylcarbonate, isocyanate and polyurethanes), and fine and specialty chemicals production (active pharmaceutical in-gredients, Taxol, Prodrugs, Agrochemicals, *etc.*). After about 20 years of effort, the majority of production is still based on phosgene. It is thus preferable to invest on a new intrinsically-safe technology as that proposed in this project. The large involvement in this project (many person/months) of a company world leader in this sector testifies that they believe that the proposed novel technology could have a major impact in this area.

HOMOGENEOUS CATALYSIS AND THE ROLE OF MULTI-PHASE OPERATIONS

Homogeneous catalysis in comparison to heterogeneous catalysis is burdened by the use of a solvent, which makes catalyst recycle and product separation costly and difficult. This is probably one of the main reasons that industry prefers heterogeneous catalysis.

Besides heterogenizing homogeneous catalysts, immobilization of the homogeneous catalysts in multiphase operation (*e.g.* two-liquid phase approach) offers promising opportunities. The two-liquid phase approach rests on the proper choice of distribution coefficients of the products with immiscible solvents. The following cases can be considered:

- The catalyst operates in a polar phase and the products form the second immiscible phase which can be "spooned off".
- Two immiscible solvents yielding two phases are used in the reactor. The catalyst remains in one phase, the products are extracted into the second phase.
- The homogeneous catalysis is carried out conventionally followed by extraction of the products with a second solvent, which is immiscible with the solvent of the phase containing the catalyst.

Examples are presented for all three cases. For case one, besides polar organic solvents, water, perfluorinated solvents, ionic liquids and supercritical CO_2 will be discussed.

Ionic liquids, having no vapour pressure and supercritical CO_2 can be used in a continuous operation. The catalyst must be soluble in the ionic liquid phase. The product will be extracted by the $scCO_2$. After separation, the CO_2 can easily be recycled. The combination of ionic liquids and compressed CO_2 provides a unique new approach of great promise.

MULTIPHASE CATALYTIC PACKED-BED REACTORS (PBRS)

Multiphase catalytic packed-bed reactors (PBRs) operate in two modes:

(1) trickle operation, with a continuous gas phase and a distributed liquid phase, and the main mass transfer resistance located in the gas, and

(2) bubble operation, with a distributed gas and a continuous liquid phase, and the main mass transfer resistance located in the liquid phase.

For three-phase reactions (gas and liquid phases in contact with a solid catalyst), the common modes of operation are trickle- or packed-bed reactors, in which the catalyst is stationary, and slurry reactors, in which the catalyst is suspended in the liquid phase. In these reactors, gas and liquid move co-currently down flow or gas is fed countercurrently upflow. Commercially, the former is the most used reactor, in which the liquid phase flows mainly through the catalyst particles in the form of films, rivulets, and droplets.

Based on the direction of the fluid flow, PBRs can then be classified as trickle-bed reactors (TBRs) with co-current gas-liquid downflow, trickle-bed reactors with countercurrent gas-liquid flow, and packed-bubble reactors, where gas and liquid are contacted in co-current upflow. To carry out the catalyst and reactor selection and process design properly, knowledge of what each reactor type can and cannot do is very important. When a fixed-bed reactor is chosen, the question frequently asked is whether to use an upflow or downflow mode of operation.

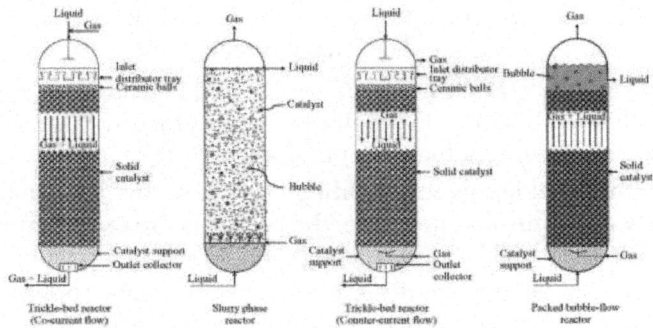

Fig. : Various types of multiphase catalytic reactors.

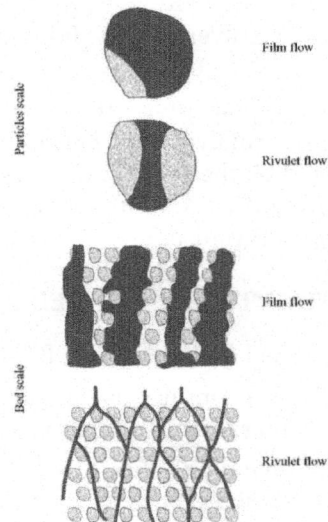

Fig. : Liquid flow texture found during the trickle-flow regime in a TBR.

In the case of catalytic packed beds with two-phase flow, such as those used for straight-run naphtha hydrodesulfurization, from a reaction engineering perspective, a large catalyst-to-liquid volume ratio and plug flow of both phases are preferred, and catalyst deactivation is very slow or negligible, which facilitates reactor modeling and design. However, for three -phase catalytic reactors such as those employed for hydrotreating of middle distillates and heavy petroleum fractions, the reaction occurs between the dissolved gas and the liquid-phase reactant at the surface of the catalyst, and the choice of upflow versus downflow operation can be based on rational considerations regarding the limiting reactant at the operating conditions of interest.

Fixed-Bed Reactors

In a TBR the catalyst bed is fixed, the flow pattern is much closer to plug flow, and the ratio of liquid to solid catalyst is small. If heat effects are substantial [*i.e.*, highly exothermic reactions such as those occurring in hydrotreating of unsaturated feeds (light cycle oil from fluid catalytic cracking units)], they can be controlled by recycling of the liquid product stream, although this may not be practical if the product is not relatively stable under reaction conditions or if very high conversion is desired, as in HDS, since recycling causes the system to approach the behavior of a continuous-stirred-tank reactor (CSTR). For such high-temperature increases, the preferred solution is quenching with hydrogen, although the use of other streams has also been reported.Even when a completely vapor-phase reaction in a fixed catalyst bed may be technically feasible, a TBR may be preferred to save energy costs due to reactant vaporization. The limiting reactant may be essentially all in the liquid phase or in both the liquid and gas phases, and the distribution of reactant and products between the gas and liquid phases may vary with conversion.

TBR with Co-current Gas-Liquid Downflow

A TBR consists of a column that may be very high (above 10 to 30 m), equipped with one or various fixed beds of solid catalysts, throughout which gas and liquid move in co-current downflow. The typical film flow texture found during a trickle-flow regime. In this mode, gas is the continuous phase and liquid holdup is lower. This operation is the one most used in practice, since there are less severe limitations in throughput than in countercur-rent operation.

For gas-limited reactions (high liquid reactant flux to the catalyst particle, low gas reactant flux to the particle), especially at partially wetted conditions, a downflow reactor is preferred, as it facilitates transport of the gaseous reactant to the catalyst. In contrast to commercial TBR, in the case of bench-scale TBR operating at equivalent space velocity, the liquid velocity and the catalyst bed length have important effects on the performance of the reactor. The principal advantages and disadvantages of TBR with downflow co-current operation are given below.

Advantages

- Recommended for gas-limited reactions
- Liquid flow approaches plug-flow behavior, which leads to high conversions

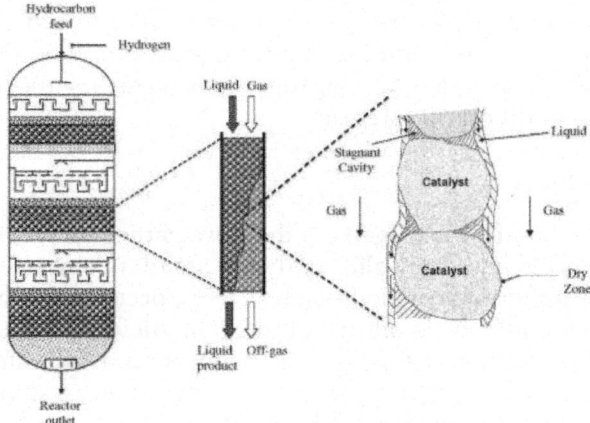

Fig. : Nonideal TBR suffering from liquid maldistribution.

- Low liquid-solid volume ratio: fewer occurrences of homogeneous side reactions
- Possibility of varying the liquid rate according to catalyst wetting and heat and mass transfer resistances
- A variety of flow regimes allowed; most flexible with respect to varying throughput demands
- The down flow mode also helps keep the bed in place, although with catalysts that are soft or deformable, this might hasten undesired cementation
- Compared with countercurrent flow operation, for co-current flow of the two phases, no limitation on the throughput arises from the phenomenon of flooding, and the quantities of the phase that can be passed depend only on the upstream pressure available because of vaporization effects
- At higher gas loadings, the texture of the liquid is modified by gas-phase friction, the liquid distribution is improved (lower liquid wall flow), and the pressure drop rises (less rapidly in co-current than in countercurrent flow)
- Easy operation with fixed adiabatic beds; for exothermic reaction systems, gas or liquid streams as quench, and the liquid and/or gas recycle limit temperature rises
- Possibility of operating at higher pressure and temperature
- Pressure drop through the bed is relatively low, thus reducing pumping costs

- Larger reactor size, and generally of simple construction, as there are no moving parts
- Lower investment and operating costs, and low catalyst loss, which is important when costly catalysts are used

Disadvantages

- Limitations on the use of viscous or foaming liquids
- Limited to reasonably fast reactions
- Lower catalyst effectiveness, due to the use of large catalyst particle size
- Particle size cannot be smaller than 1 mm because of pressure drop; risk of increasing pressure drop or obstructing catalyst pores when side reactions lead to fouling products
- Reactor-scale maldistribution, channeling, and incomplete and/or ineffective external catalyst wetting (poor contacting effectiveness) can occur with low liquid flow rates and reactor diameter/particle size ratios (<25)
- Sensitivity to thermal effects, although this drawback can be limited by recycling part of the outlet liquid or injecting cooled gas (quenching)
- Difficulties in the recovery of reaction heat
- Lower liquid holdup compared with co-current gas-liquid upflow
- Deactivation of the catalyst by deposits
- Dismantling of the reactor during catalyst replacement
- In hydrotreating (HDT) reactors, most of the bed is under the H2S and NH3 reach regime and its inhibiting effect is strongest in the region where the refractory sulfur compounds have to be converted. NH3, particularly, strongly suppresses the activity of the acidic function of the hydrocracking catalyst
- H2 partial pressure will be lowest at the HDT reactor outlet due to the combined effect of pressure drop, hydrogen consumption, and reduction of hydrogen purity as gaseous by-product yields (H-S, NH3 – and H-O) increase along the reactor
- Used in downward mode in the refining industry with less conversion; the inhibition effect of H2S and NH3 on the catalyst results in a poorer performance

TBR with Countercurrent Gas-Liquid Flow TBRs

TBR with Countercurrent Gas-Liquid Flow TBRs operating in countercurrent gas-liquid flow provide an opportunity for selective removal of by-products that may act as inhibitors (*e.g.*, in hydrodesulfurization, where hydrogen sulfide may have an inhibitory effect). The introduction of FBRs with countercurrent flow in a number of refining operations is probably either *via* redesign of existing reactors or by introduction of new technology. The goal is not an improvement in reactant (hydrogen) mass transfer, which is not rate limiting, but enhanced removal of inhibitory byproducts or in situ product separation. That is why countercurrent flow will become more prominent in the future for processes that suffer from byproduct catalyst inhibition.

A Catalytic PBR with Countercurrent Mode

A catalytic PBR with countercurrent mode is a suitable alternative to TBRs for reactions conducted over catalysts with a very large surface area-to-volume ratio. However, the main problem of the countercurrent reactor for commercial application is due to hardware limitations. There is therefore a need to develop improved hardware configurations that allow countercurrent contacting of gas and liquid in the presence of small catalyst particles. The main advantages and disadvantages of TBRs with countercurrent flow are given below.

Advantages

- Countercurrent operation is preferred over co–current when a large heat of reaction is involved
- Countercurrent operation gives a more favorable flat axial temperature profile
- Large surface area for vapor-liquid mass transfer
- High ratio of number of active sites to reactor volume
- Easy catalyst handling
- For the HDT process, the major part of the bed is in an H2S-lean regime, which protects from inhibition by H2S formed in a large part of the bed.
- H2 partial pressure is highest at the end of the bed, and temperature in this part can be lowered and more active, less sulfur-tolerant catalysts can be used in the downstream part of the bed, which will favor the chemical equilibrium for reversible reactions [*i.e.*, hydrodearomatization (HDA) reaction]. The effect of equilibrium-limited conversion and product inhibition is reduced
- The major part of the bed is in the NH3 [a by-product of hydrodenitrogenation (HDN) reaction]-lean regime, which favors the HDT reaction by protection from the inhibition of NH3 and H2S. This operation has great advantages through omitting two separate reactor stages
- The concentration of gas impurities formed during reaction is less in most parts of the bed. This favors the conversion of reactions normally limited by chemical equilibrium and enables handling more difficult feedstocks to obtain higher levels of conversion. The typical partial pressure profiles of H2S along the bed length for co-current and countercurrent operations during hydroprocessing, in which the aforementioned behavior is clearly observed

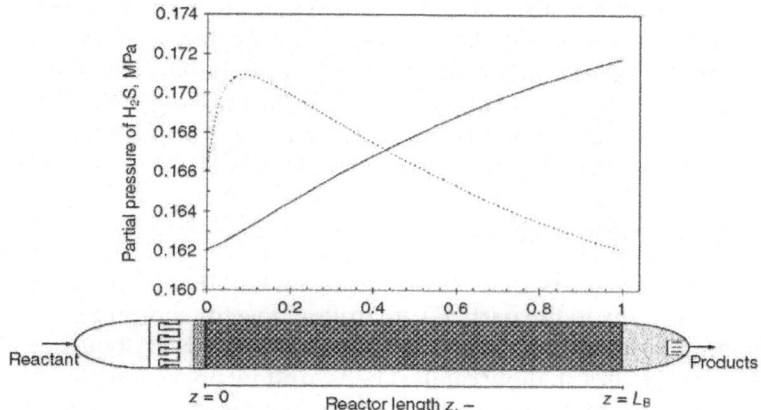

Fig. : Profiles of H_2S partial pressure along the catalytic bed in an HDT reactor
($-$, co-current; $-$, countercurrent).

- Countercurrent operation provides the highest hydrogen purity in that part of the bed where the least reactive compounds need to be converted

Disadvantages

- Presence of flooding at high liquid throughputs
- Estimation of liquid holdup, pressure drop, and mass transfer coefficients is difficult since correlations employed to calculate these parameters do not include data for the small porous catalyst packing typically used in PBRs with two-phase flow
- Limited to low velocities far below those of industrial interest, due to the occurrence of excessive pressure drop and flooding problems
- It is not possible to use smaller (1 to 5 mm) catalyst particles than those used in co-current downflow TBRs
- High axial dispersion effects in the liquid phase

Packed Bubble-Flow

Reactors with Co-current Gas-Liquid Upflow This classification includes upflow reactors, upflow co-current reactors, packed-bubble columns, upflow packed-bubble columns, and flooded fixed – bed reactors. In bubble-flow operation a continuous liquid phase, together with a dispersed gas phase, move upward co-currently through the packed bed. Such an operation would be recommended in cases where liquid reactants are treated with a relatively small amount of gas, as in the hydration of nitro compounds and olefins, or where a relatively large liquid residence time is required for the degree of conversion desired. Use of these reactors assures complete external wetting of the catalyst and high liquid holdup. In this mode the liquid is typically the continuous phase.

Bubble operation is also advantageous when the reactor diameter/particle diameter ratio is relatively small, because the liquid catalyst contact is more effective than in trickle operation. Compared with empty bubble columns, the packed bed has the advantage of reducing substantially backmixing in the flowing phases as well as the coalescence of gas bubbles. Under any conditions the wall heat transfer coefficient should also be higher than it is in trickle operation.

For Liquid-limited Reactions

For liquid-limited reactions (low liquid reactant flux to the catalyst particle, high gas reactant flux to the particle), an upflow reactor should be preferred, as it provides complete catalyst wetting and the fastest transport of the liquid reactant to the catalyst. For very shallow catalyst beds, upflow operation gives much better conversions than downflow operation under the same reaction conditions. The gas and liquid flow rates typically used in a bench-scale down-flow trickle-bed HDS reactor are such that when they are used in co-current upflow operation, a bubble flow regime will be generated.

The performance of a reactor under this hydrodynamic flow condition should be considerably different from the one obtained under trickle- flow conditions. In an upflow system the low-boiling components, which are generally more reactive, pass into the vapor phase and are swept out more rapidly than the high-boiling material, which progresses relatively slowly through the bed. This superior performance of upflow processing is attributed to the long residence time of the heavy liquid fractions, but a more important factor may be the very low liquid flow used.

When both gas and liquid flow upward, maldistribution of liquid or incomplete catalyst wetting should not be very important, particularly when the hydrodynamic conditions of bubble flow prevail within the reactor. An upflow (flooded bed) reactor, which should give good solid–iquid contacting, could be used instead of an autoclave to obtain information on the intrinsic kinetics. The main advantages and disadvantages of TBRs with co-current upflow are given below.

Advantages

- Recommended for liquid-limited reactions
- Liquid holdup is higher. The liquid holdup is larger in an upflow operation than in a downflow operation under similar conditions
- Better effective wetting
- Better thermal stability for highly exothermic reactions
- High liquid saturation
- The liquid flow can be more uniformly distributed (better distribution of liquid throughout the catalyst bed)
- The gas-liquid and liquid-solid mass transfer coefficients are larger in an upflow operation than in a downflow operation

- In backmix flow conditions, where variations in gas and liquid flow rates change the conversion, upflow operation gives better results than downflow operation under the same conditions
- Larger effective residence time
- If a catalyst gradually becomes deactivated by the deposit of polymeric or tarry materials, the upflow reactor may maintain its activity longer by washing off these deposits more effectively
- For rapid and highly exothermic reactions, heat transfer between liquid and solid may also be more effective in upflow than in downflow operation

Disadvantages

- For HDT operations, conversions of sulfur, metals, and asphaltenes decrease with an increase in gas and liquid flow rates at constant temperature and pressure. Conversion of sulfur in upflow operation is reduced faster with time than in downflow operation; however, the conversion is always highest
- Higher pump requirements in order to overcome the hydrostatic head of the liquid
- The need of some designs to avoid the fluidization of the catalyst unless the catalyst was held in place by an extra weight or suitable mechanical methods
- If limiting reactant is present in both phases, over a range of operating conditions in which catalyst pellets filled with liquid are diffusion limited, an upflow reactor would be expected to exhibit a lower reaction rate than a partially wetted TBR
- Formation of stagnant zones inside the catalyst bed
- Higher axial dispersion compared with the downflow mode of operation

Slurry-Bed Reactors

The best alternative to the use of a fixed – bed reactor with two – phase flow, either upward or downward, is a slurry-bed or ebullating-bed reactor in which the catalyst particles, which must be substantially smaller, are in motion. These reactors are sometimes termed three- phase fluidized – bed reactors or suspended-bed reactors. The main advantages and disadvantages of slurry-bed reactors are given below.

Advantages

- High heat capacity to provide good temperature control
- Potentially high reaction rate per unit volume of reactor if the catalyst is highly active

- Easy heat recovery
- Adaptability to either batch or flow processing
- Much lower pressure drop
- The catalyst may readily be removed and replaced if its working life is relatively short
- Continuous removal of solid material formed in reaction
- Because of high intraparticle diffusion rate, small particles can be used, which may allow for operating at catalyst effectiveness factors approaching unity, of special importance if diffusion limitations cause rapid catalyst degradation or poorer selectivity
- Lower external mass transfer resistance by means of a high stirring speed

Disadvantages

- Residence-time distribution patterns are close to those of a CSTR, which makes it difficult to obtain high degrees of conversion except by staging and/or increasing operation temperature
- Generation of fine particles by abrasion of the catalyst
- Catalyst removal by filtration may provoke problems with possible plugging difficulties on filters, further time of operation, and the costs of filtering systems may be a substantial portion of the capital investment
- Higher catalyst consumption than that of fixed-bed reactors
- Difficult to scale up
- The high liquid-to-solid ratio in a slurry-bed reactor allows homogeneous side reactions to become more important, if any are possible
- Potential hazard of localized overheating in the reactor because of bad fluidization
- Backmixed flow and the volume of the reactor are not fully utilized

CATALYSIS

Catalysts are substances that speed up a reaction but which are not consumed by it and do not appear in the net reaction equation. In addition, *catalysts affect the forward and reverse rates equally*; this means that catalysts have no effect on the equilibrium constant and thus on the composition of the equilibrium state. Thus a catalyst (in this case, sulfuric acid) can be used to speed up a reversible reaction such as ester formation or its reverse.

Catalysts Provide Alternative Reaction Pathways

Catalysts function by allowing the reaction to take place through an alternative mechanism that requires a smaller activation energy. This change is brought about by a specific interaction between the catalyst and the reaction components. Recall that the rate constant of a reaction is an exponential function of the activation energy, so even a modest reduction of E_a can yield an impressive increase in the rate.

Catalysts are conventionally divided into two categories: *homogeneous* and *heterogeneous*. *Enzymes*, natural biological catalysts, are often included in the former group, but because they share some properties of both but exhibit some very special properties of their own, they are treated here as a third category.

SOME COMMON EXAMPLES OF CATALYSIS

How to Burn a Sugar Cube

When heated by itself, a sugar cube (sucrose) melts at 185°C but does not burn. But if the cube is rubbed in cigarette ashes, it burns before melting owing to the catalytic action of trace metal compounds in the ashes.

Platinum as an Oxidation Catalyst

The surface of metallic platinum is an efficient catalyst for the oxidation of many fuel vapors. This property is exploited in flameless camping stoves (left). The image at the right shows a glowing platinum wire heated by the slow combustion of ammonia on its surface.

Decomposition of Hydrogen Peroxide

Hydrogen peroxide is thermodynamically unstable according to the reaction

$$2\,H_2O_2 \rightarrow 2\,H_2O + O_2 \quad \Delta G = -210\ kJ\ mol^{-1}$$

In the absence of contaminants this reaction is very slow, but a variety of substances, ranging from iodine, metal oxides, trace amount of metals, greatly accelerate the reaction, in some cases almost explosively owing to the rapid release of heat. The most effective catalyst of all is the enzyme *catalase*, present in blood and intracellular fluids; adding a drop of blood to a solution of 30% hydrogen peroxide induces a vigorous reaction.

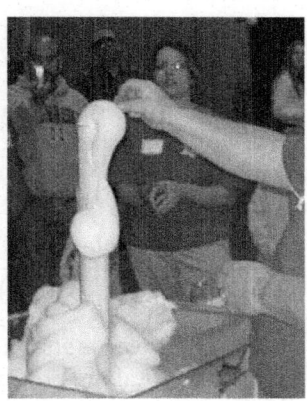

Fig. : Relative rates of H_2O_2 decomposition. Examples of rapid liberation of O_2. can result in a spectacular bubblebath if some soap is added. This same reaction has been used to power a racing car!

Each kind of catalyst facilitates a different pathway with its own activation energy. Because the rate is an exponential function of E_a (Arrhenius equation), even relatively small differences in E_a's can have dramatic effects on reaction rates. Note especially the values for catalase; the chemist is still a rank amateur compared to what Nature can accomplish through natural selection!

Catalyst	E_a kJ/mol	Relative rate
no catalyst	75	1
iodide ion	56	2140
colloidal platinum	50	24,000
catalase (enzyme)	21	2,900,000,000

How Catalytic Activity is Expressed

Changes in the rate constant or of the activation energy are obvious ways of measuring the efficacy of a catalyst. But two other terms have come into use that have special relevance in industrial applications.

Turnover Number

The turnover number (TON) is an average number of cycles a catalyst can undergo before its performance deteriorates. Reported TONs for common industrial catalysts span a very wide range from perhaps 10 to well over 10^5, which approaches the limits of diffusion transport.

Turnover Frequency

This term, which was originally applied to enzyme-catalyzed reactions, has come into more general use. It is simply the number of times the overall catalyzed reaction takes place per catalyst (or per active site on an enzyme or heterogeneous catalyst) per unit time is defined as:

$$\text{TOF} = \frac{\text{number of product molecules}}{(\text{number of active sites}) \times (\text{time})} = \frac{1}{S}\frac{dn}{dt}$$

The number of active sites S on a heterogeneous catalyst is often difficult to estimate, so it is often replaced by the total area of the exposed catalyst, which is usually experimentally measurable. TOFs for heterogeneous reactions generally fall between 10^{-2} to $10^2 s^{-1}$.

Homogeneous Catalysis

As the name implies, homogeneous catalysts are present in the same phase (gas or liquid solution) as the reactants. Homogeneous catalysts generally enter directly into the chemical reaction (by forming a new compound or complex with a reactant), but are released in their initial form after the reaction is complete, so that they do not appear in the net reaction equation.

Iodine-catalyzed Cis-trans Isomerization

Unless you are taking an organic chemistry course in which your instructor indicates otherwise, don't try to memorize these mechanisms. They are presented here for the purpose of convincing you that catalysis is not black magic, and to familiarize you with some of the features of catalyzed mechanisms. It should be sufficient for you to merely convince yourself that the individual steps make chemical sense.

You will recall that cis-trans isomerism is possible when atoms connected to each of two doubly-bonded carbons can be on the same (*cis*) or opposite (*trans*) sides of the bond. This reflects the fact that rotation about a double bond is not possible.

Conversion of an alkene between its *cis*- and *trans* forms can only occur if the double bond is temporarily broken, thus freeing the two ends to rotate. Processes that cleave covalent bonds have high activation energies, so cis-trans isomerization reactions tend to be slow even at high temperatures. Iodine is one of several catalysts that greatly accelerate this process, so the isomerization of *butene* serves as a good introductory example of homogeneous catalysis.

The mechanism of the iodine-catalyzed reaction is believed to involve the attack of iodine atoms (formed by the dissociation equilibrium ①) on one of the carbons in Step ② :

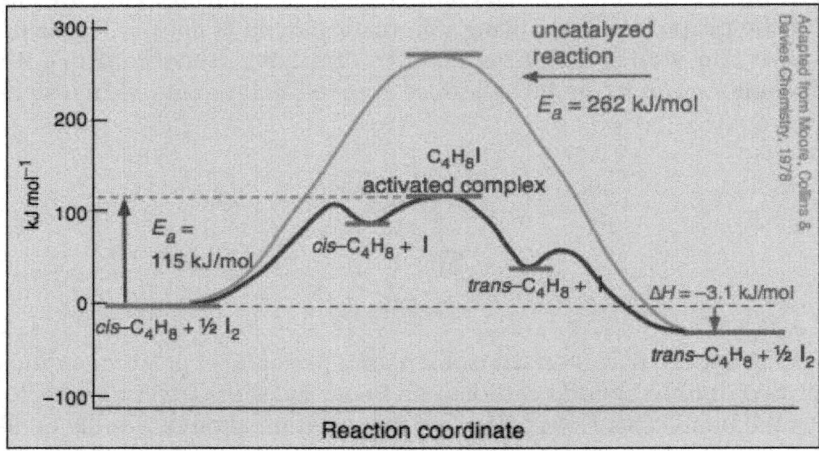

During its brief existence, the free-redical activated complex can undergo rotation about the C—C bond, so that when it decomposes by releasing the iodine (**3**), a portion of the reconstituted butene will be in its *trans*form. Finally, the iodine atom recombine into diiodine. Since processes **1** and **4** cancel out, iodine does not appear in the net reaction equation — a requirement for a true catalyst.

Acid-base Catalysis

Many reactions are catalyzed by the presence of an acid or a base; in many cases, both acids and bases will catalyze the same reaction. As one might expect, the mechanism involves the addition or removal of a proton, changing the reactant into a more kinetically labile form. A simple example is the addition of iodine to propanone

$$I_2 + (CH_3)_2\text{-}C\text{=}O \rightarrow (CH_2I)(CH_3)\text{-}C\text{=}O$$

The mechanism for the acid-catalyzed process involves several steps. The role of the acid is to provide a proton that attaches to the carbonyl oxygen, forming an unstable oxonium ion . The latter rapidly rearranges into an enol (*i.e.*, a carbon connected to both a double bond (*ene*) and a hydroxyl (*ol*) group.) This completes the catalytic part of the process, which is basically an acid-base (proton-transfer) reaction in which the role of the proton is to withdraw an electron from the ketone oxygen.

$$
\begin{array}{c}
CH_3 \\
| \\
C=O \\
| \\
CH_3
\end{array}
+ H^+(aq)
\xrightarrow[\text{①}]{\text{slow}}
\begin{array}{c}
CH_3 \\
| \\
C=O-H \\
|\;\oplus \\
CH_3
\end{array}
\xrightarrow{\text{fast}}
\begin{array}{c}
CH_2 \\
|| \\
C-O-H \\
|\quad\text{②} \\
CH_3
\end{array}
+ H^+
$$

propanone
(acetone)

In the second stage, the enol reacts with the iodine. The curved arrows indicate shifts in electron locations. An electron is withdrawn from the π orbital of the double bond by one of the atoms of the I_2 molecule. This induces a shift of electrons in the latter, causing half of this molecule to be expelled as an iodide ion. The other half of the iodine is now an iodonium ion I^+ which displaces a proton from one of the methyl groups. The resultant carbonium ion then expels the -OH proton to yield the final neutral product.

Perhaps the most well-known acid-catalyzed reaction is the hydrolysis (or formation) of an ester — a reaction that most students encounter in an organic chemistry laboratory course. This is a more complicated process involving five steps.

Oxidation-reduction Catalysis

Many oxidation-reduction (electron-transfer) reactions, including direct oxidation by molecular oxygen, are quite slow. Ions of transition metals capable of existing in two oxidation states can often materially increase the rate.

An example would be the reduction of Fe^{3+} by the vanadium ion V^{3+}:

$V^{3+} + Fe^{3+} \rightarrow V^{4+} + Fe^{2+}$

This reaction is catalyzed by either Cu^+ or Cu^{3+}, and the rate is proportional to the concentration of V^{3+} and of the copper ion, but independent of the Fe^{3+} concentration. The mechanism is believed to involve two steps:

1 $V^{3+} + Cu^{2+} \xrightarrow{k_1} V^{4+} + Cu^+$ (rate-determining)

2 $Fe^{3+} + Cu^+ \xrightarrow{k_2} Fe^{2+} + Cu^{2+}$ (very fast)

(If Cu^+ is used as the catalyst, it is first oxidized to Cu^{2+} by step 2.)

Hydrogen Peroxide, Again

Ions capable of being oxidized by an oxidizing agent such as H_2O_2 can serve as catalysts for its decomposition.

Thus H_2O_2 oxidizes iodide ion to iodate, when then reduces another H_2O_2 molecule, returning an I^- ion to start the cycle over again:

$H_2O_2 + I^- \rightarrow H_2O + IO^-$

$H_2O_2 + IO^- \rightarrow H_2O + O_2 + I^-$

Iron(II) can do the same thing. Even traces of metallic iron can yield enough Fe^{2+} to decompose solutions of hydrogen peroxide.

$$H_2O_2 + Fe^{2+} \rightarrow H_2O + Fe^{3+}$$
$$H_2O_2 + Fe^{3+} + 2H^+ \rightarrow H_2O + O_2 + Fe^{2+} + 2H^+$$

Heterogeneous Catalysts

As its name implies, a heterogeneous catalyst exists as a separate phase (almost always a solid) from the one (most commonly a gas) in which the reaction takes place. The catalytic affect arises from disruption (often leading to dissociation) of the reactant molecules brought about by their interaction with the surface of the catalyst.

Unbalanced Forces at Surfaces

You will recall that one universal property of matter is the weak attractive forces that arise when two particles closely approach each other. When the particles have opposite electric charges or enter into covalent bonding, these far stronger attraction dominate and define the "chemistry" of the interacting species. The molecular units within the bulk of a solid are bound to their neighbors through these forces which act in opposing directions to keep each under a kind of "tension" that restricts its movement and contributes to the cohesiveness and rigidity of the solid.

At the surface of any kind of condensed matter, things are quite different. The atoms or molecules that reside on the surface experience unbalanced forces which prevents them from assuming the same low potential energies that characterize the interior units. (The same thing happens in liquids, and gives rise to a variety of interfacial effects such as surface tension.)

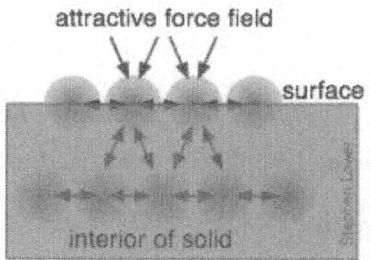

But in the case of a solid, in which the attractive forces tend to be stronger, something much more significant happens. The molecular units that reside on the surface can be thought of as partially buried in it, with their protruding parts (and the intermolecular attractions that emerge from them) exposed to the outer world. The strength of the attractive force field which emanates from a solid surface varies in strength depending on the nature of the atoms or molecules that make up the solid.

Don't confuse adsorption with absorption; the latter refers to the bulk uptake of a substance into the interior of a porous material. At the microscopic level, of course, absorption also involves adsorption. The process in which molecules in a gas or a liquid come into contact with and attach themselves to a solid surface is known as adsorption. Adsorption is almost always an exothermic process and its strength is conventionally expressed by the enthalpy or "heat" of adsorption ΔH_{ads}.

Chemisorption and Physisorption

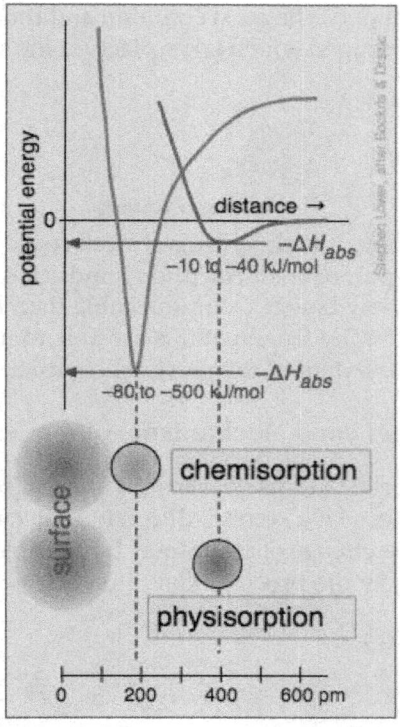

Two general categories of adsorption are commonly recognized, depending on the extent to which the electronic- or bonding structure of the attached molecule is affected. When the attractive forces arise from relatively weak van der Waals interactions, there is little such effect and ΔH_{ads} tends to be small. This condition is described as *physical adsorption* (physisorption). Physisorption of a gas to a surface is energetically similar to the condensation of the gas to a liquid, it usually builds up multiple layers of adsorbed molecules, and it proceeds with zero activation energy.

Of more relevance to catalytic phenomena is chemisorption, in which the adsorbate is bound to the surface by what amounts to a chemical bond. The resulting disruption of the electron structure of the adsorbed species "activates" it and makes it amenable to a chemical reaction (often dissociation) that could not be readily achieved through thermal activation in the gas or liquid phase. In contrast to physisorption, chemisorption generally involves an activation energy (supplied by ΔH_{ads}) and the adorbed species is always a monolayer.

MECHANISMS OF REACTIONS ON SURFACES

Dissociative Adsorption

The simplest heterogeneous process is chemisorption followed by bond-breaking as described above. The most common and thoroughly-studied of these is the dissociation of hydrogen which takes place on the surface of most transition metals.

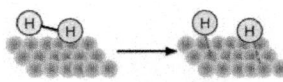

dissociative adsorbtion

The single $1s$ electron of each hydrogen atom coordinates with the d orbitals of the metal, forming a pair of chemisorption bonds (indicated by the red dashed lines). Although these new bonds are more stable than the single covalent bond they replace, the resulting hydrogen atoms are able to migrate along the surface owing to the continuous extent of the d-orbital conduction band.

The Langmuir-Hinshelwood Mechanism

Although the adsorbed atoms ("*adatoms*") are not free radicals, they are nevertheless highly reactive, so if a second, different molecular species adsorbs onto the same surface, an interchange of atoms may be possible. Thus carbon monoxide can be oxidized to CO_2 by the process illustrated below:

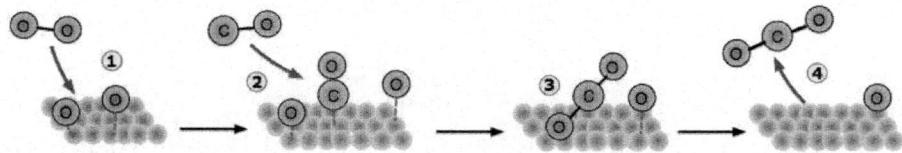

Irvin Langmuir (1881-1967) was a pioneer in surface chemistry, and as an employee of General Electric, was the first American industrial chemist to win a Nobel Prize (1932). Sir Cyril Hinshelwood (1897-1967) carried out early studies of kinetics and chain reactions, and was a joint winner of the 1956 Nobel Prize in Chemistry for this work.

In this example, only the O_2 molecule undergoes dissociation . The CO molecule adsorbs without dissociation , configured perpendicular to the surface with the chemisorption bond centered over a hollow space between the metal atoms. After the two adsorbed species have migrated near each other , the oxygen atom switches its attachment from the metal surface to form a more stable C=O bond with the carbon , followed by release of the product molecule.

The Eley-Rideal Mechanism

An alternative mechanism eliminates the second chemisorption step; the oxygen atoms react directly with the gaseous CO molecules by replacing the chemisorption bond with a new C–O bond as they swoop over the surface:

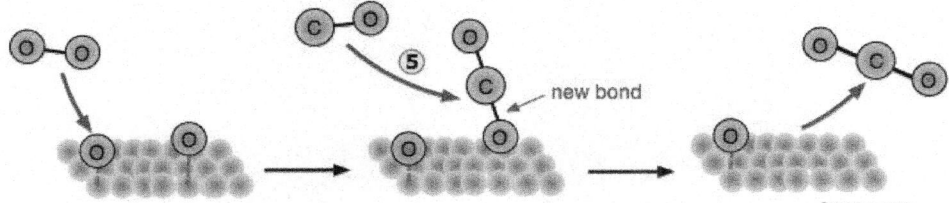

Examples of both mechanisms are known, but the Langmuir-Hinshelwood mechanism is more important in that it exploits the activation of the adsorbed reactant. In the case of carbon monoxide oxdation, studies involving molecular beam experiments support this scheme. A key piece of evidence is the observation of a short time lag between contact of a CO molecule with the surface and release of the CO_2, suggesting that CO remains chemisorbed during the interval.

The Sabatier Principle

Paul Sabatier (1854-1941) was a French chemist who developed numerous catalytic processes, particularly those involving hydrogenation of organic compounds. He was a co-winner of the 1913 Nobel Prize in Chemistry. To be effective, these processes of adsorption, reaction, and desorption must be orchestrated in a way that depends critically on the properties of the catalyst in relation to the chemisorption properties (ΔH_{ads}) of the reactants and products.

- Adsorption of the reactant onto the catalytic surface must be strong enough to perturb the bonding within the species to dissociate or activate it;
- If the resulting fragments must migrate to other locations on the surface, their chemisorption must be weak enough to allow this movement but not so small that they escape before they have a chance to react;
- The product species must have sufficiently small ΔHads values to ensure their rapid desorption from the catalyst so that surface is freed up to repeat the cycle

The importance of choosing a catalyst that achieves the proper balance of the heats of adsorption of the various reaction components is known as the *Sabatier Principle*, but is sometimes referred to as the "just-right" or "Goldilocks Principle".

In its application to catalysis, this principle is frequently illustrated by a "volcano diagram" in which the rate of the catalyzed reaction is plotted as a function of ΔH_{ads} of a substrate such as H_2 on a transition metal surface. The plot at the left shows the relative effectiveness of various metals in catalyzing the decomposition of formic acid HCOOH. The vertical axis is plotted as temperature, the idea being that the better the catalyst, the lower the temperature required to maintain a given rate.

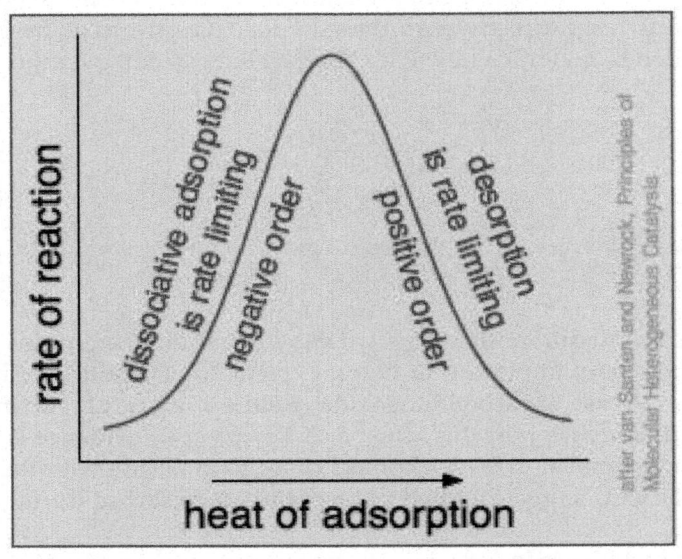

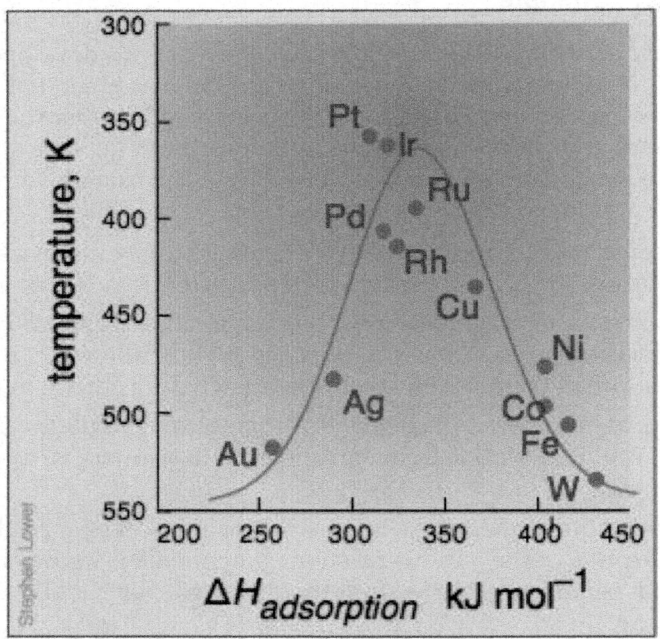

The Catalytic Cycle

This term refers to the idealized sequence of steps between the adsorption of a reactant onto the catalyst and the desorption of the product, culminating in restoration of the catalyst to its original condition. A typical catalytic cycle for the hydrogenation of propene is illustrated below.

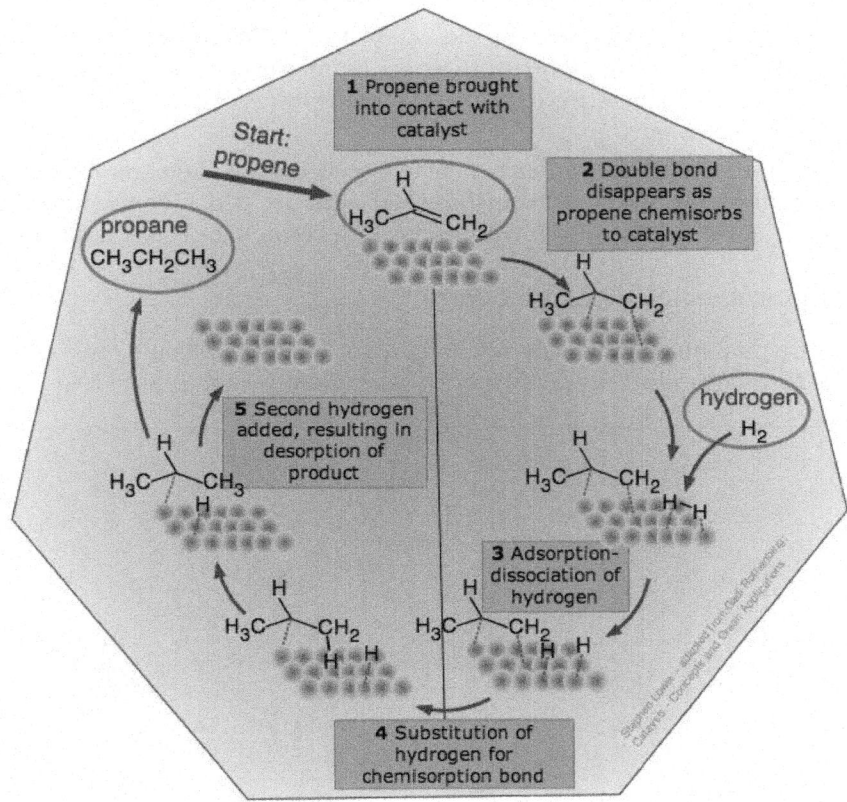

This particular reaction

$$H_3C-CH=CH_2 + H_2 \rightarrow H_3C-CH_2-CH_3$$

takes place spontaneously only in the reverse direction, but it is representative of the process used to hydrogenate the carbon-carbon double bonds in vegetable oils to produce solid saturated fats such as margarine.

Catalyst Poisoning and Breakdown

Catalyst poisoning, brought about by irreversible binding of a substance to its surface, can be permanent or temporary. In the latter case the catalyst can be regenerated, usually by heating to a high temperature. In organisms, many of the substances we know as "poisons" act as catalytic poisons on enzymes.

If catalysts truly remain unchanged, they should last forever, but in actual practice, various events can occur that limit the useful lifetime of many catalysts.

- Impurities in the feedstock or the products of side reactions can bind permanently to a sufficient number of active sites to reduce catalytic efficiency over time, "poisoning" the catalyst.

- Physical deterioration of the catalyst or of its support, often brought about by the high temperatures sometimes used in industrial processes, can reduce the effective surface area or the accessibility of reactants to the active sites.

Catalysts tend to be rather expensive, so it is advantageous if they can be reprocessed or regenerated to restore their activity. It is a common industrial practice to periodically shut down process units to replace spent catalysts.

How Heterogeneous Catalysts Work

The actual mechanisms by which adsorption of a molecule onto a catalytic surface facilitates the cleavage of a bond vary greatly from case to case.

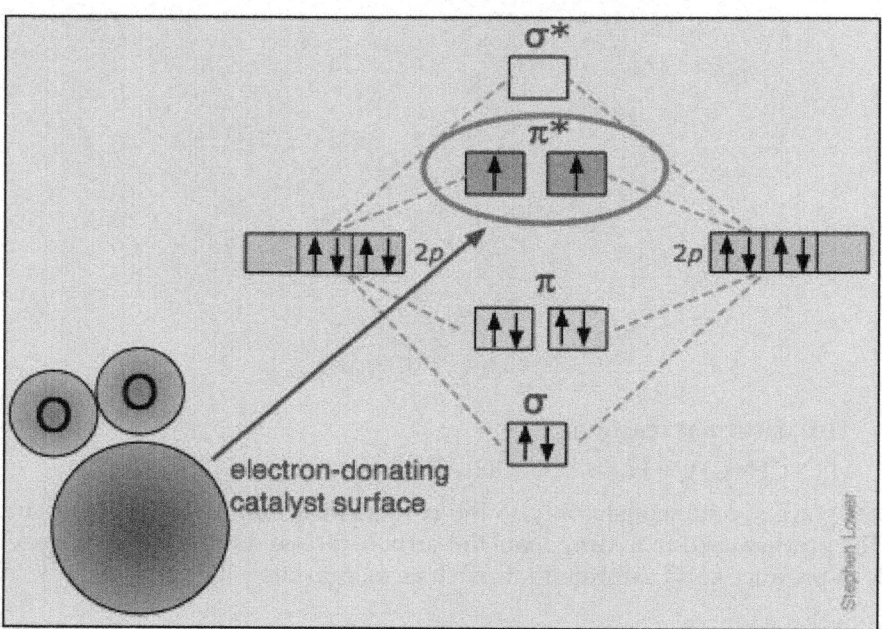

We give here only one example, that of the dissociation of dioxygen O_2 on the surface of a catalyst capable of temporarily donating an electron which enters an oxygen antibonding molecular orbital that will clearly destabilize the O–O bond. (Once the bond has been broken, the electron is given back to the catalyst.)

Types of Catalytically Active Surfaces

Heterogeneous catalysts mostly depend on one or more of the following kinds of surfaces:

Active surface type	Remarks
covalent solid	Atoms at surfaces or crystal edges in macromolecular 2- and 3-D networks such as graphite or quartz may have free valences
ionic solid	Surfaces and edges are sites of intense electric fields able to interact with ions and polar molecules
oxide	Oxides can acquire H^+ and/or OH^- groups able to act as acid- or base catalysts
metals	"Electron gas" at metal surfaces can perturb bonding in substrate molecules
d-orbitals / transition metal	Vacant d orbitals can provide a variety of coordination sites for activation.
conduction band ← band gap / semiconductor	Semiconductors (including many oxides) can supply electrons, thermally excited through reasonably small (<50 kJ) band gaps.

Some Factors Affecting Catalyst Efficacy

Since heterogeneous catalysis requires direct contact between the reactants and the catalytic surface, the area of active surface goes at the top of the list. In the case of a metallic film, this is not the same as the nominal area of the film as measured by a ruler; at the microscopic level, even apparently smooth surfaces are highly irregular, and some cavities may be too small to accommodate reactant molecules.

Consider, for example, that a 1-cm cube of platinum (costing roughly $1000) has a nominal surface area of only 6 cm^2. If this is broken up into 10^{12} smaller cubes whose sides are 10^{-6} m, the total surface area would be 60,000 cm^2, capable in principle of increasing the rate of a Pt-catalyzed reaction by a factor of 10^4. These very finely-divided (and often very expensive) metals are typically attached to an inert *supporting surface* to maximize their exposure to the reactants.

Surface topography. At the microscopic level, even an apparently smooth surface is pitted and uneven, and some sites will be more active than others. Penetration of molecules into and out of some of the smaller channels of a porous surface may become rate-limiting.

An otherwise smooth surface will always possess a variety of defects such as steps and corners which offer greater exposure and may be either the only active sites on the surface, or overly active so as to permanently bind to a reactant, reducing the active area of the surface.

In one study, it was determined that kink defects constituting just 5 percent of platinum surface were responsible for over 90% of the catalytic activity in a certain reaction.

Steric Factors

When chemisorption occurs at two or more locations on the reactant, efficient catalysis requires that the spacing of the active centers on the catalytic surface be such that surface bonds can be formed without significant angular distortion.

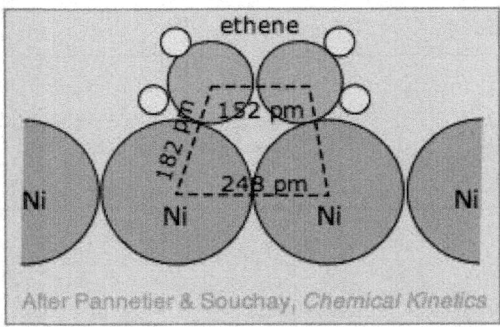

After Pannetier & Souchay, *Chemical Kinetics*

Thus activation of the ethylene double bond on a nickel surface proceeds efficiently because the angle between the C — Ni bonds and the C — C is close to the tetrahedral value of 109.5° required for carbon sp^3 hybrid orbital formation. Similarly, we can expect that the hydrogenation of benzene should proceed efficiently on a surface in which the active sites are spaced in the range of 150 and 250 pm.

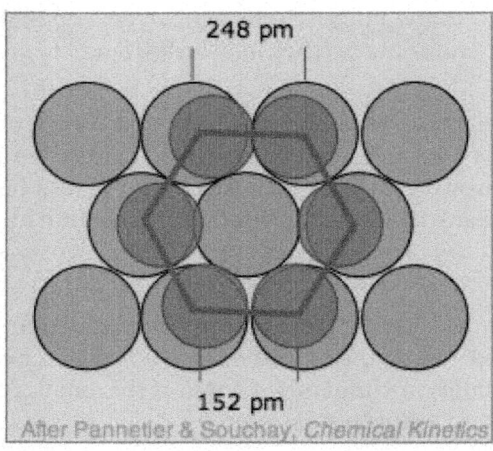

After Pannetier & Souchay, *Chemical Kinetics*

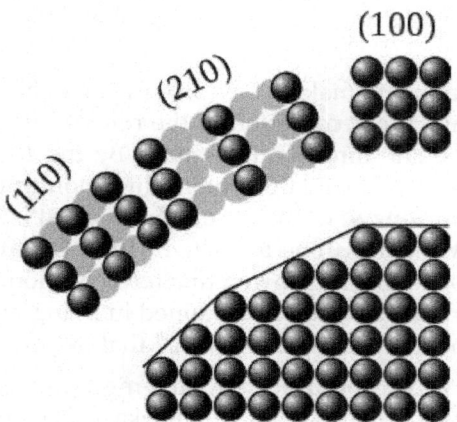

This is one reason why many metallic catalysts exhibit different catalytic activity on different crystal faces.

SOME SPECIAL TYPES OF HETEROGENEOUS CATALYSTS

Metal-cluster Catalysts

As the particle size of a catalyst is reduced, the fraction of more highly exposed step, edge, and corner atoms increases. An extreme case occurs with nano-sized (1-2 nm) metal cluster structures composed typically of 10-50 atoms. Metallic gold, well known for its chemical inertness, exhibits very high catalytic activity when it is deposited as metallic clusters on an oxide support. For example, O_2 dissociates readily on Au_{55} clusters which have been found to efficiently catalyze the oxidation of hydrocarbons.

Zeolite Catalysts

Zeolites are clay-like aluminosilicate solids that form open-framework microporous structures that may contain linked cages, cavities or channels whose dimensions can be tailored to the sizes of the reactants and products. To those molecules able to diffuse through these spaces, zeolites are in effect "all surface", making them highly efficient. This size-selectivity makes them important for adsorption, separation, ion-exchange, and catalytic applications. Many zeolites occur as minerals, but others are made synthetically in order to optimize their properties. As catalysts, zeolites offer a number of advantages that has made them especially important in "green chemistry" operations in which the number of processing steps, unwanted byproducts, and waste stream volumes are minimized.

Enzymes and Biocatalysis

This distortion of Robert Fitzgerald's already-distorted translation of the famous quatrain from the wonderful *Rubaiyat of Omar Khayyam* underlines the central role that enzymes and their technology have played in civilization since ancient times.

Fermentation and wine-making have been a part of human history and culture for at least 8000 years, but recognition of the role of catalysis in these processes had to wait until the late nineteenth century. By the 1830's, numerous similar agents, such as those that facilitate protein digestion in the stomach, had been discovered. The term "enzyme", meaning "from yeast", was coined by the German physiologist Wilhelm Kühne in 1876. In 1900, Eduard Buchner (1860-1917, 1907 Nobel Prize in Chemistry) showed that fermentation, previously believed to depend on a mysterious "life force" contained in living organisms such as yeast, could be achieved by a cell-free "press juice" that he squeezed out of yeast.

By this time it was recognized that enzymes are a form of catalyst (a term introduced by Berzelius in 1835), but their exact chemical nature remained in question. They appeared to be associated with proteins, but the general realization that enzymes *are* proteins began only in the 1930s when the first pure enzyme was crystallized, and did not become generally accepted until the 1950s. It is now clear that nearly all enzymes are proteins, the major exception being a small but important class of RNA-based enzymes known as ribozymes.

Proteins are composed of long sequences of amino acids strung together by amide bonds; this sequence defines the *primary structure* of the protein.. Their huge size (typically 200-2000 amino acid units, with total molecular weights 20,000 - 200,000) allows them to fold in complicated ways (known as secondary and tertiary structures) whose configurations are essential to their catalytic function.

Because enzymes are generally very much larger than the reactant molecules they act upon (known in biochemistry as substrates), enzymatic catalysis is in some ways similar to heterogeneous catalysis. The main difference is that the binding of a substrate to the enzyme is much more selective.

Precursors and Cofactors

Most enzymes come into being as inactive precursors (*zymogens*) which are converted to their active forms at the time and place they are needed.

- For example, the enzymes that lead to the clotting of blood are supposed to remain inactive until bleeding actually begins; a major activating factor is exposure of the blood to proteins in the damaged vessel wall.
- The enzyme that catalyzes the formation of lactose (milk sugar) in the mammary gland is formed during pregnancy, but it remains inactive until the time of birth, when hormonal changes cause a modifier unit to bind to and activate the enzyme.

Conversion to the active form may involve a simple breaking up of the protein by hydrolysis of an appropriate peptide bond or the addition of a phosphate or similar group to one of the amino acid residues.

Many enzyme proteins also require "helper" molecules, known as *cofactors*, to make them catalytically active. These may be simple metal ions (many of the trace nutrient ions of Cu, Mn, Mo, V, *etc.*) or they may be more complex organic molecules which are called *coenzymes*. Many of the latter are what we commonly refer to as *vitamins*. Other molecules, known as *inhibitors*, decrease enzyme activity; many drugs and poisons act in this way.

The Enzyme-substrate Complex

The standard model of enzyme kinetics consists of a two-step process in which an enzyme binds reversibly to its substrate S (the reactant) to form an *enzyme-substrate complex* ES:

$$E + S \underset{k_{-1}}{\overset{k_1}{\rightleftharpoons}} ES$$

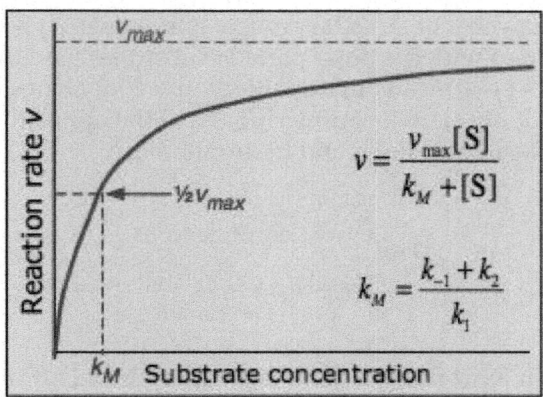

$$v = \frac{v_{max}[S]}{k_M + [S]}$$

$$k_M = \frac{k_{-1} + k_2}{k_1}$$

The enzyme-substrate complex plays a role similar to that of the activated complex in conventional kinetics, but the main function of the enzyme is to stabilize the transition state. In the second, essentially irreversible step, the product and the enzyme are released:

$$ES \xrightarrow{k_2} E + P$$

The basic kinetic treatment of this process involves the assumption that the concentrations [E] and [ES] reach steady-state values which do not change with time.

The overall process is described by the *Michaelis-Menten equation* which is plotted here. The *Michaelis constant* k_M is defined as shown, but can be simplified to the ES dissociation constant k_{-1}/k_1 in cases when dissociation of the complex is the rate-limiting step. The quantity v_{max} is not observed directly, but can be determined from k_M as shown here.

Enzymes are Proteins

In order to understand enzymes and how they catalyze reactions, it is first necessary to review a few basic concepts relating to proteins and the amino acids of which they are composed.

Amino Acids: Polar, Nonpolar, Positive, Negative

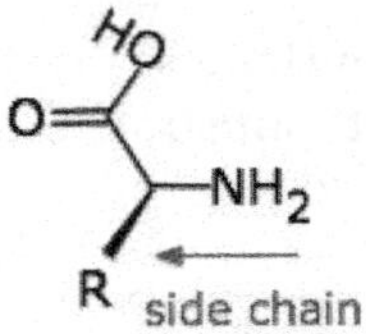

The 21 amino acids that make up proteins all possess the basic structure shown here, where R represents either hydrogen or a side chain which may itself contain additional $-NH_2$ or $-COOH$ groups. Both kinds of groups can hydrogen-bond with water and with the polar parts of substrates, and therefore contribute to the amino acid's polarity and hydophilic nature. Side chains that contain longer carbon chains and especially benzene rings have the opposite effect, and tend to render the amino acid non-polar and hydrophobic.

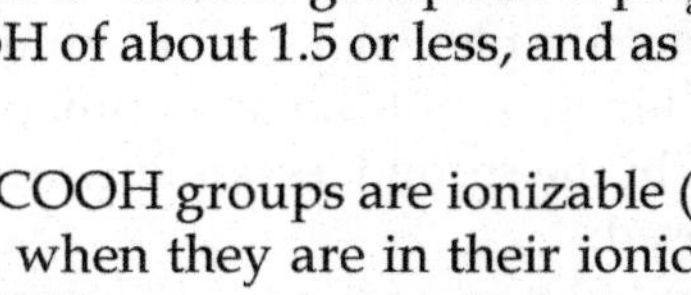

The pK_a of an ionizable group is the pH at which half of these groups will be ionized. For example, if a $-COOH$ group has a pK_a of 2.5, it will exist almost entirely in this form at a pH of about 1.5 or less, and as the negative ion $-COO^-$ at pH above about 3.5.

Both the $-NH_2$ and $-COOH$ groups are ionizable (*i.e.*, they can act as proton donors or acceptors) and when they are in their ionic forms, they will have an electric charge. The $-COOH$ groups have pKa's in the range 1.8-2.8, and will therefore be in their ionized forms $-COO^-$ at ordinary cellular pH values of around 7.4. The amino group pKa's are around 8.8-10.6, so these will also normally be in their ionized forms NH_3^+.

This means that at ordinary cellular pH, both the carboxyl and amino groups will be ionized. But because the charges have opposite signs, an amino acid that has no extra ionizable groups in its side chain will have a net charge of zero.

But if the side chain contains an extra amino or carboxyl group, the amino acid can carry a net electric charge.

For clarity, only the charges in the side chains are shown here, but of course the carboxyl and amino groups at the head end of each amino acid will also be ionized. The small green numbers are pK_a values.

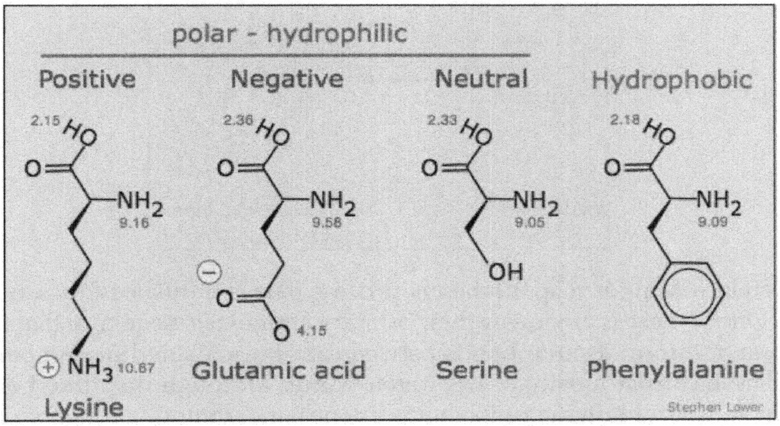

Proteins

Proteins are made up of one or more chains of amino acids linked to each other through peptide bonds by elimination of a water molecule.

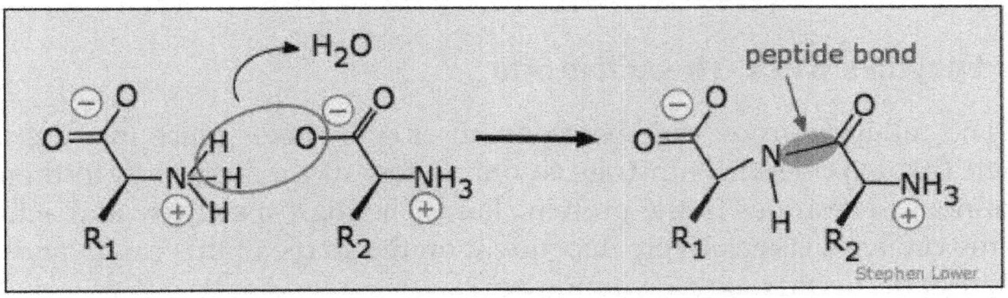

The product shown above is called a peptide, specifically it is a *dipeptide* because it contains two *amino acid residues* (what's left after the water has been removed.) Proteins are simply very long polypeptide chains, or combinations of them. (The distinction between a long polypeptide and a small protein is rather fuzzy!)

Globular Proteins

Most enzymes fall into the category of *globular proteins*. In contrast to the fibrous proteins that form the structural components of tissues, globular proteins are soluble in water and rarely have any systematic tertiary structures. They are made up of one or more amino-acid ("*peptide*") chains which fold into various shapes that can roughly be described as spherical — hence the term "globular", and the suffix "globin" that is frequently appended to their names, as in "hemoglobin".

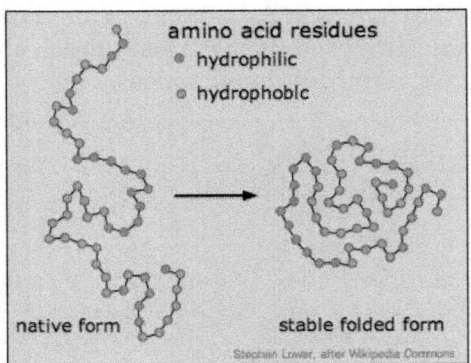

Protein folding is a spontaneous process that is influenced by a number of factors. One of these is obviously their primary amino-acid sequence that facilitates formation of intramolecular bonds between amino acids in different parts of the chain. These consist mostly of hydrogen bonds, although disulfide bonds S$-$S between sulfur-containing amino acids are not uncommon.

In addition to these intramolecular forces, interactions with the surroundings play an important role. The most import of these is the *hydrophobic effect*, which favors folding conformations in which polar amino acids (which form hydrogen bonds with water) are on the outside, while the so-called *hydrophobic* amino acids remain in protected locations within the folds.

How Enzymes Work: The Active Site

The catalytic process mediated by an enzyme takes place in a depression or cleft that exposes the substrate to only a few of the hundreds-to-thousands of amino acid residues in the protein chain. The high specificity and activity of enzyme catalysis is sensitively dependent on the shape of this cavity and on the properties of the surrounding amino acids.

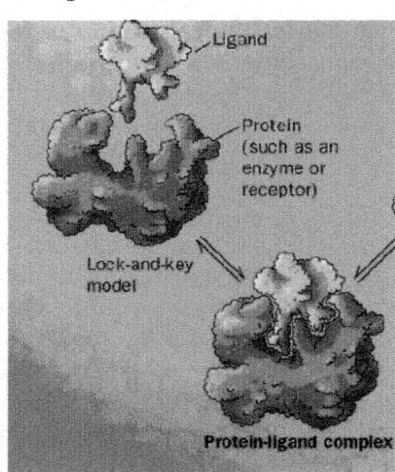

In 1894, long before it was clear that enzymes are proteins, the German chemist Emil Fischer suggested the so-called lock-and-key model as a way of understanding how a given enzyme can act specifically on only one kind of substrate molecule. This model is essentially an elaboration of the one we still use for explaining heterogeneous catalysis.

Although the basic lock-and-key model continues to be useful, it has been modified into what is now called the *induced-fit model*. This assumes that when the substrate enters the active site and interacts with the surrounding parts of the amino acid chain, it reshapes the active site (and perhaps other parts of the enzyme) so that it can engage more fully with the substrate.

One important step in this process is to squeeze out any water molecules that are bound to the substrate and which would interfere with its optimal positioning.

Within the active site, specific interactions between the substrate and appropriately charged, hydrophilic and hydrophobic amino acids of the active site then stabilize the transition state by distorting the substrate molecule in such a way as to lead to a transition state having a substantially lower activation energy than can be achieved by ordinary non-enzymatic catalysis.

Beyond this point, the basic catalytic steps are fairly conventional, with acid/base and nucleophilic catalysis being the most common.

Enzyme Regulation and Inhibition

If all the enzymes in an organism were active all the time, the result would be runaway chaos. Most cellular processes such as the production and utilization of energy, cell division, and the breakdown of metabolic products must operate in an exquisitely choreographed, finely-tuned manner, much like a large symphony orchestra; no place for jazz-improv here!

Nature has devised various ways of achieving this; we described the action of precursors and coenzymes above. Here we focus on one of the most important (and chemically-interesting) regulatory mechanisms.

Allosteric Regulation: Tweaking the Active Site

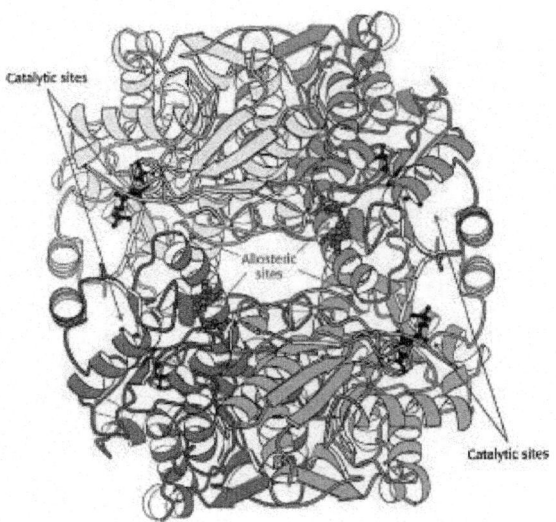

There is an important class of enzymes that possess special sites (distinct from the catalytically active sites) to which certain external molecules can reversibly bind. Although these *allosteric sites,* as they are called, may be quite far removed from the catalytic sites, the effect of binding or release of these molecules is to trigger a rapid change in the folding pattern of the enzyme that alters the shape of the active site. The effect is to enable a signaling or regulatory molecule (often a very small one such as NO) to modulate the catalytic activity of the active site, effectively turning the enzyme on or off.

In some instances, the product of an enzyme-catalyzed reaction can itself bind to an allosteric site, decreasing the activity of the enzyme and thus providing negative feedback that helps keep the product at the desired concentration. It is believed that concentrations of plasma ions such as calcium, and of energy-supplying ATP are, are regulated in this way.

Allosteric enzymes are more than catalysts: they act as control points in metabolic and cellular signaling networks

Allosteric enzymes frequently stand at the beginning of a sequence of enzymes in a metabolic chain, or at branch points where two such chains diverge, acting very much like traffic signals at congested intersections.

Enzyme Inhibition

As is the case with heterogeneous catalysts, certain molecules other than the normal substrate may be able to enter and be retained in the active site so as to competitively inhibit an enzyme's activity.

This is how penicillin and related antibiotics work; these molecules covalently bond to amino acid residues in the active site of an enzyme that catalyzes the formation of an essential component of bacterial cell walls. When the cell divides, the newly-formed progeny essentially fall apart.

Enzymes Outside the Cell

Artist Mike Perkins' conception of an immobilized enzyme within a functionalized nanoporous silica pore. The enzyme is shown as green with positively charged regions shown in red. The blue structures inside the pores represent the negatively charged functional groups added to the mesoporous silica to create a favorable chemical environment for the enzyme. The favorable environment in each pore attracts the enzyme molecule to move into the unoccupied pore. This environment stabilizes and increases the chemical reactivity of the enzyme, allowing it to convert harmful substrate materials (purple particles) into useful or harmless products (yellow and red particles).

Enzymes have been widely employed in the food, pulp-and-paper, and detergent industries for a very long time, but mostly as impure whole-cell extracts.

In recent years, developments in biotechnology and the gradual move of industry from reliance on petroleum-based feedstocks and solvents to so-called "green" chemistry have made enzymes more attractive as industrial catalysts. Compared to the latter, purified enzymes tend to be expensive, difficult to recycle, and unstable outside of rather narrow ranges of temperature, pH, and solvent composition.

HOMOGENEOUS CATALYSIS

In chemistry, homogeneous catalysis is catalysis in a solution by a soluble catalyst. Strictly speaking, homogeneous catalysis are catalytic reactions where the catalyst is in the same phase as the reactants, so homogeneous catalysis applies to reactions in the gas phase and even in a solid. Heterogeneous catalysis is the alternative to homogeneous catalysis, where the catalysis occurs at the interface of two phases, typically gas-solid. The term is used almost exclusively to describe solutions and it is often implies catalysis by organometallic compounds. The area is one of intense research and many practical applications, *e.g.*, the production of acetic acid. Enzymes are examples of homogeneous catalysts.

EXAMPLES

Acid Catalysis

The proton is the most pervasive homogeneous catalyst because water is the most common solvent. Water forms protons by the process of self-ionization of water. In an illustrative case, acids accelerate (catalyze) the hydrolysis of esters:

$$CH_3CO_2CH_3 + H_2O \rightleftharpoons CH_3CO_2H + CH_3OH$$

In the absence of acids, aqueous solutions of most esters do not hydrolyze at practical rates.

Organometallic Chemistry

Processes that utilize soluble organometallic compounds as catalysts fall under the category of homogeneous catalysis, as opposed to processes that use bulk metal or metal on a solid support, which are examples of heterogeneous catalysis. Some well-known examples of homogeneous catalysis include hydroformylation and transfer hydrogenation, as well as certain kinds of Ziegler-Natta polymerization and hydrogenation. Homogeneous catalysts has also been used in a variety of industrial processes such as the Wacker process Acetaldehyde (conversion of ethylene to acetaldehyde) as well as the Monsanto process and the Cativa process for the conversion of MeOH and CO to acetic acid.

Many non-organometallic complexes are also widely used in catalysis, *e.g.* for the production of terephthalic acid from xylene.

Other forms of Homogeneous Catalysis

Enzymes are homogeneous catalysts that are essential for life but are also harnessed for industrial processes. A well studied example carbonic anhydrase, which catalyzes the release of CO_2 into the lungs from the blood stream.

Contrast with Heterogeneous Catalysis

Homogeneous catalysis differs from heterogeneous catalysis in that the catalyst is in a different phase than the reactants. One example of heterogeneous catalysis is the petrochemical alkylation process, where the liquid reactants are immiscible with a solution containing the catalyst. Heterogeneous catalysis offers the advantage that products are readily separated from the catalyst, and heterogeneous catalysts are often more stable and degrade much slower than homogeneous catalysts. However, heterogeneous catalysts are difficult to study, so their reaction mechanisms are often unknown.

Enzymes possess properties of both homogeneous and heterogeneous catalysts. As such, they are usually regarded as a third, separate category of catalyst.

SOLID ACIDS AND BASES

Acid Catalysis

Fig. : In acid-catalyzed Fischer esterification, the proton binds to oxygens and functions as a Lewis acid to activate the ester carbonyl (top row) as an electrophile, and converts the hydroxyl into the good leaving groupwater (bottom left). Both lower the kinetic barrier and speed up the attainment of chemical equilibrium.

In acid catalysis and base catalysis a chemical reaction is catalyzed by an acid or a base. The acid is often the proton and the base is often a hydroxide ion. Typical reactions catalyzed by proton transfer areesterfications and aldol reactions. In these reactions the conjugate acid of the carbonyl group is a betterelectrophile than the neutral carbonyl group itself. Catalysis by either acid or base can occur in two different ways: specific catalysis and general catalysis.

Use in Synthesis

Acid catalysis is mainly used for organic chemical reactions. There are many possible chemical compounds that can act as sources for the protons to be transferred in an acid catalysis system. A compound such assulfuric acid, H_2SO_4, can be used. Usually this is done to create a more likely leaving group, such as converting an OH group to a H_2O^+ group, which can then be eliminated as water (H_2O). Acids specifically used for acid catalysis include hydrofluoric acid (in the alkylation process), phosphoric acid, toluenesulfonic acid, polystyrene sulfonate, heteropoly acids, zeolites and graphene oxide.

With carbonyl compounds such as esters, synthesis and hydrolysis go through a tetrahedral transition state, where the central carbon has an oxygen, an alcohol group, and the original alkyl group. Strong acids protonate the carbonyl, which makes the oxygen positively charged, so that it can easily receive the double bond electrons when the alcohol attacks the carbonyl carbon. This enables ester synthesis and hydrolysis. The reaction is an equilibrium between the ester and its cleavage to carboxylic acid and alcohol. On the contrary, strong bases deprotonate the attacking alcohol or amine, which also promotes the reaction. However, bases also deprotonate the acid, which is irreversible. Therefore, in a strongly basic, aqueous environment, esters only hydrolyze.

Solid Acid Catalysts

In industrial scale chemistry, many processes are catalysed by "solid acids." As heterogeneous catalysts, solid acids do not dissolve in the reaction medium. Well known examples include zeolites, alumina, and various other metal oxides. Such acids are used in cracking. A particularly large scale application is alkylation, *e.g.* the combination of benzene and ethylene to give ethylbenzene. Many alkylamines are prepared by amination of alcohols.

Fig. : Zeolite, ZSM-5 is widely used as a solid acid catalyst.

KINETICS

Specific Catalysis

In specific acid catalysis taking place in solvent S, the reaction rate is proportional to the concentration of the protonated solvent molecules SH^+. The acid catalyst itself (AH) only contributes to the rate acceleration by shifting the chemical equilibrium between solvent S and AH in favor of the SH^+ species.

$$S + AH \rightarrow SH^+ + A^-$$

For example in an aqueous buffer solution the reaction rate for reactants R depends on the pH of the system but not on theconcentrations of different acids.

$$\text{rate} = -\frac{d[R1]}{dt} = k[SH^+][R1][R2]$$

This type of chemical kinetics is observed when reactant R_1 is in a fast equilibrium with its conjugate acid R_1H^+ which proceeds to react slowly with R_2 to the reaction product; for example, in the acid catalysed aldol reaction.

General Catalysis

In general acid catalysis all species capable of donating protons contribute to reaction rate acceleration. The strongest acids are most effective. Reactions in which proton transfer is rate-determining exhibit general acid catalysis, for example diazonium coupling reactions.

$$\text{rate} = -\frac{d[R1]}{dt} = k_1[SH^+][R1][R2] + k_2[AH^1][R1][R2] + k_3[AH^2][R1][R2] + \ldots$$

When keeping the pH at a constant level but changing the buffer concentration a change in rate signals a general acid catalysis. A constant rate is evidence for a specific acid catalyst.

REDOX CATALYSIS

Understanding and Exploiting Enzymatic Redox Catalysis

Redox reactions are the foundation of life. Reactions in which a substrate becomes more oxidized or more reduced are known as 'redox' reactions. These are the basis for much of the energy used to support life: reduction of O_2 to 2 H_2O in respiration, oxidation of sugars to CO_2 and countless chemotrophic modes of metabolism. Redox enzymes accelerate these reactions enough for them to be biologically useful and support life (sugar sitting on your table in air does not oxidize significantly in a year, but you get energy from eating, and oxidizing, sugar in a few minutes). Some of the most demanding reactions in biology are catayzed by redox-active enzymes, for example reductive cleavage of the triple bond of N_2, which is the second strongest bond known. Yet nitrogenase cleaves N_2 at standard temperature and pressure. Many redox-active enzymes use inorganic or organic cofactors to actually execute the chemistry, which in many cases would not be accessible *via* amino acids alone. However having incorporated reactive cofactors, the protein must be able to control the activated intermediates that occur during turnover, and ensure that only a specific substrate has access to the cofactor, and a single specific reaction follows. Thus, instead of mediating wholesale, random oxidation of all the many susceptible molecules that make up a cell, redox enzymes permit utilization of specific reactions as sources of energy, in parallel with separate utilization of other similar molecules to build and run the cell, and thus provide the basis for controlled burning of selected fuels only, as opposed to an all-consuming wild-fire.

Enzymes are needed to make redox reactions useful and to control them. We seek to understand how enzymes can both accelerate difficult reactions by many orders of magnitude (factors of up to $\times 10^{19}$), yet on the other hand retain very tight control over what compounds react and what course the reaction takes. These tasks are especially challenging for redox enzymes, since they imply control over the outcome of movements of electrons, and electrons are exceedingly small, fast moving, fundamental particles, while proteins are relatively large molecules, whose structures can be inherently dynamic and flexible. Thus we truly work at the interfacebetween physics and some of the most important questions underlying the possibility of life. Our research addresses fundamental elements of enzymatic redox catalysis, and enzyme catalysis in general.

We are currently concentrating on three enzymes, superoxide dismutase (SOD) and nitroreductase (NR). These provide complementary perspectives on how redox catalysis can be accomplished, and keep us thinking about how our studies of a particular system can elucidate recurring themes in redox catalysis *in general.* Thus, SOD mediates single-electron chemistry based on a bound metal ion (Fe or Mn), and NR exploits an organic cofactor (flavin) in two-electron chemistry. SOD consumes superoxide radicals and thus protects against oxidative damage and forestalls aging. NR metabolizes toxic nitrated aromatics such as TNT and has potential utility in bioremediation, weapons detection and cancer therapy.

Spectroscopic methods enable us to directly observe the transitions or orbitals of the valence electrons that participate in the reaction. EPR is used to monitor redox-active metal ions directly and thus elucidate the nature of the orbitals accommodating the valence electrons. NMR spectroscopy is used to directly observe the flavin cofactor to elucidate the valence orbitals, and define the geometry and protonation of the system, as well as the hydrogen bonds that tune its reactivity. Since proton transfer is tightly coupled to biological electron transfer, and protons can be directly observed by NMR, NMR is also invaluable for identifying the amino acid residues which participate, and evaluating the energies involved. NMR, EPR and a variety of other spectroscopic methods are used to study substrate binding modes, substrate binding energies, substrate activation and active site residues involved, and NMR enables us to correlate movements throughout the protein with chemical activity of the active site. Our spectroscopic studies are complemented by X-ray crystallography, mechanistic and structural studies, and are aided by the full battery of modern molecular biological methods including site directed mutagenesis, directed evolution, mass spectrometry and specific isotopic labeling.

REDOX CATALYSIS FOR A SUSTAINABLE ENERGY INFRASTRUCTURE

The main research theme in my group is to understand and mimic bioinorganic multi-electron processes that are relevant to our future energy infrastructure. Reduction of protons generates hydrogen that can be used as a chemical fuel. Alternatively to gaseous hydrogen, the reduction of carbon dioxide can afford a liquid carbon based fuel. In case of both reactions protons and electrons are necessary which are generated by oxidation of water to produce dioxygen. We focus in particular on molecular catalysts that allow for specific structural modifications in order to tune and control the observed catalytic activity.

Bio-inspired and Ligand Assisted Small Molecule Activation

In nature the role of the protein environment of the catalytic site has several roles. Depending on the electron donating or electron withdrawing character of the ligands coordinated directly to the transition metal active site, the electronic properties of the active site is tuned. Also the geometry of the catalytic site is perfectly oriented for catalysis. These roles of the ligand environment have been well established in coordination chemistry. In addition the protein environment is actively involved in activation of the substrate (*e.g.* H^+, CO_2, N_2, O_2 or H_2O) *via* hydrogen bonding networks, and shuttling protons from or to the active site. The latter events are critical for efficient multi-electron catalytic processes, yet have hardly been explored. In my group molecular catalysts are studied that are employed with ligands that bear functionalities in the second coordination sphere. These functionalities are positioned as such that they assist in activation of the substrate and can shuttle protons away from the catalytic site in case of oxidation reactions and to the active site in case of the reduction reactions.

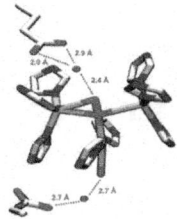

The active dioxygen reduction site of Laccase. Three copper atoms coordinate and reduce dioxygen, while protons are transferred to the catalytic site via nearby acetate functionalities via a hydrogen bonding network.

Enzyme Mimics

Since natural multi-electron redox catalysts operate very efficiently, such systems form a major inspiration for development of new catalytic systems. In addition fundamental studies on these enzyme mimics will indirectly reveal how catalysis in nature may occur. Currently, my group is interested in model systems of the oxygen evolving center of photosystem II, and the active sites of the dioxygen reduction enzymes Laccase and cytochrome c oxidase.

Electrochemistry and in Situ Spectroscopy

To catalyze redox half reactions we rely on electrochemical techniques to study the rates of the catalytic reaction as a function of applied potential and the number of active sites involved. Our focus lies with fundamental understanding of these systems and with unraveling the mechanisms wherein these bio-inspired molecular redox catalysts operate. To determine the structure of the catalytic species that are present in the reaction mixture at a particular applied potential we use Raman spectroscopy, IR spectroscopy and UV spectroscopy coupled to electrochemistry. In combination with reaction kinetics by means of various electrochemical techniques gives us a clear picture how catalysis proceeds. Typically several catalytic intermediates are transient species and cannot be observed experimentally. To shed light on these intermediate species of the catalytic cycle we rely on density functional theory. In our approach we use experimental data to calibrate our computational model.

Synthesis and Structure Activity Studies

A major advantage of molecular catalysts is that these can be varied to a large degree in order to radically improve the catalytic activity of the targeted multi-electron redox reaction. For example implementation of a proton delivery or removal system may seriously affect the reaction rates. An important research question is how and to which extend a particular ligand modification affect the rate determining step, or bottleneck, of the catalytic reaction. Another important parameter is the potential that is required to achieve a particular catalytic rate. We aim to conduct the chemical reaction of choice at the mildest conditions possible, as this will result in the highest efficiency in terms of energy consumption. This latter is the major crux in the formation of chemical fuels from renewable energy and the subsequent consumption of these chemical fuels.

ORGANOMETALLIC CATALYSTS OXIDIZE WATER TO PRODUCE HYDROGEN CHEAPLY AND RENEWABLY

These multi-electron redox catalysts may facilitate a cost-effective oxidation of water to produce hydrogen as an alternative fuel source. Hydrogen is attractive as a fuel source, especially for vehicles, because it is clean and renewable. However, a large barrier to its use as an alternative fuel source is the high cost of producing renewable, clean hydrogen. Currently, most of the hydrogen produced in the United States comes from natural gas, the cheapest source. Ecofriendly methods such as electrolysis from water and photochemical water splitting cost too much for widespread use. The United States has spent billions of dollars on efforts to improve hydrogen production and isolation techniques. Researchers at the University of Florida have developed an organometallic catalyst that may hasten and initiate the electrochemical and photochemical production of water, potentially lowering costs and removing a barrier to widespread hydrogen production. Because hydrogen has the potential to power many types of devices, the possible market applications for this inexpensive catalyst are extensive.

Application

Catalysts that may produce hydrogen *via* water splitting as a renewable energysource

Advantages

- Utilizes common resources such as water and metal ions, increasing the probability that a renewable supply of hydrogen is available to meet increasing energy demands
- Lowers the cost and increases the speed for producing hydrogen, removing a barrier to widespread adoption
- Helps produce hydrogen via a "green" method, safeguarding the environment from damaging pollutants

Technology

These redox catalysts are composed of large organic scaffolds that support several metal centers, arranged around a central cavity. These metals provide a confined site in which redox reactions involving water are able to occur. This means that the large organometallic catalyst provides an initiation site for the oxidation of water, to produce hydrogen. It also provides a site for the reduction of hydrogen, to produce water and electricity. This catalysis speeds the electrochemical process, providing amore efficient production of hydrogen, or a more efficient fuel cell.

CASCADE AND DOMINO CATALYTIC REACTIONS

Cascade Reaction

A cascade reaction or tandem reaction or domino reaction is a consecutive series of intramolecular organic reactions which often proceed *via* highly reactive intermediates. It allows the organic synthesis of complex multinuclear molecules from a single acyclic precursor. The substrate contains many functional groups that take part in chemical transformations one at a time. Often a functional group is generated in situ from the previous chemical transformation. The definition includes the prerequisite *intramolecular* in order to distinguish this reaction type from a multi-component reaction. In this sense it differs from the definition of a biochemical cascade. The main advantages of a cascade reaction in organic synthesis are that the reaction is often fast due to its intramolecular nature, the reaction is also clean, displays high atom economy, does not involve workup and isolation of many intermediates, and adds much complexity in effectively one step.

A cascade reaction is sometimes called a *living reaction* because it shares some characteristics with a living polymerization. In cascade reactions one can identify an initiation site, a relay moiety and a termination moiety. Examples of cascade reactions are numerous (*e.g.* the Aldol-Tishchenko reaction) and especially so in alkyne chemistry (theBanert cascade to name just one) or polyolefin polycycloisomerizations. Other alkyne coupling reactions are classified based on common features such as type of compound synthesised, for instance the spiro mode cascade:

or the linear-fused mode cascade, through application of the intramolecular Heck reaction:

or the zipper mode cascade.

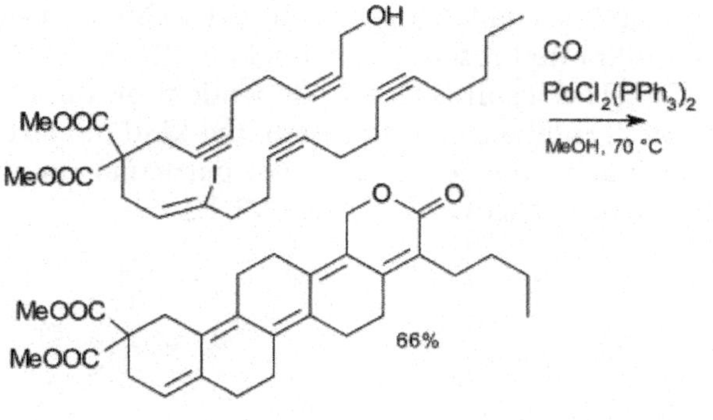

66%

Other cascade reactions are included in Diels-Alder reactions, oxirane ring-opening reactions, and Pauson–Khand reactions.

An example of an oxirane cascade reaction is given by the synthesis of certain polyether ladder polymers:

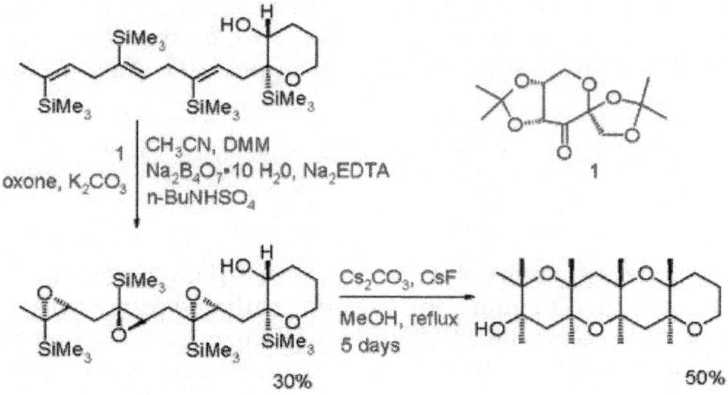

This type of ladder compounds are found in marine lifeforms such as red tide. The tri-epoxide is prepared from a triene through asymmetric Shi epoxidation and oxone as primary oxidizing agent. The hydroxyl group in the tri-epoxide is activated as a nucleophile by the presence of the base caesium carbonate The bulky trimethylsilyl groups make such that the polyether is formed with the correct stereochemistry and they are removed in situ by caesium fluoride.

MULTICOMPONENT CATALYTIC REACTIONS

Multicomponent Reactions

Multicomponent Reactions (MCRs) are convergent reactions, in which three or more starting materials react to form a product, where basically all or most of the atoms contribute to the newly formed product. In an MCR, a product is assembled according to a cascade of elementary chemical reactions. Thus, there is a network of reaction equilibria, which all finally flow into an irreversible step yielding the product. The challenge is to conduct an MCR in such a way that the network of pre-equilibrated reactions channel into the main product and do not yield side products. The result is clearly dependent on the reaction conditions: solvent, temperature, catalyst, concentration, the kind of starting materials and functional groups. Such considerations are of particular importance in connection with the design and discovery of novel MCRs.

1-CR   2-CR   6-CR

Multicomponent Reactions with Carbonyl Compounds

Some of the first multicomponent reactions to be reported function through derivatization of carbonyl compounds into more reactive intermediates, which can react further with a nucleophile. One example is the Mannich Reaction:

Fig. : Mannich Reaction.

Obviously, this reaction only proceeds if one carbonyl compound reacts faster with the amine to give an imine, and the other carbonyl compound plays the role of a nucleophile. In cases where both carbonyl compounds can react as the nucleophile or lead to imines with the same reaction rate, preforming the intermediates is an alternative, giving rise to a standard multistep synthesis.

Carbonyl compounds played a crucial role in the early discovery of multicomponent reactions, as displayed by a number of name reactions:

Fig. : Biginelli Reaction.

Fig. : Bucherer-Bergs Reaction.

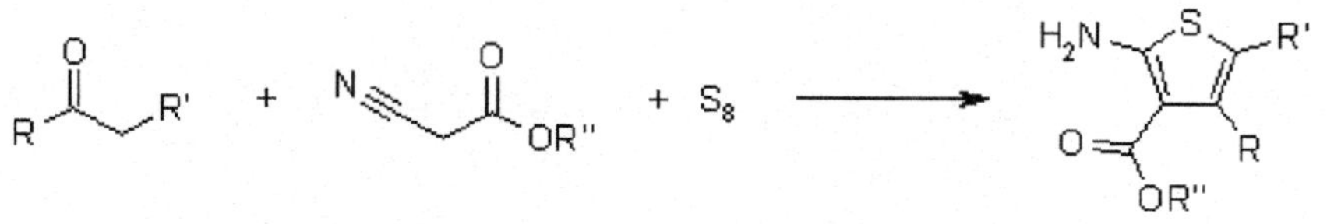

Fig. : Gewald Reaction

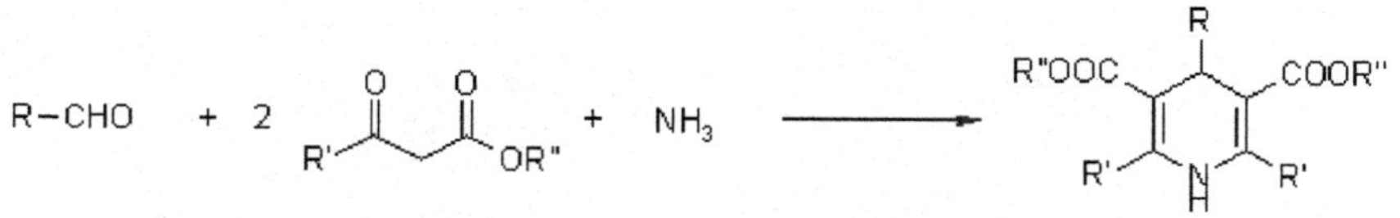

Fig. : Hantzsch Dihydropyridine (Pyridine) Synthesis

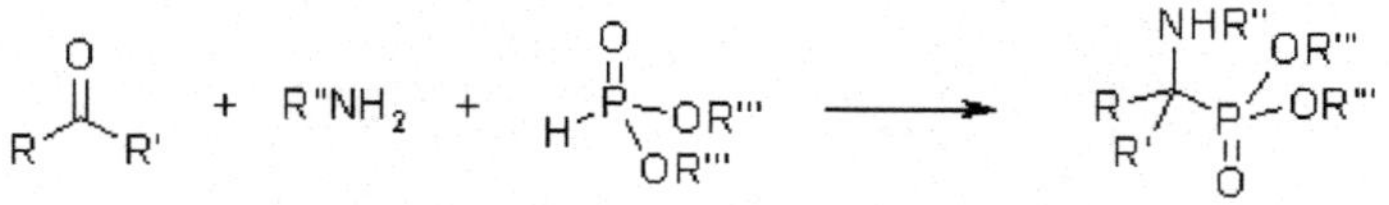

Fig. : Kabachnik-Fields Reaction

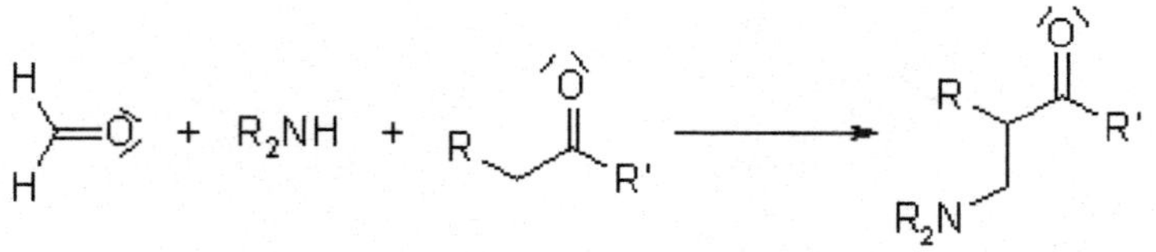

Fig. : Mannich Reaction

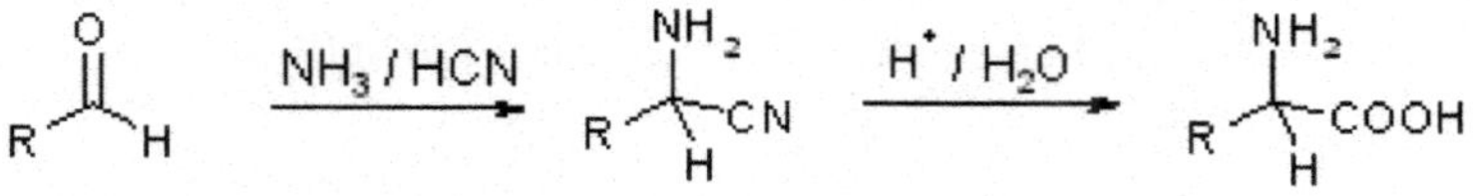

Fig. : Strecker Synthesis

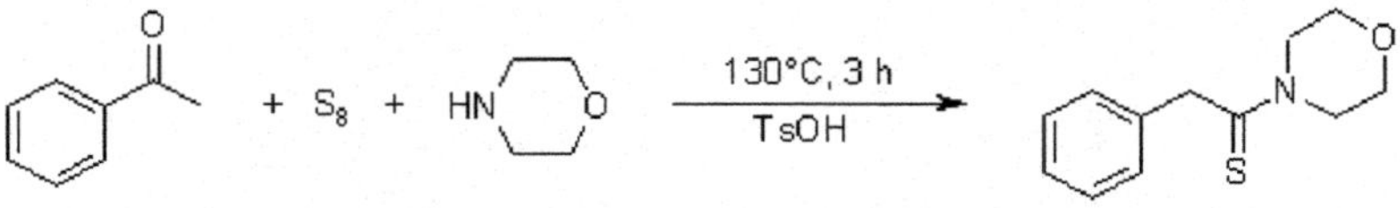

Fig. : Kindler Thioamide Synthesis

Isocyanide-based Multicomponent Reactions

Isocyanides play a dual role as both a nucleophile and electrophile, allowing interesting multicomponent reactions to be carried out. One of the first multicomponent reactions to use isocyanides was the Passerini Reaction. The mechanism shows how the isocyanide displays ambient reactivity. The driving force is the oxidation of C^{II} to C^{IV}, leading to more stable compounds.

Fig. : Passerini Reaction.

This interesting isocyanide chemistry has been rediscovered, leading to an overwhelming number of useful transformations. One of these is the Ugi Reaction:

Fig. : Ugi Reaction.

Both the Passerini and Ugi Reactions lead to interesting peptidomimetic compounds, which are potentially bioactive. The products of these reactions can constitute interesting lead compounds for further development into more active compounds. Both reactions offer an inexpensive and rapid way to generate compound libraries. Since a wide variety of isocyanides are commercially available, an equivalently diverse spectrum of products may be obtained.

Variations in the starting compounds may also lead to totally new scaffolds, such as in the following reaction, in which levulinic acid simultaneously plays the role of a carboxylic acid and a carbonyl compound:

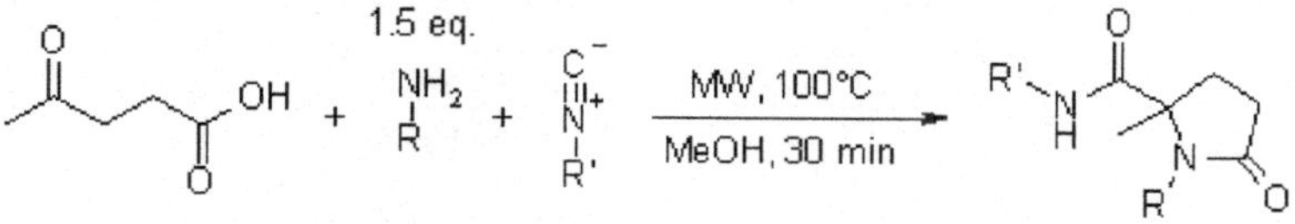

But how can multicomponent reactions be discovered? It's sometimes a simple matter of trial and error. Some very interesting MCRs have even been discovered by preparing libraries from 10 different starting materials. By analyzing the products of each combination (three-, four-, up to ten-component reactions), one is able to select those reactions that show a single main product. HPLC and MS are useful analytical methods, because the purity and mass of the new compounds help to decide rapidly whether a reaction might be interesting to investigate further.

MULTICOMPONENT REACTIONS

Display all Abstracts

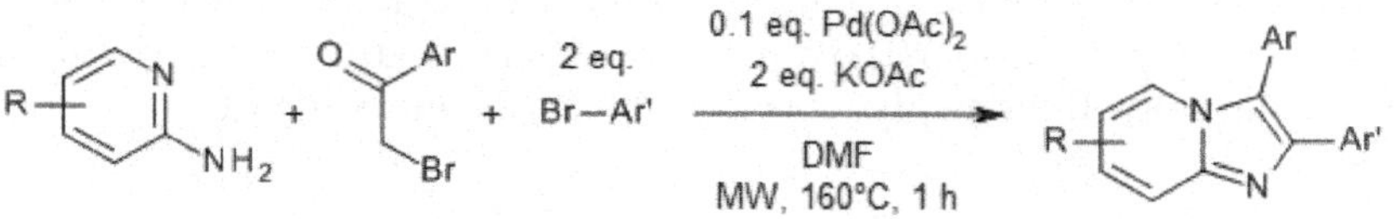

Microwave irradiation enables an expeditious one-pot, ligand-free, Pd(OAc)$_2$-catalyzed, three-component reaction for the synthesis of 2,3-diarylimidazo[1,2-*a*]pyridines. This methodology offers high availability of commercial reagents and great efficiency in expanding molecule diversity.

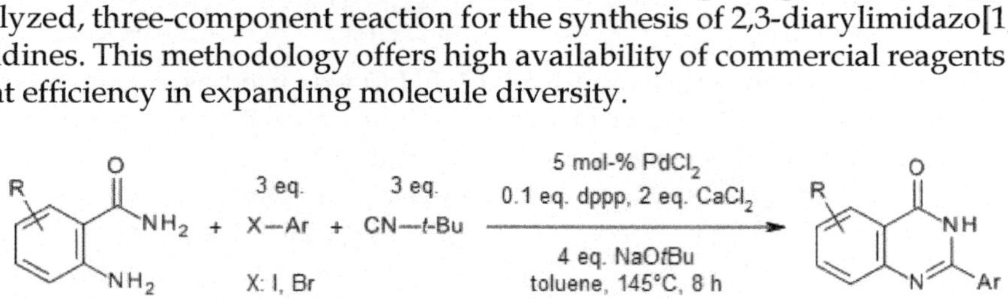

An operational simple palladium-catalyzed three-component reaction of readily available 2-aminobenzamides, aryl halides, and tert-butyl isocyanide efficiently constructs quinazolin-4(3*H*)-ones in good yields *via* a palladium-catalyzed isocyanide insertion/cyclization sequence.

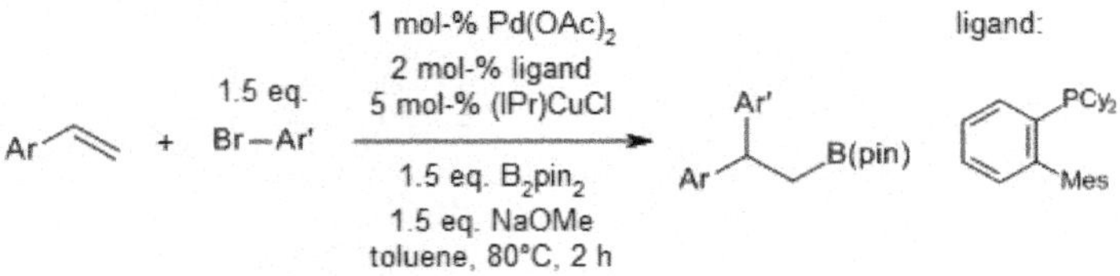

Arylboration of vinylarenes and methyl crotonate with aryl halides and bis(pinacolato)diboron by cooperative Pd/Cu catalysis gives 2-boryl-1,1-diarylethanes and α-aryl-β-boryl ester in a regioselective manner. The reaction is compatible with various functionalities and can be scaled-up to a gram scale.

A copper(II)-catalyzed intermolecular three-component oxyarylation of allenes using arylboronic acids as a carbon source and TEMPO as an oxygen source proceeded under mild conditions with high regio- and stereoselectivity and functional group tolerance.

An efficient synthesis of 2H-indazole derivatives in a one-pot three-component reaction of 2-chloro- and 2-bromobenzaldehydes, primary amines and sodium azide is catalyzed by copper(I) oxide nanoparticles (Cu$_2$O-NP) under ligand-free conditions in polyethylene glycol (PEG 300) as a green solvent.

The odorless and stable solid Na$_2$S$_2$O$_3$·5H$_2$O was used as a convenient and environmentally friendly source of sulfur in a Pd-catalyzed cross-coupling of aryl halides and alkyl halides to deliver aromatic sulfides.

Microwave-assisted conditions enabled a simple, rapid, one-pot synthesis of arylaminomethyl acetylenes in very good yields using arylboronic acids, aqueous ammonia, propargyl halides, copper(I) oxide and water as the solvent within ten minutes.

A multicomponent protocol enables the synthesis of highly substituted imidazole derivatives in excellent yield from various α-azido chalcones, aryl aldehydes, and anilines in the presence of erbium triflate as a catalyst.

A three-component reaction of alkynes, elemental sulfur, and aliphatic amines allows a general, straightforward, and atom-economical synthesis of thioamides.

A highly practical copper-catalyzed intermolecular cyanotrifluoromethylation of alkenes provides a general and straightforward way to synthesize various useful CF_3-containing nitriles, which can be used for the preparation of pharmaceutically and agrochemically important compounds.

A facile formation of C-N, C-O, and C-S bonds from ynals, pyridin-2-amines, and alcohols or thiols enables a transition-metal-free three-component reaction for the construction of imidazo[1,2-a]pyridines.

Unusual N-Acyl-N,O-acetals are present in a number of bioactive natural products and can act as a synthetic precursor to unstable reactive N-acylimines. Various N-acyl-N,O-acetals can be prepared under mild conditions mediated by titanium ethoxide (Ti(OEt)$_4$). The method also offers a new strategy to make other O-alkyl-N,O-acetals.

An operationally simple and rapid copper-catalyzed three-component synthesis of trisubstituted N-aryl guanidines involving cyanamides, arylboronic acids, and amines is performed in the presence of K_2CO_3, a catalytic amount of $CuCl_2 \cdot 2H_2O$, bipyridine, and oxygen (1 atm).

Heating a solution of an aldehyde, an aromatic amine, and a nitroalkane in 20% water-methanol at 60°C for five hours enables an environmentally benign three-component, one-pot synthesis of 2-nitroamines in the absence of a catalyst.

A Lewis acid palladium-catalyzed reaction of amides, aryl aldehydes, and arylboronic acids enables a practical and general synthesis of α-substituted amides from simple, readily available building blocks.

ORGANOCATALYSIS

In organic chemistry, the term organocatalysis (a portmanteau of the terms "organic" and "catalyst") refers to a form ofcatalysis, whereby the rate of a chemical reaction is increased by an organic catalyst referred to as an "organocatalyst" consisting of carbon, hydrogen, sulfur and other nonmetal elements found in organic compounds. Because of their similarity in composition and description, they are often mistaken as a misnomer for enzymes due to their comparable effects on reaction rates and forms of catalysis involved.

Organocatalysts which display secondary amine functionality can be described as performing either enamine catalysis (by forming catalytic quantities of an active enamine nucleophile) or iminium catalysis (by forming catalytic quantities of an activated iminium electrophile). This mechanism is typical for covalent organocatalysis. Covalent binding of substrate normally requires high catalyst loading (for proline-catalysis typically 20-30 mol%). Noncovalent interactions such as hydrogen-bonding facilitates low catalyst loadings (down to 0.001 mol%).

Organocatalysis offers several advantages. There is no need for metal-based catalysis thus making a contribution to green chemistry. In this context, simple organic acids have been used as catalyst for the modification of cellulose in water on multi-ton scale. When the organocatalyst is chiral an avenue is opened to asymmetric catalysis, for example the use of proline inaldol reactions is an example of chirality and green chemistry.

Regular achiral organocatalysts are based on nitrogen such as piperidine used in the Knoevenagel condensation. DMAP used in esterfications and DABCO used in the Baylis-Hillman reaction. Thiazolium salts are employed in the Stetter reaction. These catalysts and reactions have a long history but current interest in organocatalysis is focused onasymmetric catalysis with chiral catalysts, called asymmetric organocatalysis or enantioselective organocatalysis. A pioneering reaction developed in the 1970s is called the Hajos–Parrish–Eder–Sauer–Wiechert reaction. Between 1968 and 1997, there were only a few reports of the use of small organic molecules as catalysts for asymmetric reactions (the Hajos–Parrish reaction probably being the most famous), but these chemical studies were viewed more as unique chemical reactions than as integral parts of a larger, interconnected field.

99%, 93% ee

In this reaction, naturally occurring chiral proline is the chiral catalyst in an Aldol reaction. The starting material is an achiral triketone and it requires just 3% of proline to obtain the reaction product, a ketol in 93% enantiomeric excess. This is the first example of an amino acid-catalyzed asymmetric aldol reaction.

Many chiral organocatalysts are an adaptation of chiral ligands (which together with a metal center also catalyze asymmetric reactions) and both concepts overlap to some degree.

Organocatalyst Classes

Organocatalysts for asymmetric synthesis can be grouped in several classes:

- Biomolecules: proline, phenylalanine. Secondary amines in general. The cinchona alkaloids, certain oligopeptides.
- Synthetic catalysts derived from biomolecules.
- Hydrogen bonding catalysts, including TADDOLS, derivatives of BINOL such as NOBIN, and organocatalysts based on thioureas
- Triazolium salts as next-generation Stetter reaction catalysts

Examples of asymmetric reactions involving organocatalysts are:

- Asymmetric Diels-Alder reactions
- Asymmetric Michael reactions
- Asymmetric Mannich reactions
- Shi epoxidation
- Organocatalytic transfer hydrogenation

Proline

Proline catalysis has been reviewed.

Imidazolidinone Organocatalysis

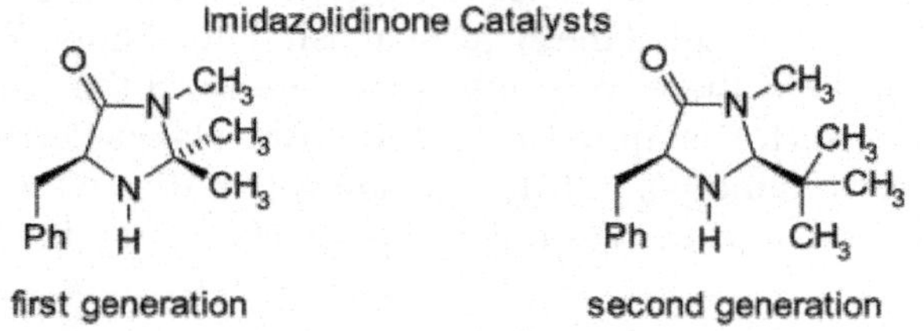

Imidazolidinone Catalysts

first generation   second generation

A certain class of imidazolidinone compounds (also called MacMillan organocatalysts) are suitable catalysts for many asymmetric reactions such as asymmetric Diels-Alder reactions. The original such compound was derived from the biomolecule phenylalanine in two chemical steps (amidation with methylamine followed bycondensation reaction with acetone) which leave the chirality intact:

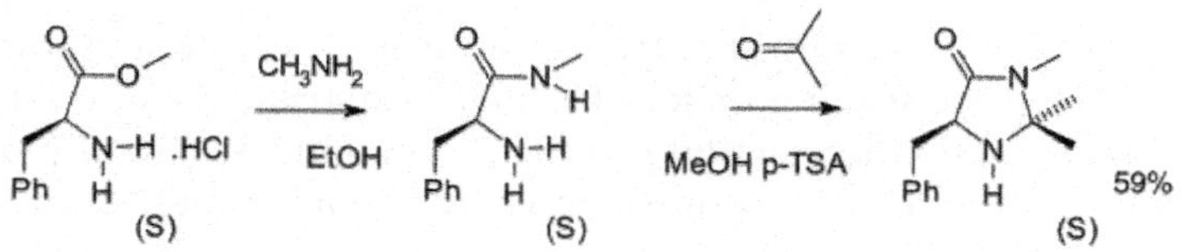

This catalyst works by forming an iminium ion with carbonyl groups of α,β-unsaturated aldehydes (enals) and enones in a rapid chemical equilibrium. This iminium activation is similar to activation of carbonyl groups by a Lewis acid and both catalysts lower the substrate's LUMO:

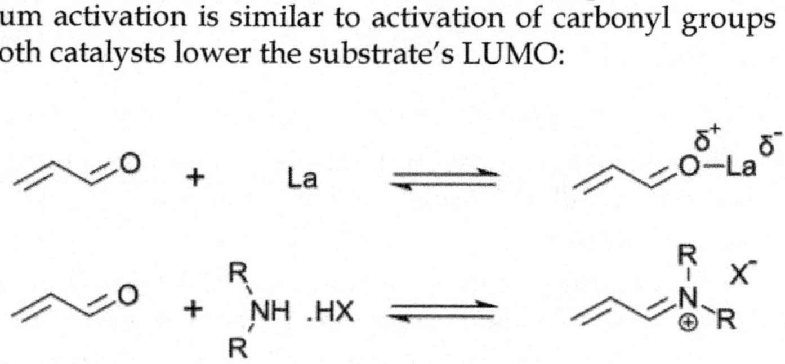

The transient iminium intermediate is chiral which is transferred to the reaction product *via* chiral induction. The catalysts have been used in Diels-Alder reactions, Michael additions, Friedel-Crafts alkylations, transfer hydrogenations and epoxidations.

One example is the asymmetric synthesis of the drug warfarin (in equilibrium with the hemiketal) in a Michael addition of 4-hydroxycoumarin and benzylideneacetone:

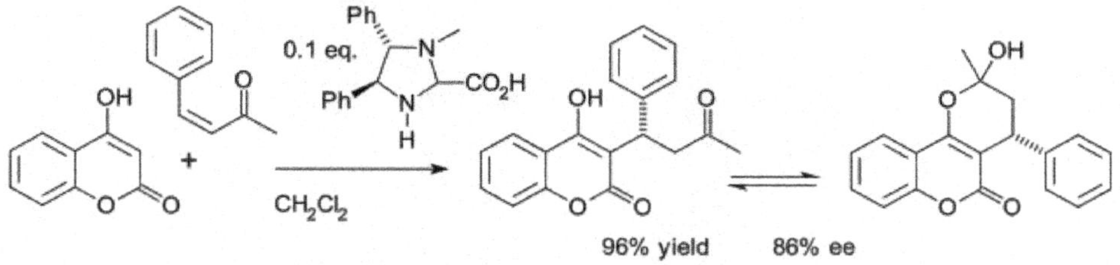

96% yield   86% ee

A recent exploit is the vinyl alkylation of crotonaldehyde with an organotri-fluoroborate salt:

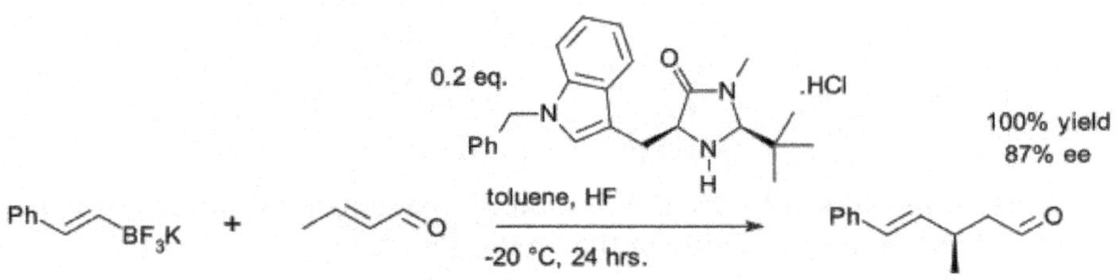

100% yield
87% ee

Organocatalysis

Organocatalysis uses small organic molecules predominantly composed of C, H, O, N, S and P to accelerate chemical reactions. The advantages of organocatalysts include their lack of sensitivity to moisture and oxygen, their ready availability, low cost, and low toxicity, which confers a huge direct benefit in the production of pharmaceutical intermediates when compared with (transition) metal catalysts.

In the example of the Knoevenagel Condensation, it is believed that piperidine forms a reactive iminium ion intermediate with the carbonyl compound:

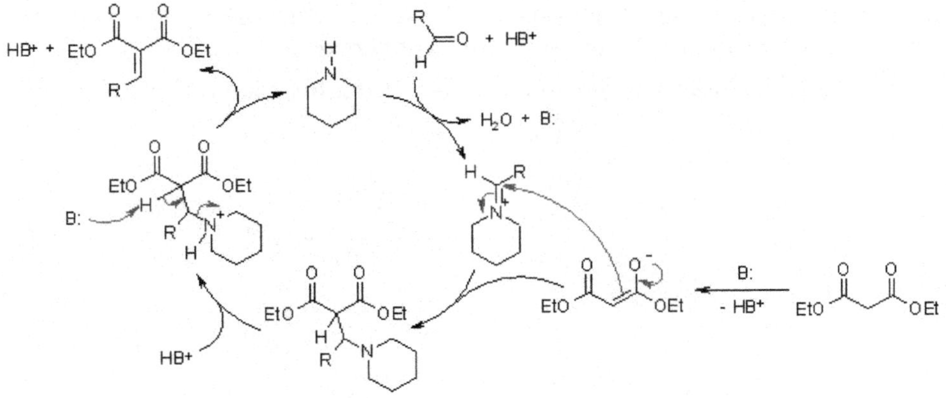

Another organocatalyst is DMAP, which acts as an acyl transfer agent:

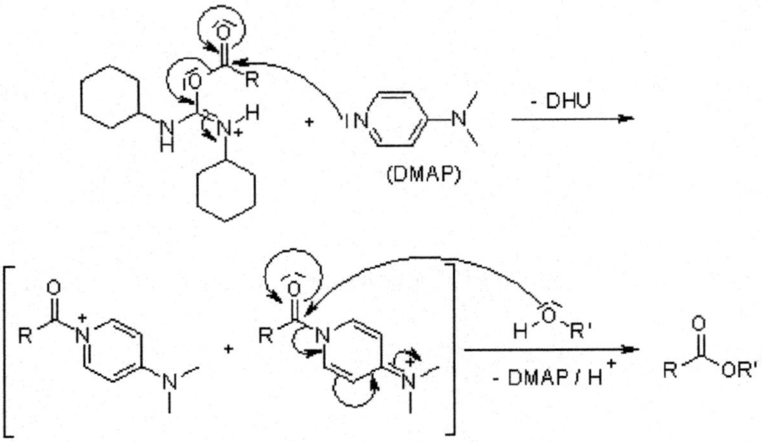

Fig. : Steglich Esterification

Thiazolium salts are versatile umpolung reagents (acyl anion equivalents), for example finding application in the Stetter Reaction:

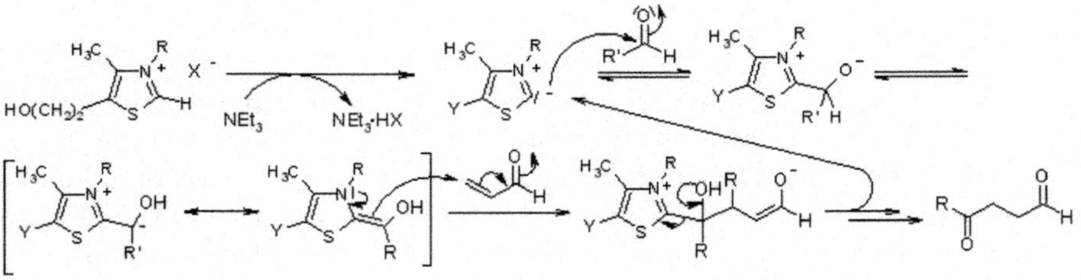

All of these organocatalysts are able to form temporary covalent bonds. Other catalysts can form H-bonds, or engage in pi-stacking and ion pair interactions (phase transfer catalysts). Catalysts may be specially designed for a specific task - for example, facilitating enantioselective conversions.

An early example of an enantioselective Stetter Reaction is shown below:

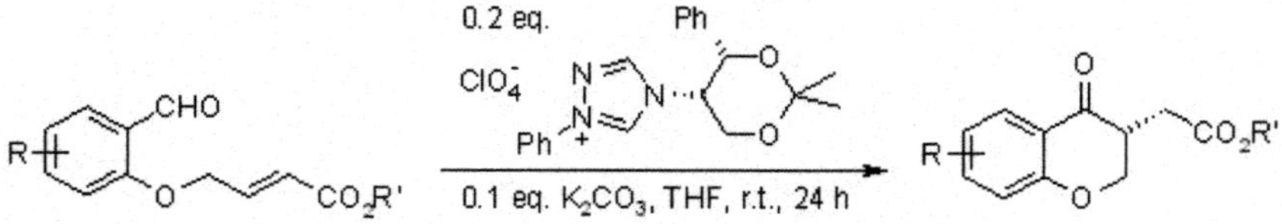

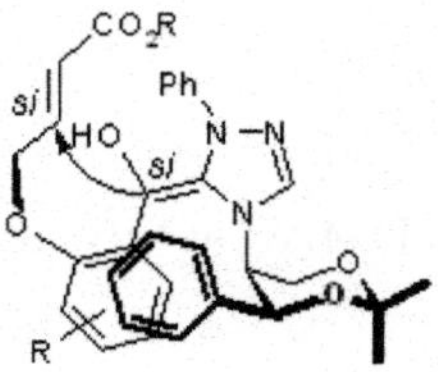

Fig. : model explaining the facial selectivity

Enantioselective Michael Addition using phase transfer catalysis:

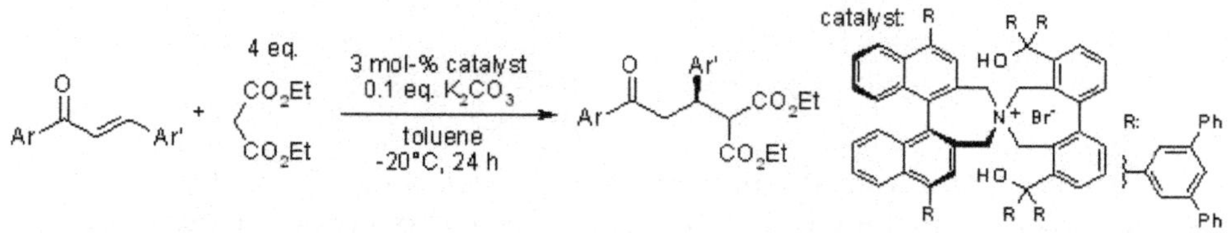

The first enantioselective organocatalytic reactions had already been described at the beginning of the 20th century, and some astonishing, selective reactions such as the proline-catalyzed synthesis of optically active steroid partial structures by Hajos, Parrish, Eder, Sauer and Wiechert had been reported in 1971. However, the transition metal-based catalysts developed more recently have drawn the lion's share of attention.

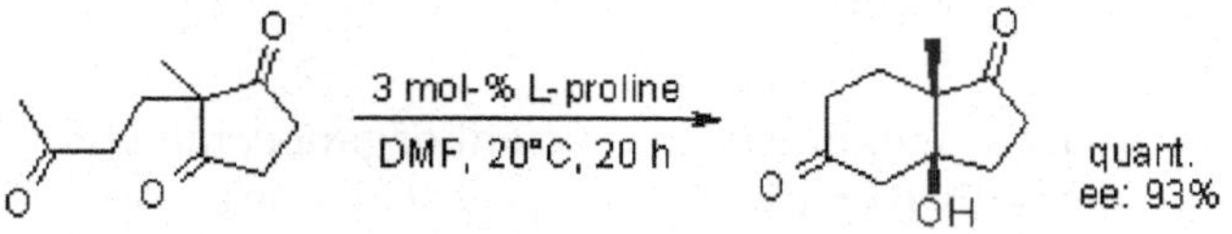

Fig. : Hajos-Parrish-Eder-Sauer-Wiechert reaction (example).

The first publications from the groups of MacMillan, List, Denmark, and Jacobson paved the way in the year 1990. These reports introduced highly enantioselective transformations that rivaled the metal-catalyzed reactions in both yields and selectivity. Once this foundation was laid, mounting interest in organocatalysis was reflected in a rapid increase in publications on this topic from a growing number of research groups.

Proline-derived compounds have proven themselves to be real workhorse organocatalysts. They have been used in a variety of carbonyl compound transformations, where the catalysis is believed to involve the iminium form. These catalysts are cheap and readily accessible:

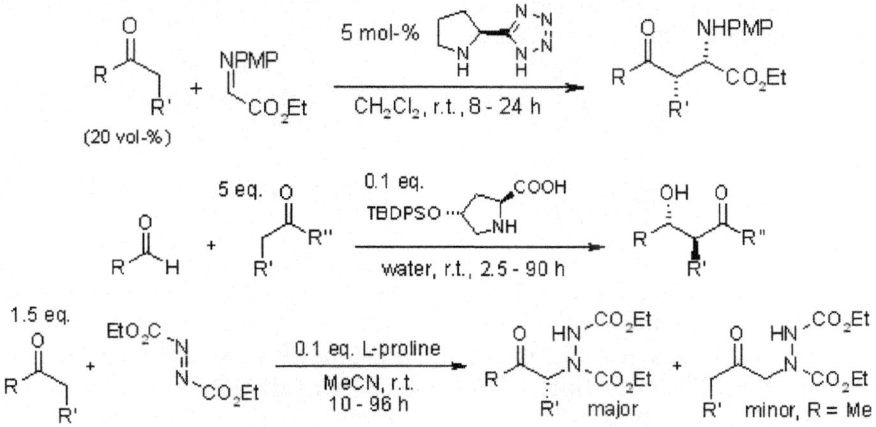

A readily synthesized chiral sulfinamide based organocatalyst enables an asymmetric ring-opening (ARO) reaction of *meso* epoxides with anilines in high yields of with excellent enantioselectivity at room temperature. A probable mechanism for the catalytic ARO reaction is envisaged by ^{1}H and ^{13}C NMR experiments.

2,2,2-trifluoroacetophenone is an efficient organocatalyst for a cheap, mild, fast, and environmentally friendly epoxidation of alkenes. Various olefins, mono-, di-, and trisubstituted, are epoxidized chemoselectively in high to quantitative yields utilizing low catalyst loadings and H_2O_2 as a green oxidant.

In a synergistical combination of photoredox catalysis and organocatalysis for the direct β-alkylation of saturated aldehydes, photon-induced enamine oxidation provides an activated β-enaminyl radical intermediate, which readily combines with a wide range of Michael acceptors to produce β-alkyl aldehydes in a highly efficient manner. This redox-neutral, atom-economical C–H functionalization can be achieved both inter- and intramolecularly.

The asymmetric Payne oxidation of *N*-sulfonyl aldimines catalyzed by a *P*-spiro chiral triaminoiminophosphorane enables a practical synthesis of optically active *N*-sulfonyl oxaziridines with high efficiency and an excellent level of enantioselectivity. The versatility of this method was demonstrated by the diastereoselective kinetic oxidation of racemic α-chiral *N*-sulfonyl imines.

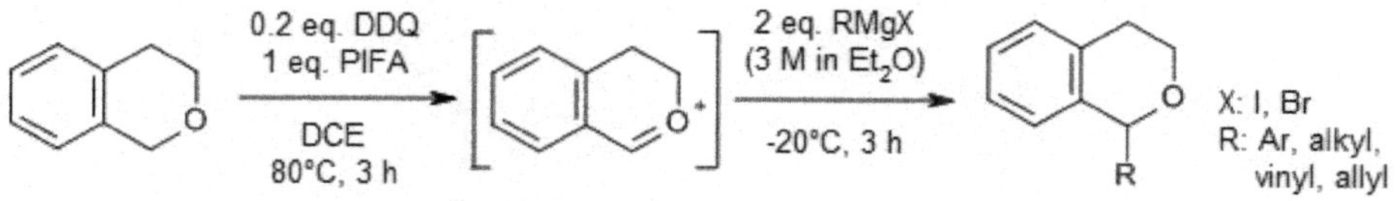

0.2 eq. DDQ
1 eq. PIFA

DCE
80°C, 3 h

2 eq. RMgX
(3 M in Et$_2$O)

-20°C, 3 h

X: I, Br
R: Ar, alkyl,
vinyl, allyl

The use of [bis(trifluoroacetoxy)iodo]benzene as stoichiometric oxidant and 2,3-dichloro-5,6-dicyano-1,4-benzoquinone as organocatalyst enables a convenient oxidation of isochromans. A further reaction with grignard reagents or amides affords the corresponding isochroman derivatives.

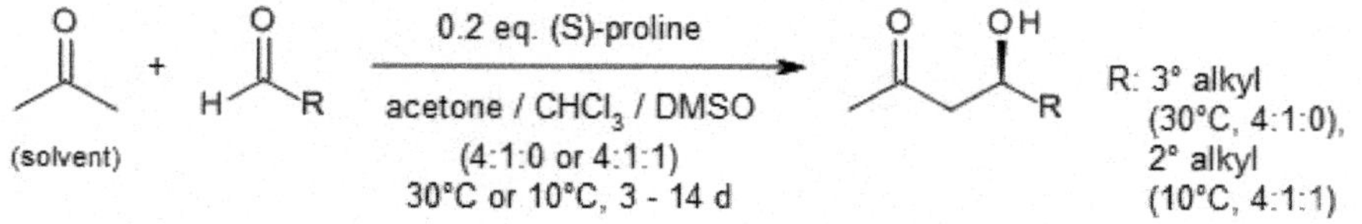

(solvent)

H   R

0.2 eq. (S)-proline

acetone / CHCl$_3$ / DMSO
(4:1:0 or 4:1:1)
30°C or 10°C, 3 - 14 d

R

R: 3° alkyl
(30°C, 4:1:0),
2° alkyl
(10°C, 4:1:1)

Challenges in controlling the proline-catalyzed asymmetric aldol reaction between aliphatic aldehydes and acetone include side reactions such as aldol condensation and self-aldolization. Recently optimized conditions enable high yields of the desired addition product and good to excellent enantioselectivities.

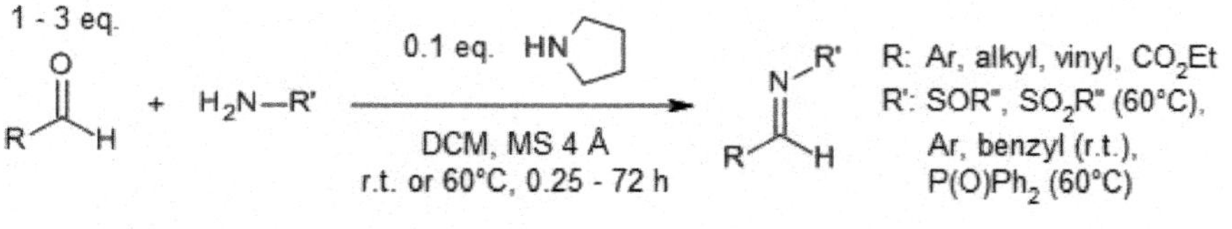

1 - 3 eq.

R   H   +   H$_2$N—R'

0.1 eq. HN

DCM, MS 4 Å
r.t. or 60°C, 0.25 - 72 h

R   H

R: Ar, alkyl, vinyl, CO$_2$Et
R': SOR", SO$_2$R" (60°C),
Ar, benzyl (r.t.),
P(O)Ph$_2$ (60°C)

Pyrrolidine as an organocatalyst enables a general and efficient biomimetic method for the synthesis of aldimines from aldehydes and compounds bearing an amino group. This nucleophilic catalysis proceeds *via* iminium activation with outstanding yields in the absence of acids and metals under simple conditions. The method has been applied to the synthesis of *N*-sulfinyl, *N*-sulfonyl imines, *N*-phosphinoyl, *N*-alkyl, and *N*-aryl imines.

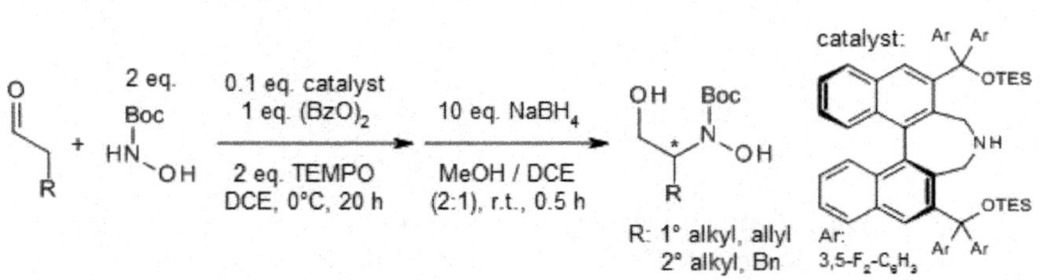

catalyst:

R   +   HN   Boc   OH

2 eq.
Boc

0.1 eq. catalyst
1 eq. (BzO)$_2$

2 eq. TEMPO
DCE, 0°C, 20 h

10 eq. NaBH$_4$

MeOH / DCE
(2:1), r.t., 0.5 h

OH   Boc
N   OH
R

R: 1° alkyl, allyl
2° alkyl, Bn

Ar:
3,5-F$_2$-C$_6$H$_3$

OTES
NH
OTES

A highly regio- and enantioselective hydroxyamination of aldehydes with in situ generated nitrosocarbonyl compounds from hydroxamic acid derivatives was realized by combining TEMPO and BPO as oxidants in the presence of a binaphthyl-modified amine catalyst.

After NHC-catalyzed activation of acetic esters, the catalytically generated triazolium enolate intermediates serve as two-carbon nucleophiles that undergo highly enantioselective reactions with enones and α,β-unsaturated imines to give α-unsubstituted δ-lactones and lactams, respectively.

Visible light activates diarylketone catalysts to abstract a benzylic hydrogen atom selectively, which enables an operationally simple direct fluorination of benzylic C-H groups in the presence of a fluorine radical donor. 9-Fluorenone catalyzes benzylic C-H monofluorination, while xanthone catalyzes benzylic C-H difluorination.

An N-Heterocyclic carbene (NHC)-catalyzed highly enantioselective lactonization of modified enals with enolizable aldehydes proceeds *via* α,β-unsaturated acylazolium intermediates and yields synthetically important 4,5-disubstituted dihydropyranones.

A chiral disulfonimide efficiently catalyzes an asymmetric Mannich reaction of silyl ketene acetals with N-Boc-amino sulfones with excellent yields and enantioselectivities. The chiral catalyst also promotes the in situ generation of the N-Boc imine intermediates. Kinetic studies confirm a stepwise mechanism.

A facile and mild photooxidation of alcohols gives carboxylic acids and ketones using easily handled 2-chloroanthraquinone as an organocatalyst under visible light irradiation in an air atmosphere.

An efficient enantioselective hydrazination/cyclization cascade reaction of α-substituted isocyanoacetates to azodicarboxylates is catalyzed by a squaramide catalyst derived from *Cinchona* alkaloid. The reaction affords optically active 1,2,4-triazolines in excellent yields and very good enantioselectivities under mild conditions.

A chiral cyclopropenimine catalyzes Mannich reactions between glycine imines and *N*-Boc-aldimines with high levels of enantio- and diastereocontrol. The reactivity of this catalyst is substantially greater than that of a widely used thiourea cinchona alkaloid-derived catalyst. Various aryl and aliphatic *N*-Boc-aldimines are effective substrates for this transformation.

5-Ethyl-3-methyl-2',4':3',5'-di-*O*-methylenedioxy-riboflavinium perchlorate, which is readily derived from commercially available vitamin B₂, exhibits high catalytic activity for the oxidation of organic sulfides under an oxygen atmosphere with the assistance of hydrazine hydrate as a reductant. This is an inexpensive, convenient, and environmentally benign method for the selective oxidative transformation of sulfides into sulfoxides.

A highly efficient organocatalytic asymmetric sulfa-Michael addition of thiols to hexafluoroisopropyl α,β-unsaturated esters performs well over a broad scope of α,β-unsaturated esters and diversified thiols. Introducing electron-withdrawing hexafluoroisopropyl ester is crucial to enhancing the electrophilicity of unsaturated esters as Michael acceptors.

An organocatalytic iodination of activated aromatic compounds using 1,3-di-iodo-5,5-dimethylhydantoin (DIH) as the iodine source with thiourea catalysts in acetonitrile is applicable to a number of aromatic substrates with significantly different steric and electronic properties. The iodination is generally highly regioselective and provides high yields of isolated products.

An efficient and highly enantioselective Payne-type oxidation of *N*-sulfonyl imines exhibits broad substrate generality and unique chemoselectivity based on the combined use of hydrogen peroxide and trichloroacetonitrile under the catalysis of P-spiro chiral triaminoiminophosphorane.

Solvent-Free Enantioselective Friedländer Condensation with Wet 1,1'-Binaphthalene-2,2'-diamine-Derived Prolinamides as Organocatalysts

N-heterocyclic carbene (NHC)-catalyzed C-C bond cleavage of carbohydrates as formaldehyde equivalents generates acyl anion intermediates for Stetter reaction *via* a retro-benzoin-type process. The renewable nature of carbohydrates, accessible from biomass, further highlights the practical potential of this fundamentally interesting catalytic activation.

A valine-derived, electron-deficient triazolium salt catalyzes a high yielding, highly chemo- and enantioselective cross-benzoin reaction between aliphatic aldehydes and α-ketoesters. Diastereoselective reduction of the products gives access to densely oxygenated compounds with high chemo- and diastereoselectivity.

Using cinchona alkaloid-derived primary amines as catalysts and aqueous hydrogen peroxide as the oxidant, highly enantioselective Weitz-Scheffer-type epoxidation and hydroperoxidation reactions of α,β-unsaturated carbonyl compounds take place. Acyclic enones, cyclic enones, and α-branched enals can be converted. Intermediates have been characterized by MS and NMR. DFT calculations explain the activation of H_2O_2.

An efficient catalytic and highly enantioselective protonation of silyl ketene imines is catalyzed by the chiral phosphoric acids TRIP or STRIP in the presence of a stoichiometric amount of methanol as the proton source and silyl acceptor. Various substituted racemic silyl ketene imines have been transformed into highly enantioenriched nitriles.

ENANTIOSELECTIVE ORGANOCATALYSIS

Catalysis of reactions using an organic molecule as the catalytic species has been a significantly overlooked area in recent years. However, organocatalysis often has notable advantages over conventional transition metal catalysis. For example, organocatalytic reactions seldom suffer from the requirement for inert conditions frequently necessary in metal chemistry. Furthermore, environmental factors favour the use of non-toxic small organic molecules and they are often less expensive than the rare metal species commonly used in synthesis.

Organic Catalysts

We are developing new enantioselective organocatalytic processes that generate highly functionalised molecules from simple starting materials in a single step.

Enantioselective Organocatalytic Cyclopropanation *via* Ammonium Ylides

The cyclopropane motif is a common feature in complex molecule synthesis and in medicinal chemistry due to a unique combination of reactivity and structural properties. These properties have made the preparation of cyclopropanes an attractivetarget for new methodology development. Despite the many processes for the synthesis of functionalized cyclopropanes there are surprisingly few general catalytic enantioselective methods. Of the methods available the carbenoid mediated reactions are most often utilized however, there are isolated examples of catalytic ylide based enantioselective cyclopropanations.

Recently, we described a new cyclopropanation process using a reaction that was mediated by a stoichiometric quantity of a nucleophilic tertiary amine through the formation of an ammonium ylide. Here, we report the development of a new enantioselective organocatalytic cyclopropanation reaction *via* ammonium ylides that produces a range of functionalized molecules with excellent diastereo- and enantioselectivity.

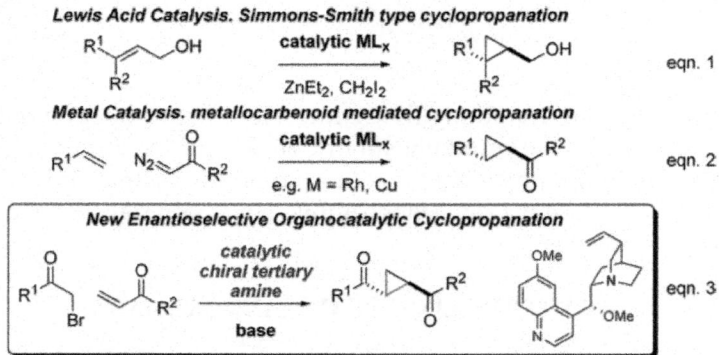

Methods of Enantioselective Catalytic Cyclopropanation

An organocatalytic cyclopropanation process *via* ammonium ylides has a number of advantages over its metal counterparts. There are no toxic transition metals involved in the reaction and the starting materials are readily available and conveniently handled. Furthermore, the number of known chiral amines represents a significant pool from which potential catalysts can be selected.

In this system an A-bromo carbonyl compound **1** undergoes S_N2 displacement with the tertiary amine catalyst **3** to form a quaternary ammonium salt **I**. Deprotonation with mild base forms the ylide **II** that undergoes conjugate addition to alkene **2**, forming **III**. Finally, 3-*exo*-tet cyclization generates the cyclopropane **4** and reforms the catalyst.

Proposed Catalytic Cycle

The scope of this process is broad with more than 10 examples giving ee > 90% as either enantiomer. We are currently exploring applications in natural product synthesis and chemical biology using cyclopropanes as the key component. In summary, we have developed a new enantioselective organocatalytic cyclopropanation reaction. Importantly the cyclopropanes can be produced as either enantiomer using the quinine or quinidine series of cinchona alkaloid catalysts. The reaction is applicable to a range of substrates with a variety of versatile functional groups.

Organocatalytic Intramolecular Cyclopropanation

Catalytic processes that form functionalized cyclic molecules represent a key transformation for synthetic organic chemistry. Recently, a number of organocatalytic processes have emerged that often provide excellent levels of both enantio- and diastereocontrol for the synthesis of cyclic molecules.

We have developed a *new organocatalytic intramolecular cyclopropanation reaction* that forms synthetically versatile [n.1.0]-bicycloalkanes using a nucleophilic tertiary amine catalyst.

The synthesis of these bicycloalkanes is most commonly carried out by inter- or intramolecular metal catalyzed carbene transfer of diazo-compounds to electron rich alkenes. Other than these methods there are few general alternatives to form [n.1.0]-bicycloalkenes. [n.1.0]-Bicycloalkanes offer many exciting applications in complex molecule synthesis due to the high levels of stereochemistry and latent reactivity inherent within their structure. Therefore, new complementary methods for their catalytic enantioselective synthesis are very important. We have identified an interesting organocatalytic intramolecular cyclopropanation process that is stereoselective and produces highly functionalized bicycloalkanes containing three stereocentres, two rings and three levels of orthogonal functionality in a single step from linear building blocks.

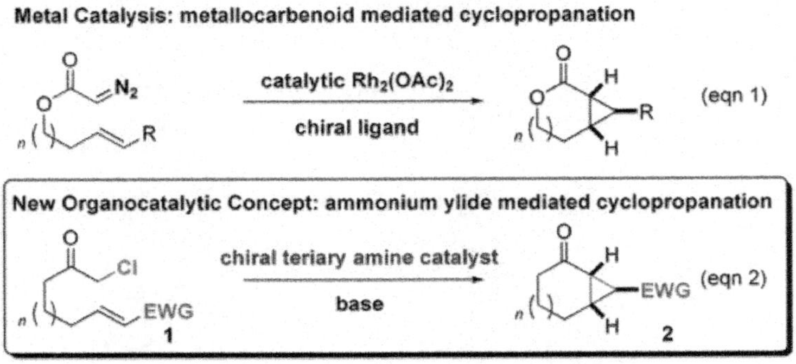

Metal Catalysis: metallocarbenoid mediated cyclopropanation

(eqn 1)

New Organocatalytic Concept: ammonium ylide mediated cyclopropanation

(eqn 2)

Methods for Catalytic Intramolecular Cyclopropanation

In this approach an α-chloroketone with a tethered electron deficient alkene reacts through a catalytically generated ammonium ylide to form the bicyclic structure 2. This organocatalytic strategy precludes the use of highly sensitive diazo compounds. It should also offer a wider substrate scope due to compatibility with the metal-free catalyst, and produce bicycloalkanes with higher levels of functionality. Furthermore, there are many readily available chiral tertiary amines from which an enantioselective process can be developed.

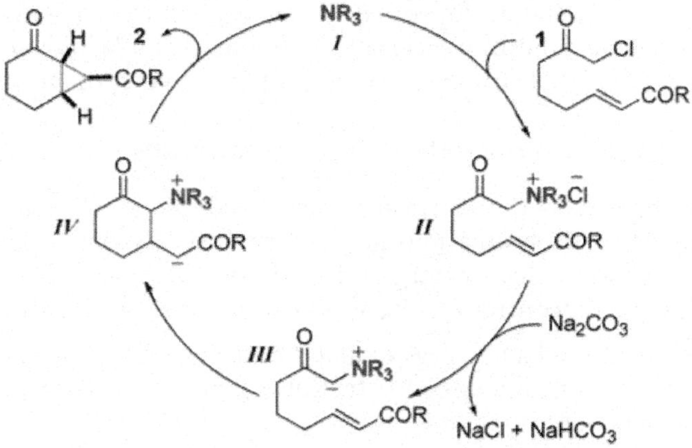

Proposed Catalytic Cycle

A proposed catalytic cycle is shown. The amine catalyst *I* displaces the chloride in 1 giving the quaternary ammonium salt *II*. *Deprotonation forms the ammonium ylide III and intramolecular conjugate addition forms IV and finally the bicycloalkane 2 is generated through displacement of the ammonium group, concurrently re-generating catalyst I.*

The scope of this new reaction was broad when DABCO was used as the catalyst with excellent yields and diastereoselectivity for a variety of substrates. However, DABCO was selected as the catalyst because of structural similarity to the cinchona alkaloids making it a racemic model for an enantioselective reaction. On replacement of DABCO with chiral cinchona alkaloid catalyst the reaction produced the bicycloalkanes in good yield and excellent enantioselectivity (95%). Importantly both enantiomers are accessible using the pseudoenantiomeric quinine/quinidine catalysts. To the best of our knowledge this excellent result represents *the first enantioselective organocatalytic intramolecular cyclopropantion reaction*.

entry	catalyst loading	time / h	additive	yield %	e.e. %
1	20 mol% A	96	–	67	64 (+)
2	20 mol% A	24	0.4 eq NaBr	61	94 (+)
3	20 mol% A	24	0.4 eq NaI	64	95 (+)
4	20 mol% B	24	0.4 eq NaBr	48	94 (−)

In summary, we have developed an organocatalytic intramolecular cyclopropanation reaction for the formation of synthetically versatile [n.1.0]-bicycloalkanes as single diastereoisomers. This powerful catalytic process effects the controlled formation of three stereocenters, two carbon-carbon bonds and two rings in a single transformation. The reaction is enantioselective with a catalytic amount of chiral amine and can form either enantiomer. We are currently exploring the scope of the catalytic enantioselective process and applications towards the synthesis of complex molecules.

Organocatalysis

Various studies suggest a close connection between global climate change and the emission of anthropogenic greenhouse gases. By far largest part of this emission is attributed to carbon dioxide (CO_2). Apart from the goal of reducing CO_2 emissions, its use as synthetic building block is the central point of the overall CO_2 management strategy. The atom economic and efficient utilization of CO_2 as synthetic building block is closely connected to the effective activation of this very stable molecule.

The aim of our work is the development of novel metal free catalysts, so called organocatalysts, for the synthesis of industrially relevant products with CO_2 as a C1-building block. The aim is that combining those catalysts with metal-catalyzed or enzyme-catalyzed procedures in (sequential) one pot reactions leads to innovative and sustainable catalytic systems with high selectivity and energy efficiency respectively. These alternative methods, taking steps in the upstream and downstream phases, are targeted at changing and extending the raw material base, utilizing CO_2. Subjects under study include transformations and products of much interest to industry and large CO_2-fixation potential.

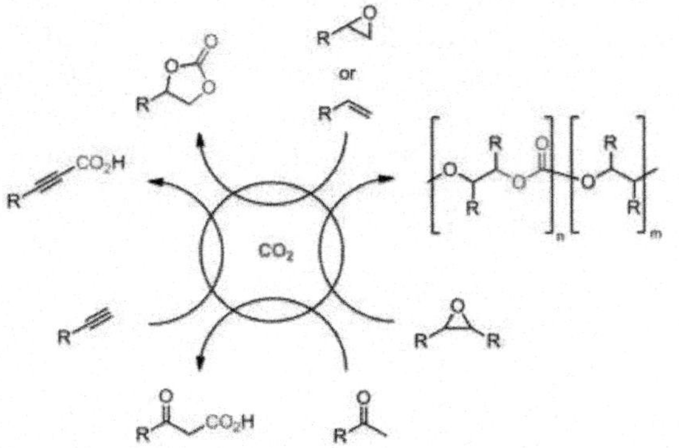

Fig. : Utilization of CO_2 *via* direct Fixation.

Below a short outline about two selected sub-projects will be given:

Synthesis of Cyclic Carbonates

So far we prepared a series of bifunctional phosphonium salts and evaluated their potential as catalysts under various reaction conditions in the synthesis of cyclic carbonates from epoxides and carbon dioxide. Based on this catalyst screening we determined structure activity relationships and identified promising candidates. Moreover, the reaction parameters were optimized for a model reaction. Even under mild conditions (90°C, $p(CO_2)$= 10 bar) quantitative conversions were obtained after 2–3 h for 15 different epoxides. The reaction was also performed on a multi gram scale and monitored by in-situ FTIR.

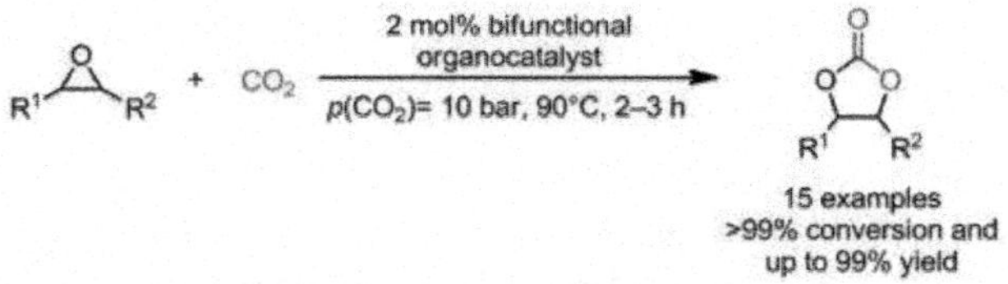

15 examples
>99% conversion and
up to 99% yield

Fig. : Conversionen of epoxides with CO_2 yielding cyclic carbonates.

Synthesis of Polycarbonates

Bis-urea derivatives are known as oxo-anion receptors. We prepared a series of those compounds and employed them as catalysts in the copolymerisation of CO_2 with epoxides. The intention in using the urea derivatives was to stabilize the anionic intermediates in polymer chain growth and thus to support the incorporation of carbon dioxide into the polymer chain. The aim is for this to lead to a polymer with well-defined structure and material characteristics. We obtained copolymers with >80% carbonate linkage and molecular weights of >35,000 g·mol–1 in excellent yields: results depended on the catalytic system and the reaction conditions.

Fig. : Copolymerisation of epoxides with CO_2 yielding polycarbonates.

Novel Organocatalyzed Processes based on Phosphorous Containing Catalysts

Phosphines are an important class of ligands in metal complexes. Due to their outstanding role in homogeneous catalysis, there have been numerous publications reported concerning the synthesis of achiral as well as chiral phosphine-based ligands. In addition many of them are commercially available. However phosphines and their derivatives are not only employed as ligands but are also important reagents in organic synthesis. For example, they mediate the conversion of alcohols to halides (Appel reaction), the olefination of ketones and aldehydes (Wittig reaction) and the synthesis of imines *via* Staudinger reaction. In contrast to their tremendous importance as ligands and reagents, the application of phosphines as organocatalysts typically has a somewhat secondary role.

We are exploring the application of phosphorus-based organic compounds as catalysts in a variety of reactions. Therefore we are utilizing their Lewis basic and Lewis acidic properties, respectively. On the one hand our aim is to develop asymmetric versions of the reactions indicated, by employing chiral phosphines or derivatives. A second aim is to develop catalytic (asymmetric) variations of methods which hitherto employ phosphines in stoichiometric amounts, *e.g.* Wittig Reactions.

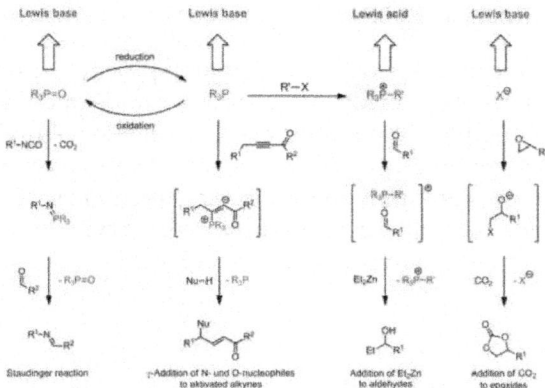

Fig. : Lewis-basic and Lewis-acidic activity of phosphines and derivatives, respectively, as well as selected possible applications.

Asymmetric Intramolecular γ-addition to Activated Alkynes

A variety of O-, S-, and N-functionalized activated alkynes were prepared and cyclised in the phosphine catalyzed intramolecular γ-Addition. For the first time we successfully converted N-derivatives in this reaction. In an intense screening of chiral phosphine catalysts products with up to 84% *ee* were obtained in an intense screening of chiral phosphine catalysts.

X= O, S, NPG

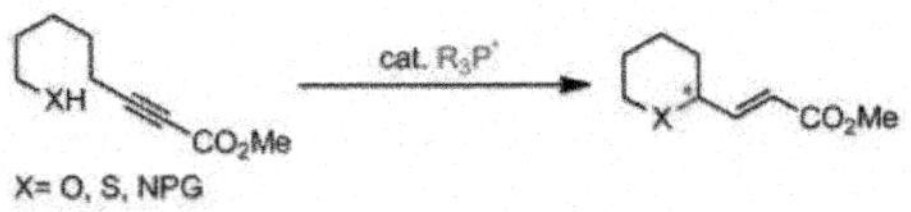

Fig. : Asymmetric intramolecular γ-addition to activated alkynes.

Addition of Et2Zn to Aldehydes

Alongside the application of phosphines as Lewis basic catalysts we are also interested in the utilization of phosphonium salts as potential Lewis acidic organocatalysts. In this context we studied the addition of diethyl zinc to aldehydes. It is known that this reaction can be catalyzed by Lewis acids. We showed that simple Bu_4PCl is an efficient catalyst for this reaction. However, when chiral phosphonium salts were used, only racemic products were obtained and it was observed that the anion exerted a strong influence. Further investigations revealed that diethyl zinc is activated by the Lewis basic counter ion and the reaction can be catalyzed by a combination of simple alkaline metal salts and crown ethers.

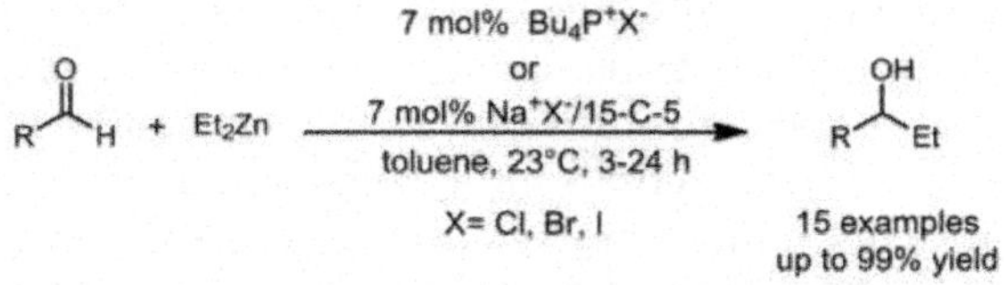

7 mol% $Bu_4P^+X^-$
or
7 mol% Na^+X^-/15-C-5
toluene, 23°C, 3-24 h

X= Cl, Br, I

15 examples
up to 99% yield

Fig. : Addition of diethyl zinc to aldehydes.

Catalytic Wittig Reaction

In our efforts to convert phosphine mediated processes into catalytic variants we investigated the Wittig reaction. So far we have developed a microwave-assisted catalytic variant employing commercially-available tributyl phosphineoxide, as well as a thermal method utilizing an easily-accessible phospholane oxid as a pre-catalyst. The right combination of reducing agent, base and solvent recently enabled us to convert over 20 substrates. In some cases yields and selectivity levels were significantly higher than previously reported.

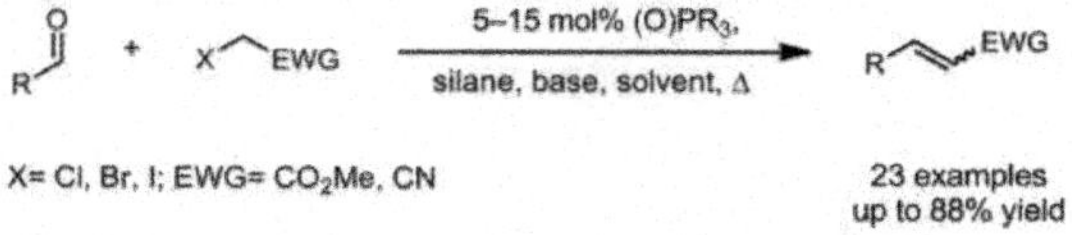

5–15 mol% $(O)PR_3$,
silane, base, solvent, Δ

X= Cl, Br, I; EWG= CO_2Me, CN

23 examples
up to 88% yield

Fig. : Catalytic Wittig reaction.

Metall-Hydrid-Mediated Tishchenko Reaction

During our studies concerning the catalytic Wittig reaction we observed the dimerization of benzaldehyde yielding benzyl benzoate (Tishchenko reaction) when sodium hydride was used as the base in our model reaction. Based on this result we developed an efficient method for the dimerization of aromatic aldehydes. The reaction was also produced on a large scale and easily monitored by in situ FT-IR spectroscopy. A mechanism was postulated and confirmed by labelling and capture experiments. Thus, we developed an efficient method for the dimerization of aromatic as well as hetero-aromatic and aliphatic aldehydes.

$$2 \quad R \overset{O}{\underset{}{\overset{\|}{\diagup}} H \quad \xrightarrow[\text{Solvent, } \Delta]{\text{1–5 mol\% MH}} \quad R \overset{O}{\underset{O}{\overset{\|}{\diagup}} R}$$

R = Ar, Het-Ar, Alkyl
M= Li, Na, K, [Al]

30 examples
up to 95% yield

Fig. : Metallhydride mediated Tishchenko reaction.

Chapter 3

MEMBRANE TECHNOLOGIES AT THE SERVICE OF SUSTAINABLE

DEVELOPMENT THROUGH PROCESS INTENSIFICATION MEMBRANE TECHNOLOGY

Membrane technology covers all engineering approaches for the transport of substances between two fractions with the help of permeable membranes. In general, mechanical separation processes for separating gaseous or liquid streams use membrane technology.

Applications

Fig. : Ultrafiltration for a swimming pool.

Membrane separation processes operate without heating and therefore use less energy than conventional thermal separation processes such as distillation, sublimation or crystallization. The separation process is purely physical and both fractions (permeate and retentate) can be used. Cold separation using membrane technology is widely used in the food technology, biotechnology and pharmaceuticalindustries. Furthermore, using membranes enables separations to take place that would be impossible using thermal separation methods. For example, it is impossible to separate the constituents of azeotropic liquids or solutes which form isomorphic crystals by distillation or recrystallization but such separations can be achieved using membrane technology. Depending on the type of membrane, the selective separation of certain individual substances or substance mixtures is possible. Important technical applications include the production of drinking water by reverse osmosis (worldwide approximately 7 million cubic metres annually), filtrations in the food industry, the recovery of organic vapours such as petro-chemical vapour recovery and the electrolysis for chlorine production.

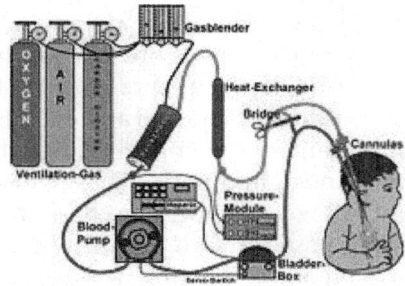

Fig. : Venous-arterial extracorporeal membrane oxygenation scheme.

In waste water treatment, membrane technology is becoming increasingly important. With the help of ultra/microfiltration it is possible to remove particles, colloids and macromolecules, so that waste-water can be disinfected in this way. This is needed if waste-water is discharged into sensitive waters especially those designated for contact water-sports and recreation.

About half of the market is in medical applications such as use in artificial kidneys to remove toxic substances by hemodialysis and asartificial lung for bubble-free supply of oxygen in the blood.

The importance of membrane technology is growing in the field of environmental protection (NanoMemPro IPPC Database). Even in modern energy recovery techniques membranes are increasingly used, for example in fuel cells and in osmotic power plants.

Current Market and Forecast

The global demand for membrane modules was estimated at approximately 15.6 billion USD in 2012. Driven by new developments and innovations in material science and process technologies, global increasing demands, new applications, and others, the market is expected to grow around 8% annually in the next years. It is forecast to increase to 21.22 billion USD in 2016 and reach 25 billion in 2018.

Mass Transfer

Two basic models can be distinguished for mass transfer through the membrane:

- The solution-diffusion model and
- The hydrodynamic model.

In real membranes, these two transport mechanisms certainly occur side by side, especially during ultra-filtration.

Solution-diffusion Model

In the solution-diffusion model, transport occurs only by diffusion. The component that needs to be transported must first be dissolved in the membrane. The general approach of the solution-diffusion model is to assume that the chemical potential of the feed and permeate fluids are in equilibrium with the adjacent membrane surfaces such that appropriate expressions for the chemical potential in the fluid and membrane phases can be equated at the solution-membrane interface. This principle is more important for *dense* membranes without natural pores such as those used for reverse osmosis and in fuel cells. During the filtration process a boundary layer forms on the membrane. Thisconcentration gradient is created by molecules which cannot pass through the membrane. The effect is referred as concentration polarization and, occurring during the filtration, leads to a reduced trans-membrane flow (flux). Concentration polarization is, in principle, reversible by cleaning the membrane which results in the initial flux being almost totally restored. Using a tangential flow to the membrane (cross-flow filtration) can also minimize concentration polarization.

Hydrodynamic Model

Transport through pores – in the simplest case – is done convectively. This requires the size of the pores to be smaller than the diameter of the two separate components. Membranes which function according to this principle are used mainly in micro- and ultrafiltration. They are used to separate macromolecules from solutions, colloids from adispersion or remove bacteria. During this process the not-passing particles or molecules form a pulpy mass (filter cake) on the membrane, and this blockage of the membrane hampers the filtration. This blockage can be reduced by the use of the cross-flow method (cross-flow filtration). Here, the liquid to be filtered flows along the front of the membrane and is separated by the pressure difference between the front and back of the membrane into retentate (the flowing concentrate) on the front and permeate (filtrate) on the back. The tangential flow on the front creates a shear stress that cracks the filter cake and reduces the fouling.

Membrane Operations

According to the driving force of the operation it is possible to distinguish:

1. Pressure driven operations
   - microfiltration
   - ultrafiltration
   - nanofiltration
   - reverse osmosis

2. Concentration driven operations
   - dialysis
   - pervaporation
   - forward osmosis
   - artificial lung
   - gas separation
3. Operations in an electric potential gradient
   - electrodialysis
   - membrane electrolysis *e.g.* chloralkali process
   - electrodeionization
   - electrofiltration
   - fuel cell
4. Operations in a temperature gradient
   - membrane distillation

Membrane Shapes and Flow Geometries

There are two main flow configurations of membrane processes: cross-flow (or) tangential flow and dead-end filtrations. In cross-flow filtration the feed flow is tangential to the surface of membrane, retentate is removed from the same side further downstream, whereas the permeate flow is tracked on the other side. In dead-end filtration the direction of the fluid flow is normal to the membrane surface. Both flow geometries offer some advantages and disadvantages. Generally, dead-end filtration is used for feasibility studies on a laboratory scale. The dead-end membranes are relatively easy to fabricate which reduces the cost of the separation process. The dead-end membrane separation process is easy to implement and the process is usually cheaper than cross-flow membrane filtration.

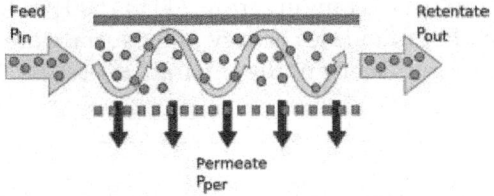

Fig. : Cross-flow geometry.

The dead-end filtration process is usually a batch-type process, where the filtering solution is loaded (or slowly fed) into the membrane device, which then allows passage of some particles subject to the driving force. The main disadvantage of a dead end filtration is the extensive membrane fouling andconcentration polarization. The fouling is usually induced faster at higher driving forces. Membrane fouling and particle retention in a feed solution also builds up a concentration gradients and particle back flow (concentration polarization). The tangential flow devices are more cost and labor-intensive, but they are less susceptible to fouling due to the sweeping effects and high shear rates of the passing flow. The most commonly used synthetic membrane devices (modules) are flat sheets/ plates, spiral wounds, and hollow fibers.

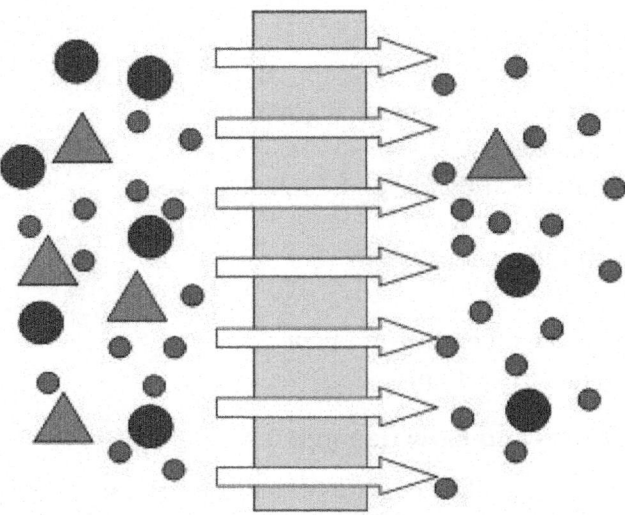

Fig. : Dead-end geometry.

Flat plates are usually constructed as circular thin flat membrane surfaces to be used in dead-end geometry modules. Spiral wounds are constructed from similar flat membranes but in the form of a "pocket" containing two membrane sheets separated by a highly porous support plate. Several such pockets are then wound around a tube to create a tangential flow geometry and to reduce membrane fouling. hollow fiber modules consist of an assembly of self-supporting fibers with dense skin separation layers, and a more open matrix helping to withstand pressure gradients and maintain structural integrity. The hollow fiber modules can contain up to 10,000 fibers ranging from 200 to 2500 μm in diameter; The main advantage of hollow fiber modules is very large surface area within an enclosed volume, increasing the efficiency of the separation process.

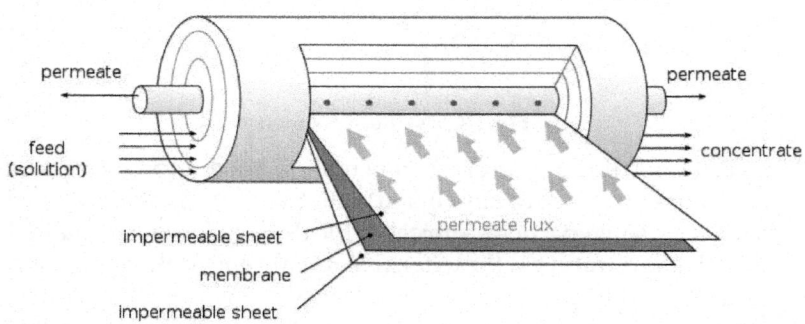

Fig. : Spiral wound membrane module.

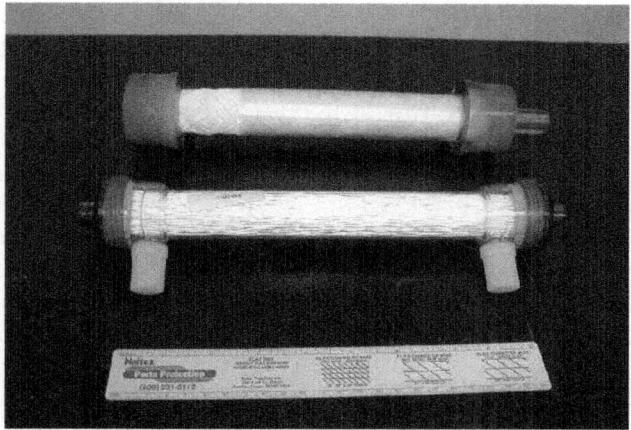

Fig. : Hollow fiber membrane module.

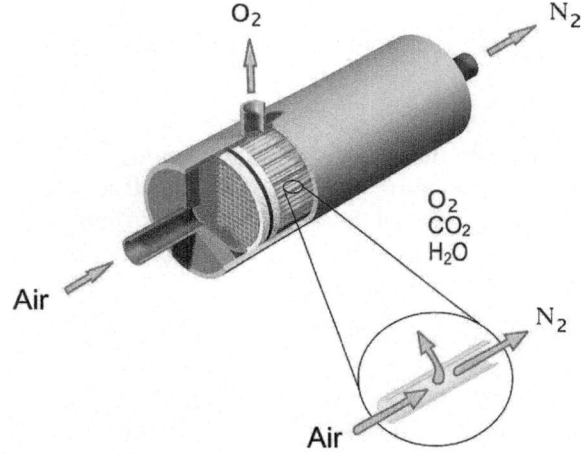

Fig. : Separation of air into oxygen and nitrogen through a membrane.

Membrane Performance and Governing Equations

The selection of synthetic membranes for a targeted separation process is usually based on few requirements. Membranes have to provide enough mass transfer area to process large amounts of feed stream. The selected membrane has to have high selectivity (rejection) properties for certain particles; it has to resist fouling and to have high mechanical stability. It also needs to be reproducible and to have low manufacturing costs. The main modeling equation for the dead-end filtration at constant pressure drop is represented by Darcy's law:

$$\frac{dV_p}{dt} = Q = \frac{\Delta p}{\mu} A \left(\frac{1}{R_m + R} \right)$$

where V_p and Q are the volume of the permeate and its volumetric flow rate respectively (proportional to same characteristics of the feed flow), μ is dynamic viscosity of permeating fluid, A is membrane area, R_m and R are the respective resistances of membrane and growing deposit of the foulants. R_m can be interpreted as a membrane resistance to the solvent (water) permeation. This resistance is a membrane intrinsic property and is expected to be fairly constant and independent of the driving force, Δp. R is related to the type of membrane foulant, its concentration in the filtering solution, and the nature of foulant-membrane interactions. Darcy's law allows for calculation of the membrane area for a targeted separation at given conditions. The solute sieving coefficient is defined by the equation:

$$S = \frac{C_p}{C_f}$$

where C_f and C_p are the solute concentrations in feed and permeate respectively. Hydraulic permeability is defined as the inverse of resistance and is represented by the equation:

$$L_p = \frac{J}{\Delta p}$$

where J is the permeate flux which is the volumetric flow rate per unit of membrane area. The solute sieving coefficient and hydraulic permeability allow the quick assessment of the synthetic membrane performance.

Membrane Separation Processes

Membrane separation processes have a very important role in the separation industry. Nevertheless, they were not considered technically important until the mid-1970s. Membrane separation processes differ based on separation mechanisms and size of the separated particles. The widely used membrane processes include microfiltration,ultrafiltration, nanofiltration, reverse osmosis, electrolysis, dialysis, electrodialysis, gas separation, vapor permeation, pervaporation, membrane distillation, and membrane contactors. All processes except for pervaporation involve no phase change. All processes except (electro)dialysis are pressure driven. Microfltration and ultrafiltration is widely used in food and beverage processing (beer microfiltration, apple juice ultrafiltration), biotechnological applications and pharmaceutical industry (antibiotic production, protein purification), water purification and wastewater treatment, the microelectronics industry, and others. Nanofiltration and reverse osmosis membranes are mainly used for water purification purposes. Dense membranes are utilized for gas separations (removal of CO_2 from natural gas, separating N_2 from air, organic vapor removal from air or a nitrogen stream) and sometimes in membrane distillation. The later process helps in the separation of azeotropic compositions reducing the costs of distillation processes.

Cut-offs of different liquid filtration techniques							
Micrometer logarithmic scaled	0,001	0,01	0,1	1	10	100	1000
Angstroms logarithmic scaled	1	10	100	1000	10^4	10^5	10^6 10^7
Molecular weight (Dextran in kD)	0,5	50	7.000				

Ranges of membrane based separations.

Pore Size and Selectivity

The pore sizes of technical membranes are specified differently depending on the manufacturer. One common distinction is by *nominal pore size*. It describes the maximum pore size distribution and gives only vague information about the retention capacity of a membrane. The exclusion limit or "cut-off" of the membrane is usually specified in the form of *NMWC* (nominal molecular weight cut-off, or *MWCO*, molecular weight cut off, with units in Dalton). It is defined as the minimum molecular weight of a globular molecule that is retained to 90% by the membrane. The cut-off, depending on the method, can by converted to so-called D_{90}, which is then expressed in a metric unit. In practice the MWCO of the membrane should be at least 20% lower than the molecular weight of the molecule that is to be separated.

Filter membranes are divided into four classes according to pore size:

Pore size	Molecular mass	Process	Filtration	Removal of
> 10		"Classic" filter		
> 0.1 µm	> 5000 kDa	microfiltration	< 2 bar	larger bacteria, yeast, particles
100-2 nm	5-5000 kDa	ultrafiltration	1-10 bar	bacteria, macromolecules, proteins, larger viruses
2-1 nm	0.1-5 kDa	nanofiltration	3-20 bar	viruses, 2- valent ions
< 1 nm	< 100 Da	reverse osmosis	10-80 bar	salts, small organic molecules

The form and shape of the membrane pores are highly dependent on the manufacturing process and are often difficult to specify. Therefore, for characterization, test filtrations are carried out and the pore diameter refers to the diameter of the smallest particles which could not pass through the membrane.

The rejection can be determined in various ways and provides an indirect measurement of the pore size. One possibility is the filtration of macromolecules (often dextran,polyethylene glycol or albumin), another is measurement of the cut-off by gel permeation chromatography. These methods are used mainly to measure membranes for ultrafiltration applications. Another testing method is the filtration of particles with defined size and their measurement with a particle sizer or by laser induced breakdown spectroscopy (LIBS). A vivid characterization is to measure the rejection of dextran blue or other colored molecules. The retention of bacteriophage and bacteria, the so-called "bacteriachallenge test", can also provide information about the pore size.

Nominal pore size	micro-organism	ATCC root number
0.1 μm	*Acholeplasma laidlawii*	23206
0.3 μm	*Bacillus subtilis spores*	82
0.5 μm	*Pseudomonas diminuta*	19146
0.45 μm	*Serratia marcescens*	14756
0.65 μm	*Lactobacillus brevis*	

To determine the pore diameter, physical methods such as porosimetry (mercury, liquid-liquid porosimetry and Bubble Point Test) are also used, but a certain form of the pores (such as cylindrical or concatenated spherical holes) is assumed. Such methods are used for membranes whose pore geometry does not match the ideal, and we get "nominal" pore diameter, which characterizes the membrane, but does not necessarily reflect its actual filtration behavior and selectivity.

The selectivity is highly dependent on the separation process, the composition of the membrane and its electrochemical properties in addition to the pore size. With high selectivity, isotopes can be enriched (uranium enrichment) in nuclear engineering or industrial gases like nitrogen can be recovered (gas separation). Ideally, even racemics can be enriched with a suitable membrane.

When choosing membranes selectivity has priority over a high permeability, as low flows can easily be offset by increasing the filter surface with a modular structure. In gas phase filtration different deposition mechanisms are operative, so that particles having sizes below the pore size of the membrane can be retained as well.

NEW PARADIGM FOR AFRICAN AGRICULTURE SEES SUSTAINABLE INTENSIFICATION IN A NEW LIGHT

The new report from the Montpellier Panel – a panel of international experts led by Professor Sir Gordon Conway of Agriculture for Impact – provides innovative thinking and examples into the way in which the techniques of sustainable intensification are being used by smallholderfarmers in Africa to address the continent's food and nutrition crisis.

Over recent years, the term "Sustainable Intensification" – producing more-outputs with more efficient use of all inputs on a durable basis, while reducing environmental damage and building resilience, natural capital and the flow of environmental services – has come to take on a highly charged and politicised meaning, becoming synonymous with big, industrial agriculture. As we strive to feed a population expected to reach nine billion by 2050 sustainably, the risk is that we may lose sight of the term's scientific value and its potential relevance to all types of agricultural systems, including for smallholder farmers in Africa.

"It is clear that we need to boost the harvest of food and fibre from any given area of land," says Dr Camilla Toulmin, director of the International Institute for Environment and Development (IIED), and one of the Montpellier Panel's members.

"But rather than doing this in conventional unsustainable ways, which meanmore pollution, less biodiversity and more climate change, we can choose to intensify farming in a sustainable way with fewer adverse impacts. This means scientists and local farmers working together, building on tradition, and apply-ing solutions at a local scale. Many of these solutions exist — they involve better use of soils, water and ecological systems, as well as diverse crop mixes, such as grains, legumes and livestock. They also need secure land rights, and support from policymakers and the development community to help them to spread."

The report begins by examining the process and elements of Intensification itself, before considering how we then ensure that the intensification is sustain-able, and concludes with practical solutions in action today across the African continent, that underline the positive impacts the framework can produce if scaled up more effectively.

Examples of sustainable intensification in action referenced in the report include:

- Microdosing of fertilisers in Niger, Mali and Burkina Faso, using the cap of a soda bottle to measure precise amounts of nutrients for each seed hole
- Planting of Faidherbia trees, a leguminous tree which curiously sheds its leaves in the wet season – providing a natural nutrient source to crops, such as maize, planted underneath and allowing for sunlight to pass through during the growing season
- Conservation Farming in Zambia as a replacement for the traditional long fallow system of the region
- New Rice for Africa (NERICA), a cross-fertilisation between Asian and African rice species, resulting in Uganda being able to reduce its rice imports by half and an increase in farmers' incomes
- Farmers' cooperative associations, such as Faso Jigi in Mali which assists smallholder producers of cereals and shallots in marketing their products

and receiving higher prices because the association offers centralisation of stocks, better quality of storage facilities and accessibility.

The report outlines four principles essential in delivering the ambitious objectives of sustainable intensification. These are:

- Prudent, in the use of inputs, particularly those which are scarce, are expensive and/or encourage natural resource degradation and environmental problems;
- Efficient, in seeking returns and in reducing waste and unnecessary use of scarce inorganic and natural inputs;
- Resilient, to future shocks and stresses that may threaten the natural and farming systems;
- Equitable, in that the inputs and outputs of intensification are accessible and affordable amongst beneficiaries at the household, village, regional or national level to ensure the potential to sustainably intensify is an opportunity for all.

Recommendations

Sustainable intensification offers a pathway towards the goal of producing morefood with less impact on the environment, intensifying food production while ensuring the natural resource base on which agriculture depends is sustained, and indeed improved, for future generations.

The Montpellier Panel recommends that Governments, in Africa and the developed countries and in partnerships with the private sector and NGOs, recognise and act on the paradigm of sustainable intensification through:

- Adoption of appropriate policies and plans at national and local levels
- Increased financial support for global and domestic research and innovation to develop and identify suitable technologies and processes
- Scaling up and out of appropriate and effective technologies and processes
- Increased investment in rural agricultural market systems and linkages that support the spread and demand for sustainable intensification
- Greater emphasis on ensuring that inputs and credit are accessible and that rights to land and water are secure for African smallholder farmers
- Building on and sharing the expertise of African smallholder farmers in their practice of Sustainable Intensification.

EMERGING TECHNOLOGIES & PROCESS INTENSIFICATION

General Information

Food is vital and the processing of food has been essential since the beginning of humankind as known today. In the last 50 years, revolutionary developments have been taking place in the food processing industry and new technologies are emerging continuously towards safer food, with better taste and with an each time more relevant nutritional content. In the same way, new challenges emerge in the sector coming from a growing global population and its highly diverse demands. The security of the food supply is fundamental and one of the main treats to this supply lies on the dependency on non-renewable energy sources that additionally contribute to an acceleration of the climate change on which the base of food lies.

In this project, a collection of emerging technologies is presented with a potentially large range of application in the food processing industry. Non thermal emergent technologies (high pressure processing, ultrasound, *etc.*) are found with a low level of intervention on food and with a broad range of application enabling great potential towards minimal processing of food using less energy and potentially diversifying the energy sources. All the emergent technologies found have an edge on conventional technologies that may enable their substitution or important synergies towards a faster processing, higher quality or the development of new food products.

New technologies can lead towards different critical directions in the development of an industrial system (Process intensification or emergent technologies), therefore the implementation and management of the implantation of the technology is fundamental in order to achieve the set goals.

For each typical process, a brief description of the applications of different emerging technologies is made. For each technology there is a link to the detailed sheet containing an overview of the technology starting with the brief historical origin of the technology and its main advantages and disadvantages. Then the scientific base of the technology, the natural principles and phenomena involved. Following, the description of the application on the different unit operations is made. Finally, there is an energy potential section about the findings regarding energy savings and change in the energy system.

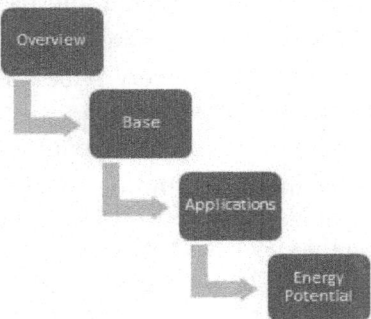

Fig. : Information content of technology description.

The development of the industrial ecosystem is in great part the development of its technology. It is proposed two ways to proceed beyond the best available techniques in the processing industry, process Intensification and Emerging Technologies implementation.

Emerging Technologies (ET)

Emerging technologies are those technical innovations in which technologies considered previously from distinct fields are converging towards a stronger inter-connections and similar goals. An emerging technology can be the case of a mature technology on certain field that finds new applications in another field. The development of the emerging technology in the new field can capitalize on the maturity of the same technology in the other field creating an effective close to market situation. All the emerging technologies in this work has this characteristic explicitly pointed out in the first section of the technology format.

In the case of emerging technologies for the food processing industry there are two main categories, thermal and non-thermal. The thermal ET is a technology that involve directly thermal energy in order to achieve the goal of the process. The effect of this category of emerging technology is related to improve limiting factor of the production process or fully overcome the limiting factors.

Non-thermal Emerging Technologies fulfil the unit operation goal using a different forms of energy or procedures that do not involve thermal energy directly, usually replacing heat leading often to lower process temperature levels. They include membrane processes, micro or radio frequency waves, pulsating method, inductive and resistive heating methods, ultrasound, ultraviolet light and other irradiation technologies, high pressure processing technologies, *etc.* As a result, thermal degradation of the product due to high temperatures and an improved solar thermal integration can be further reached, since the panels work more efficiently at lower temperatures

Process Intensification (PI)

The goal of PI is to achieve optimal function of the process. Four basic principles are taken into account:

- Maximize the effectiveness of intra-and intermolecular events
- Give each molecule the same processing experience
- Optimize the driving forces on every scale and maximize the specific areas to which those driving forces apply
- Maximize the synergistic effects from events and partial processes.

By the use of emerging technologies limiting factors can be overcome making the space for a great improvement in process intensification. This leads to more efficient processes that exploit the synergies between processes and targeted process control, producing much more with much less.

Examples of new technologies applied to process intensification are:

- A micro-structured device can be used for reaction, heat exchange, mixing, separation (microchannel reactor)
- Microwave irradiation can be used for reaction, product engineering, polymer processing (*e.g.* curing, welding), food processing (*e.g.* pasteurization, drying)

- A Rotating Packed Bed can be used for reaction, distillation, absorption, stripping and nano-product formation (precipitation)

Integrated Technology Development

Very often emerging technologies and process intensification are vaguely separated or even mixed. Given the insights from the systems approach for industrial development and sustainability, the two main factors for the development of a system can be associated to emerging technologies and process intensification respectively, this is to the development of an industrial system in the same way that general development of a system.

Comparing the two possibilities for technology development (process intensification and emerging technologies) with the main parameters of systems sustainability development in the literature (efficiency and resilience), it can be observe that processes intensification is oriented towards efficiency and emerging technologies have a higher potential to build resilience in the system.

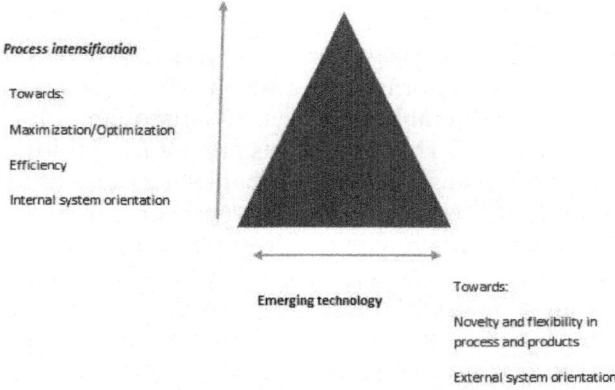

Fig. : Technology development in the processing industry.

The traditional focus of technology development is oriented towards improving the efficiency of the production processes with little or no formal space for resilience. As the system sustainability studies have shown, the relevance of the resilience has a major role in keeping a natural system running and reaching the "windows of vitality". The technology development based on emerging technologies may have a major role in the developing of a sustainable industry.

Emerging technologies can enable results that do not lead to direct efficiency improvements. These results can lead to improvements in the flexibility of the production processes, to the development of new products, to an easier incorporation of renewable energy in the production system and/or to the creation of connections with other fields. This results of emerging technologies can be associated with increased levels of diversity and connectedness in the production system. From a system perspectives in this same way, the management of emerging technologies can lead to a more resilient system.

The collection of technologies presented in this section can be used for processes intensification or emerging technologies. It depends on the management of the implementation of them.

FROM THE DEFINITIONS TO THE FUNCTION: A FEW FUNDAMENTAL IDEAS

Expert Roundtable on Critical Research Priorities in Sustainable Development

The eminent biologist E.O. Wilson describes humankind as passing through a harrowing bottleneck in the 21st century. In the 200 years since the Industrial Revolution, human population has increased six-fold and economic activity an estimated 50-fold. Since then scientific and technological advances have enabled much of the world's population to make vast gains in economic prosperity, health and well being. Yet the sheer volume of people and the intensity of their activities are leaving a trail of environmental degradation that poses future risks of unprecedented complexity.

To note just a few of the risks, critical ecosystems such as forests that hold top-soils and wetlands that purify fresh water are being degraded or destroyed, leaving many people vulnerable to floods, collapsed agricultural productivity and unsafe drinking water. These problems are not new but they are occurring at an unprecedented global scale. Rapid urbanization and inadequate planning is having deleterious effects on city-dwellers and surrounding ecosystems. The vast oceans are proving equally vulnerable, with major fisheries declining more rapidly than we thought possible even a few years ago. While world leaders once hoped globalization would pull the poorest people out of poverty, hundreds of millions are falling in a precipitously downward spiral of deeper impoverishment and environmental degradation. Human impact on the long-term global climate is threatening to disrupt fragile Earth systems in future decades unless we can create novel energy sources, technologies and/or regulatory systems that can halt continued changes to the planet's atmosphere.

These are the type of risks that the study of *sustainable development* examines and the challenges to which we seek solutions. No solutions can be possible unless we develop a clear understanding of the phenomenon at work through research studies of the basic processes. Virtually all the key areas of risk are the outcome (often unintended) of human and natural systems interactions. Consider as a simple case the impact of natural disasters such as storms, droughts, and earthquakes. It is now well documented from studies by the UNDP and World Bank that the impacts of geophysically or meteorologically identical extreme events produce very different mortality and economic outcomes. Development status is the primary determinant because poorer communities are both physically and socially more vulnerable than prosperous ones. For some countries development is massively impeded by disasters while others have been able to recover and re-build in a way that positively influences growth, at least in some sectors of the economy. However, no developed country has been impacted by a major disaster in several decades and the consequences of, for instance, a magnitude 9+ earthquake in Tokyo can only be speculated upon. Disasters are the extreme case, of course, but they underscore the complex dynamics of natural and human system interactions and the level of social exposure that some communities face.

A program of research in sustainable development at NSF would provide a critical stimulus to advance our understanding of the urgent issues that face our planet and its peoples. Research is needed both to understand the fundamental social and natural phenomena at work in determining human futures and to device solutions that will achieve sustainable development. The research and the construction of solutions will involve studies within and across virtually all the major disciplines sponsored at NSF. We believe that much of the needed research can be characterized as basic in nature because there are fundamental processes in play that are not well understood and require the full suite of tools of basic research to uncover. The research is also outcome driven in the sense that there is a well defined goal of devising ways to achieve sustainable development and may be analogous to Donald Stokes' "use inspired" basic research, lying in the so-called Pasteurs' Quadrant of scientific research. Coming up with pathways to achieve sustainable development will certainly require interactions across the basic and applied sciences. In many instances engineering programs and public health interventions will play a key role.

We therefore propose a focused Roundtable discussion that would bring together a diverse expert group comprising the leading innovators in the social, natural and applied sciences in a one day meeting to define the key elements of research programs in sustainable development at NSF.

Sustainable Development

There has been ongoing debate about the meaning of the term sustainable development since the Brundtland Commission definition of 1987. It is generally agreed though that when we speak about sustainable development we are referring to the future and security of the Earth and the lives of people who live on it. To simplify, generally speaking we can separate the two concepts contained in the term sustainable development, and say that by *development* we specifically mean the challenges of spreading social, political and economic opportunity to the entire global community, particularly the poorest of the poor. By prefacing with *sustainable*, we refer to the objective of managing the world's development in a manner consistent with the continued healthy functioning of the Earth's ecosystems, oceans, atmosphere and climate. The two concepts are deeply connected across many levels and scales and there are innumerable caveats and nuances of meaning lying beneath the simplistic definitions given here – for instance, is sustainable development possible for some groups now only by reducing the development options for others now and in the future? Keeping the separation of the two components to the term we can highlight different issues that might serve as a vehicle for guiding discussion at the proposed Roundtable.

Development Challenges and Research Priorities

Studies of development apply to all levels of human prosperity but, as our meaning suggests, we make special emphasis on poor societies and for very compelling reasons. Poor countries pose urgent and specific challenges that are quite different from those for countries where market economies are functioning and where growth, at least for the moment, can be assured. The history of global economic development now makes very clear that growth in one part of the world, no matter how aggressive, will not simply sweep countries in other parts of the world along in its wake, but will reflect development inequalities within regions, countries and districts. Vast areas of the world have been sidelined so that today, despite astonishing prosperity in some places, nearly half of humanity still lives on less than two dollars per day and one sixth still lives in chronic poverty surviving on less than a dollar a day [dollars adjusted to PPP – purchasing power parity – and hence represent truly comparable values]. The figures describing the harshest conditions of deprivation, to the extent that they are available, are appalling – at least 25,000 people die daily from poverty-related causes and the number may be as high as 40,000. Most of these deaths occur in sub-Saharan Africa and the causes of virtually all have known forms of prevention or cure. Life and death there go on much as they have done for centuries.

Progress in improving the lives of those who suffer the many harsh deprivations of poverty throughout the world has proven to be agonizingly slow in large part because understanding the root causes of poverty and the pathways to improving the condition of the world's poorest people has proven to be an extremely refractory problem. A key question that has emerged from analyses of poor country settings and performance concerns the fundamentals of poor country economies as they differ from those of growing economies. Poor countries may not simply be waiting their turn, stalled at the initial phase of economic growth as described by neo-classical economic growth theory. Instead they may be caught up in a so-called Poverty Trap, a stable state or equilibrium in which the conditions of poverty become self-perpetuating. Rather than following a classic Solow-Swan growth function in which growth is initially rapid and then decreases with time, the starting conditions of poor countries may cause growth to be initially negative. Determining factors include the low marginal productivity of capital in settings where basic infrastructure is lacking, savings rates that can be very low or negative, and rapid population growth with very low capital-labor ratios. Economies cannot move out of these trapping conditions without external interventions and may remain in persistent decline and evolve toward a low level stable equilibrium of poverty. In other words, poverty itself becomes the cause of further poverty just as growth fuels further growth in expanding economies.

A further key insight is that the factors that lead to low growth and the potential for poverty traps include, and may in many instances be dominated by, a suite of complex co-dependencies between human well-being and Earth's natural systems. Progress in poor settings then is more than a matter of economic and other public policies, governance and appropriate institutions, but also of profoundly important interactions between human and natural systems. It is the quest for basic human needs in these settings that is often the principal cause of severe environmental degradation, depletion of forest resources, fertile soils and other natural assets, while causing extreme vulnerability to climate variations and other natural extremes, possibly triggering conflicts. Though resulting *from* poverty these outcomes are equally *determinants* of poverty: the cruel backlash of a desperate struggle to survive or break the bonds of poverty. These co-dependent relationships are thought to be multi-interactive and cross many scales. They are likely to be non-linear with emergent properties that can include the potential for extremely rapid declines. Systemic shocks like major natural catastrophes may plunge weakly growing economies deep into poverty trap situations.

For countries in these situations questions of sustainability enter quite differently from those in well developed or emerging economies – the current conditions may be self perpetuating and hence stable, but deeply undesirable both in terms of human well being and for the associated degradation in the environment. Of considerable concern is that global climate change may cause sufficient stress on marginal economies that they will descend into conditions of scarcity that could trigger widespread poverty traps and conflict situations from which they will be unable to emerge.

Sustainability Challenges and Research Priorities

For the poorest countries the urgent need is to light the flame of economic growth. For growing economies, especially the rapidly growing economies of India (doubling every 10 years) and China (doubling every 7 years), the issue is how to keep the flame burning without it causing destruction to the very environment that provides the fuel for growth.

Sir Partha Dasgupta, the Cambridge economist opened his critique of Jared Diamond's recent book *Collapse* in the London Book Review by asking the following questions:

"Are our dealings with nature sustainable? Can we expect world economic growth to continue for the foreseeable future? Should we be confident that our knowledge and skills will increase in ways that will lessen our reliance on nature despite our growing numbers and rising economic activity?"

These questions neatly encapsulate the core issues of sustainable development for growing economies. Diamond's well-known view is that the answers to them are generally "no" and he cites many examples from the past and a number from the present in his well known popular book *Collapse* to support his case. Dasgupta, in a withering criticism of what he sees as Diamond's single minded neo-Malthusian treatment of our future prospects, along with Ronald Bailey's equally caustic review in Reason Magazine, sets out the other side of the discussion of key issues in the sustainability of humankind's current trajectory. They argue that the past holds numerous examples of how many societies have prospered very well despite such factors as high population densities, aggressive use of forest and other natural resources and all the other maladies that brought some societies to their knees. For every negative example, economists are able to find a strong counter-example. A key factor is that, thus far, modern societies have met every sustainability challenge with technological innovations that have allowed economies to grow while actually lessening the environmental impact of economic activities.

The growth of economies though, may not be properly assessed and development may be incorrectly calculated, even in strongly growing economies. Dasgupta points out the incomplete accounting that measures such as GNP provide. An economy's productive base consists of its capital assets and its institutions; ecological economists argue that estimating wealth should include not only the value of manufactured assets but also 'human' capital (knowledge, skills health), natural capital (ecosystems, minerals, fossil fuels), and institutions (government, civil society, the rule of law). GNP does not include depreciation of natural capital. The list of unaccounted for natural capital assets is vast including fresh water, ocean fisheries, soils, forests, wetlands, mangroves, coral reefs, atmospheric quality. So long as depreciation of natural capital (a rather formal way of referring to depletion and/or destruction of the environment) is not included in estimates of economic growth, such estimates will be deeply misleading in evaluating the sustainability of current trajectories.

Knowledge, skills, institutions and manufactured capital can in principal substitute for nature's resources, so that loss of some natural capital can be compensated by investment in other forms of capital. As Robert Solow has put it "If it is very easy to substitute other factors for natural resources, then there is, in principle, no problem. The world can, in effect, get along without natural resources." In other words, nature can be thought of as expendable – just use it up and it can be substituted for other forms of capital.

Mounting evidence from almost every sector and almost every corner of the planet implicates the pressure of human economic activity as moving natural capital rapidly toward depletion or such severe degradation that their use is impossible in a time frame well before substitution can be achieved, be cost effective, desirable or even possible. Continued growth even at the present rates, particularly in the rapidly expanding economies of China and India, would lead to massive resource depletions and widespread degradations of virtually every natural asset on the planet. Add to this pressure the legitimate aspiration for poorer countries to become much less poor on a planet with an additional 3 billion inhabitants and by 2050 the economic throughput would be at least five times the current level. Technical innovation that allows substitution and more efficient uses of natural assets may come to the rescue, but many observers fear that we are on the cusp of a sustainability crisis at present, with little or no time left to innovate.

Development and Sustainability in the face of Climate Change

The challenges of the rich and poor world, of development and of sustainability, converge in the face of the truly global threats created by global warming. Across the planet societies have established themselves in a manner that is in equilibrium with local conditions. That equilibrium can be quite precarious as it is in the harshest regions of the world in the high Arctic and in sub-Saharan Africa. In such places people's lives already hang in a balance that depends on sparse fragile ecosystems or very low rainfall, or as in the exploding peri-urban slums of mega-cities in the developing world, so many of which are in coastal settings vulnerable to sea level rise and increased storm frequency. Such unstable equilibria can be unsettled by even small changes to ambient conditions resulting in massive disruptions to already stressed life support systems. Loss of arable land or critical ecosystems could lead to massive internal and cross-border migrations, and the potential for conflict is just one dire outcome.

In places where the equilibrium is relatively robust and natural variations such as hurricanes or dry periods can presently be managed without major economic setbacks, the large-scale shift in base conditions like sea level rise, storm intensity, the movement or loss of ecosystems, hydrological systems changes, are also anticipated to be massively disruptive. The very high latitudes, though much less populated, will suffer the most severe, direct perturbations to extremely fragile ecosystems. Although the effects of climate change are acknowledged to be quite variable across the planet with some regions even expected to benefit, the rapid re-arrangement of natural systems and the associated re-distribution of prosperity is unprecedented in human history and overlays on a planet whose sustainability is already at risk even without this additional stress.

Proposed Roundtable

We propose that NSF sponsor a Roundtable with around 25 invited participants who are recognized leaders in the natural, engineering, health and social sciences and who can help shape a vision for a sustainable development program area at NSF. Rather than a set of formal presentations we envisage a structured deliberation among this relatively small expert group over a day at NSF's offices in Virginia. Key participants or groups (perhaps four or five) would be asked to prepare a short white paper to be distributed ahead of the Roundtable that would assess the state of understanding of key issues in sustainable development from their particular perspective. They would also be asked to suggest a suite of research themes that should be explored to significantly advance our understanding of those issues. One possible format for the Roundtable would have the day lead off by a key note address setting the overall objectives of the Roundtable, followed over the course of the day with four sessions, say, in which the white papers were presented by their authors and discussed by the whole group. The day might conclude with a sessions designed to draw together the elements that were presented during the day into a coherent research plan.

By assembling a small group of the world's recognized leading thinkers from a diverse suite of backgrounds we would animate a new set of dialogues on the inherently cross-disciplinary field of sustainable development that would identify the importance of the subject, the urgency of the issues, and the intellectual challenges they present.

Many specific questions arise around the subject of sustainable development and we list just a sub-set in the attached Appendix. None of these questions have adequate answers and all require efforts in fundamental inter-disciplinary inquiry to establish answers. In many instances the lack of suitable answers stems from the fact that research into these questions has to date been based largely out of specific disciplines. The problems themselves derive from the multiple ways in which human societies interact with Earth's systems and hence solutions to them will lie in studies at the nexus of the social and natural sciences. A research program that is both broad in the scope of disciplines it engages and focused in its attention to the key issues in sustainable development is urgently required. The proposed Roundtable would take a significant step toward defining such a program.

Among the individuals who could participate in the Roundtable if available are the economists Prof. Jeffrey Sachs and Prof. Joseph Stiglitz from Columbia University, both of whom are world renowned for their contributions to the sustainable development (and co-directors of the new PhD in Sustainable Development in the School of International and public affairs at Columbia). Others who we could consider in the role of white paper authors are: Paul Collier who has made seminal contributions on development issues including the burdens of conflict and is author of a new book "The Bottom Billion"; in Public Health, Dr. Barry R. Bloom, Dean of the Harvard School of Public Health; Prof. E. O. Wilson would be an exceptional contributor on the subject of biodiversity; we have several times referred to the thinking of Sir Partha Dasgupta, Frank Ramsey Professor of Economics at Cambridge University; Prof. Joel Cohen from Rockefeller University is a world authority on population and a major contributor to ideas in sustainable development; Dr. R.K. Pachauri, Chairman of the Intergovernmental Panel on Climate Change (IPCC).

There are many others we could consider and these names are identified to give a sense of the caliber of participant we would hope to engage. Our immediate step in making plans for the meeting are to develop a small steering committee at Columbia whose task it would be to take on two interconnected tasks: one is to set out an agenda in some detail, the other is to determine a list of invited participants and organize invitations. Professor John Mutter will chair the steering committee and will keep NSF staff up to date on the panning. Management staff of the Earth Institute at Columbia would support the steering committee. We will establish an intranet site to support the work and ensure that the progress of the planning is available to appropriate parties. The committee would meet mostly by phone and/or email and would meet at NSF a half day before the Roundtable to ensure that the meeting is set up satisfactorily and would stay a half day after to begin the reporting. The committee would be responsible for providing published output from the Roundtable in Web documents, publications and print materials.

Our overall objective is to define the elements of a new research program at NSF that addresses the fundamental challenges of sustainable development across a broad front. We would produce a report for the NSF that would outline the recommendations for the new research program and publish summary discussions in vehicles such as EOS for the natural scientists and social science publications that would reach a wide range of colleagues in many fields.

DIMENSIONS OF SUSTAINABLE DEVELOPMENT

One of the most important outcomes of Our Common Future was the realisation that environment and development issues are inextricably linked and therefore worrying about either environment or development on its own was inappropriate. The World Commission concluded that:

Environment and development are not separate challenges. Development cannot subsist on a deteriorating environmental resource base; the environment cannot be protected when growth leaves out of account the costs of environmental destruction. These problems cannot be treated separately by fragmented institutions and policies. They are linked in a complex system of cause and effect.

Source: World Commission on Environment and Development (1987) Our Common Future, Oxford University Press, Oxford, p. 37.

The WCED therefore argued for an approach to development that would take into account the relationship between ecological, economic, social and technological issues. The WCED called this approach 'sustainable development', defining it as:

... development that meets the needs of the present without compromising the ability of future generations to meet their own needs.

Source: World Commission on Environment and Development (1987) Our Common Future, Oxford University Press, Oxford, p. 43.

The ultimate goal of sustainable development is to improve the quality of life for all members of a community and, indeed, for all citizens of a nation and the world – while ensuring the integrity of the life support systems upon which all life, human and non-human, depends.

There is sometimes confusion about the meanings of 'sustainable development' and 'sustainability' and the relationship between them. A report on Education for Sustainable Development in New Zealand proposed the following explanation:

Sustainability is the goal of sustainable development – an unending quest to improve the quality of peoples' lives and surroundings, and to prosper without destroying the life-supporting systems on which current and future generations of humans depend. Like other important concepts, such as equity and justice, sustainability can be thought of as both a destination and a journey.

Four Dimensions of Sustainable Development

Sustainable development requires simultaneous and balanced progress in four dimensions that are totally interdependent:

- Social
- Economic
- Ecological
- Political

You may have found it difficult to identify these different dimensions in the photograph. This is because there are always close linkages among them. Similarly, decisions or actions in one area always affect the others.

For example, if economic development is to be sustainable:

- it cannot neglect environmental constraints or be based on the destruction of natural resources;
- it cannot succeed without the parallel development of social resources;
- it will require transformation of the existing industrial base and the development and diffusion of more Earth-friendly technologies;
- it must consider the needs of all species and their rights to enjoy the same quality of life and share of resources;
- it must support fairness between all people so that everyone can enjoy the same standard of living and the same level of access to resources and quality of life; and
- it must consider the needs of future generations.

A Dynamic Balance

The special contribution of the concept of sustainable development is that it emphasises respect for cultural values and, thus, does not see economic indicators as the sole measure of development. Rather, sustainable development represents the balanced integration of social and environmental objectives with economic development. These three aspects of sustainable development – society, environment and economics – were named as the three pillars of sustainable development at the World Summit on Sustainable Development in Johannesburg in 2002.

In relation to Education for Sustainable Development, these three pillars of sustainable development involve:

Society

An understanding of social institutions and their role in change and development, as well as the democratic and participatory systems which give opportunity for the expression of opinion, the selection of governments, the forging of consensus and the resolution of differences.

Environment

An awareness of the resources and fragility of the physical environment and the affects on it of human activity and decisions, with a commitment to factoring environmental concerns into social and economic policy development.

Economy

Skills to earn a living as well as a sensitivity to the limits and potential of economic growth and its impact on society and on the environment, with a commitment to assess personal and societal levels of consumption out of concern for the environment and for social justice.

However, politics and culture are also a key dimension of sustainable development, which influence the interactions of and between the three pillars. They are concerned with the values we cherish, the ways in which we perceive our relationship with others and with the natural world, and with how we make decisions. The values, diversity, knowledge, languages and worldviews associated with culture and politics strongly influence the way issues of sustainable development are decided and, thus, provide it with local relevance.

As a result of the close relationships between the four these dimensions of sustainable development, achieving this goal requires a dynamic balance between:

- Production and consumption;
- Ecology and economics;
- Development and conservation;
- Culture and ecology; and
- Democracy and economics.

However, the particular nature of the balance between these factors will vary between the developing countries of the South and the industrial countries of the North.

Analyse these five aspects of a dynamic balance for sustainable development from the viewpoints of both the South and the North.

Many commentators on sustainable development often refer to what they call "the triple bottom line" of economic sustainability, social sustainability, and ecological sustainability. In such cases, the focus is only on three dimensions of sustainable development – social, economic and environmental.

Sometimes these are referred to as the "3 E's" of sustainable development – Equity, Economy, and Ecology.

Different Definitions and Values

The definition of sustainable development proposed in the Brundtland Report and Agenda 21 has been adopted in many countries. However, the idea of 'development that meets the needs of the present without compromising the ability of future generations to satisfy their needs' has been interpreted in many different ways. In fact, several hundred different definitions of sustainable development now exist.

While these definitions seek to make the broad definition from the Brundtland Report and Agenda 21 more concrete, many tend to reflect different emphases in the social, economic, ecological and political dimensions of sustainable development. Being able to identify these different emphases in discussions about sustainable development is an important critical thinking skill. You can practise this skill by analysing the emphases in five different definitions of sustainable development.

Goals for Sustainable Human Development

The General Assembly of the United Nations met from 6-8 September, 2000 in a special Millennium Summit where a special United Nations Millennium Declaration was adopted. This Declaration emphasised the social aspects of sustainable development and the importance of overcoming poverty and inequality. It declared that sustainable human development is central to world peace and future progress:

> We recognise that, in addition to our separate responsibilities to our individual societies, we have a collective responsibility to uphold the principles of human dignity, equality and equity at the global level. As leaders we have a duty therefore to all the world's people, especially the most vulnerable and in particular, the children of the world, to whom the future belongs.

The Millennium Declaration was based upon six fundamental values that underly sustainable human development:

Freedom

Men and women have the right to live their lives and raise their children in dignity, free from hunger and from the fear of violence, oppression or injustice. Democratic and participatory governance based on the will of the people best assures these rights.

Equality

No individual and no nation must be denied the opportunity to benefit from development. The equal rights and opportunities of women and men must be assured.

Solidarity

Global challenges must be managed in a way that distributes the costs and burdens fairly in accordance with basic principles of equity and social justice. Those who suffer or who benefit least deserve help from those who benefit most.

Tolerance

Human beings must respect one other, in all their diversity of belief, culture and language. Differences within and between societies should be neither feared nor repressed, but cherished as a precious asset of humanity. A culture of peace and dialogue among all civilizations should be actively promoted.

Respect for Nature

Prudence must be shown in the management of all living species and natural resources, in accordance with the precepts of sustainable development. Only in this way can the immeasurable riches provided to us by nature be preserved and passed on to our descendants. The current unsustainable patterns of production and consumption must be changed in the interest of our future welfare and that of our descendants.

Shared Responsibility

Responsibility for managing worldwide economic and social development, as well as threats to international peace and security, must be shared among the nations of the world and should be exercised multilaterally. As the most universal and most representative organization in the world, the United Nations must play the central role.

In partnership with the World Bank, the International Monetary Fund and the OECD, the United Nations agreed on a set of Millennium Development Goals inspired by these fundamental values. The goals came from the agreements and resolutions of the world conferences organised by the United Nations in the first half of the 1990s.

The eight Millennium Development Goals are to:

Eradicate extreme poverty and hunger:

- Reduce by half the proportion of people living on less than one dollar a day;
- Reduce by half the proportion of people who suffer from hunger.

Achieve universal primary education:

- Ensure that all boys and girls complete a full course of primary schooling.

Promote gender equality and empower women:

- Eliminate gender disparity in primary and secondary education preferably by 2005, and at all levels by 2015.

Reduce child mortality:

- Reduce by two thirds the mortality rate among children under five.

Improve maternal health:

- Reduce by three quarters the maternal mortality ratio.

Combat HIV/AIDS, malaria and other diseases:

- Halt and begin to reverse the spread of HIV/AIDS;
- Halt and begin to reverse the incidence of malaria and other major diseases.

Ensure environmental sustainability:

- Integrate the principles of sustainable development into country policies and programmes; reverse loss of environmental resources;
- Reduce by half the proportion of people without sustainable access to safe drinking water;
- Achieve significant improvement in the lives of at least 100 million slum dwellers by 2020.

Develop a global partnership for development:

- Develop further an open trading and financial system that is rule-based; predictable and non-discriminatory. Includes a commitment to good governance, development and poverty reduction – nationally and internationally;
- Address the least developed countries' special needs. This includes tariff- and quota-free access for their exports; enhanced debt relief for heavily indebted poor countries; cancellation of official bilateral debt; and more generous official development assistance for countries committed to poverty reduction;
- Address the special needs of landlocked and small island developing States;
- Deal comprehensively with developing countries' debt problems through national and international measures to make debt sustainable in the long term;
- In cooperation with the developing countries, develop decent and productive work for youth;
- In cooperation with pharmaceutical companies provide access to affordable essential drugs in developing countries;

- In cooperation with the private sector, make available the benefits of new technologies – especially information and communication technologies.

The eight Millennium Development Goals are interdependent and action is needed on all of them simultaneously if they are to be achieved. Many countries and regions are making progress towards them:

The goals for international development address that most compelling of human desires – a world free of poverty and free of the misery that poverty breeds … Each of the goals addresses an aspect of poverty. They should be viewed together because they are mutually reinforcing. Higher school enrolments, especially for girls, reduce poverty and mortality. Better basic health care increases enrolment and reduces poverty. Many poor people earn their living from the environment. So progress is needed on each of the seven goals.

The goals will not be easy to achieve, but progress in some countries and regions shows what can be done. China reduced its number in poverty from 360 million in 1990 to about 210 million in 1998. Mauritius cut its military budget and invested heavily in health and education. Today all Mauritians have access to sanitation, 98% to safe water, and 97% of births are attended by skilled health staff. And many Latin American countries moved much closer to gender equality in education.

The message: if some countries can make great progress towards reducing poverty in its many forms, others can as well. But conflict is reversing gains in social development in many countries in Sub-Saharan Africa. The spread of HIV/AIDS is impoverishing individuals, families and communities on all continents. And sustained economic growth – that vital component for long-run reductions in poverty – still eludes half the world's countries. For more than 30 of them, real per capita incomes have fallen over the past 35 years. And where there is growth, it needs to be spread more equally.

So, the goals can be met. But it will take hard work.

Development Goals and Indicators

Development indicators, such as the Millennium Development Goals, can be used at all levels – local, national and global.

Match different goals and indicators to clarify your understanding of sustainable human development.

Development indicators are important tools for monitoring changes in sustainable human development. They can help identify where successes are being achieved and where further action is needed.

However, we need to be cautious in our use of indicators. For example:

Indicators are Numbers

They summarise all the influences that affect the human experiences of love, happiness, health, *etc.*, as numbers. This makes it easy to forget that indicators actually apply to people.

Indicators Should not be Used in Isolation

No one indicator is 'proof' of anything – at least by itself. It is only when used in combination with other indicators that reliable patterns can start to be identified.

As a result, the Human Development Index has been developed to measure the average achievements in a country in three basic dimensions of human development: longevity, knowledge and a decent standard of living. As a composite index, the Human Development Index thus contains three variables: life expectancy; educational attainment (adult literacy and combined primary, secondary and tertiary enrolment); and real GDP per capita.

Indicators are Averages

They summarise all the differences in a country into a simple number. Thus, they can disguise differences in development between different regions in a country (e.g. urban – rural), people of different ages, people at different income levels, the life experiences of people from different ethnic groups, etc. And, unless gender-sensitive, indicators can disguise the different experiences of males and females.

The importance of gender in development is recognised by the way development goals and indicators now focus explicitly on women and girls. The Gender-Related Development Index, for example, measures achievements in the same dimensions and variables as the Human Development Index, but takes account of inequality in achievement between women and men. The greater the gender disparity in basic human development, the lower a country's Gender-Related Development Index compared with its Human Development Index.

Alternative Measures of Development

The adequacy of current measures of economic performance, in particular those based on GDP figures, has been a concern for a long time. Moreover, there are broader concerns about the relevance of these figures as measures of societal well-being, as well as measures of economic, environmental, and social sustainability.

There are a number of other alternative tools for measuring economics human development than those already mentioned above. Some of them include:

- Gross National Income
- Gross Hapiness Index
- Genuine Progress Index

President Sarkozy (France) launched a Commission in 2008 to look at a broad range of issues in measuring progress. The aim of the Commission was to to identify the limits of GDP as an indicator of economic performance and social progress, to consider additional information required for the production of a more relevant picture, to discuss how to present this information in the most appropriate way, and to check the feasibility of measurement tools proposed by the Commission. Their work was not focused on France, nor on developed countries.

The Commission was led by by Professor Joseph E. Stiglitz, Columbia University, Professor Amartya Sen, Harvard University, and Professor Jean-Paul Fitoussi, Institut d'Etudes Politiques de Paris, and was made up of renowned experts from universities, governmental and intergovernmental organisations, in several countries (USA, France, United Kingdom, India). The Commission's report was released in September 2009.

The Cost of Progress

How much would achieving the Millennium Development Goals cost? Can the world afford sustainable human development?

Yes, it can.

In fact, the cost of making significant progress on eighteen different actions needed to achieve these goals would be less than one-third of the $780 billion the world spends on military activities each year.

See a chart of this estimate in relation to annual global military expenditure.

Review details of what this alternative expenditure could achieve:

- Eliminate starvation and malnourishment
- Provide health care and AIDS control
- Provide shelter
- Provide clean, safe water
- Eliminate illiteracy
- Provide clean, safe energy through the use of renewable sources
- Retire developing nations debt
- Stabilise population
- Prevent soil erosion
- Stop deforestation
- Stop ozone depletion
- Prevent acid rain
- Prevent global warming
- Provide clean, safe energy through greater efficiency
- Remove landmines
- Refugee relief
- Eliminating nuclear weapons
- Build democracy

UNESCO acknowledges the support of the osEarth Inc. (OSE) for providing this section of Teaching and Learning for a Sustainable Future. OSE promotes the activities of the World Game Institute (WGI), a non-profit research and education organisation.

Activity 4: The World Summit on Sustainable Development

Ten years after the Rio Earth Summit, in 2002, the United Nations General Assembly organised a World Summit on Sustainable Development in Johannesburg, South Africa. The purpose of the World Summit was to synthesise the conclusions and agreements of the series of international conferences of the 1990s, review progress on progress towards achieving the goals of Agenda 21, and make plans for cooperative efforts to progress sustainable human development.

The conditions under which the World Summit took place were generally positive. For example, it was a time of increasing recognition that many countries had not paid sufficient attention to their commitments in Agenda 21, increased importance of the need to address global poverty, and positive commitment to cooperative action on the Millennium Development Goals. However, it was also a time of rising uncertainty about global security and the negative impacts of globalisation. These conditions produced both optimism and caution as government and non-government delegates engaged in two weeks of intense discussion.

As largest ever gathering of world leaders, the World Summit clearly showed that sustainable development is of key concern around the world today. In addition to government delegations, over 21 000 participants from nearly 200 government, intergovernmental and non-governmental organizations, the private sectors and the scientific community attended.

Summit Outcomes

The outcomes of the World Summit included:

- A Political Declaration agreed by all the Presidents, Prime Ministers and heads of government delegations of the world who were present.
- A Plan of Implementation for agreed improvements in sanitation, energy, trade, health, education, human rights, biodiversity, climate change and so on.
- Many action plans and agreements under which different United Nations agencies, governments, corporations, industry associations, professional and scientific organisations, trade unions and/or non-government organizations agreed to work in partnership to attain the goals in the Plan of Implementation.

Read a summary of the Key Outcomes and Commitments of the World Summit.

As one observer noted:

Through signing the 54 page Summit Implementation Plan, those present committed to using and producing chemicals in ways that do not harm, reducing biodiversity loss by 2010; restoring fisheries to their maximum sustainable yields by 2015, establishing a representative network of marine protected areas by 2012, and implementing a Global Programme for the Protection of the Marine Environment – the most important outcomes were commitments made to half the proportion of people without access to sanitation and safe drinking water by 2015. The US, EU and others committed over a billion dollars to bring this about. Similar financial commitments and type II partnerships were made to improve access to energy. The shifting of commitments towards socio-development issues such as poverty, health and sanitation was seen by delegates as the key successes of this Summit.

Source: Tilbury, D. (2003) The World Summit, sustainable development and environmental education, Australian Journal of Environmental Education, Vol. 19, p. 110.

However, in the years immediately following the World Summit – perhaps before there has been time to see too many improvements – there has been growing concern that the promise of the World Summit are being achieved too slowly. Some feel that too much attention was paid at the Summit to the symptoms of global issues and problems and that their root causes were not sufficiently addressed. Others are concerned that the threats to sustainable development from globalisation, unfair trading practices and low levels of international development assistance were not dealt with strongly enough.

Projects and Progress

This does not mean that exciting sustainable development projects are not being undertaken in all parts of the world. In fact, more people than ever before are working towards a sustainable future. Many projects for integrating cultural concerns and indigenous perspectives into sustainability planning are under way. And new laws and policies are being developed to secure environmental and human health, promote gender equity, protect and conserve biodiversity, oceans, fisheries and the world's freshwater supplies, and to support sustainable agriculture, sustainable tourism and sustainable community development – and to improve education's role in enhancing individual and national capacity for sustainable development.

There are case studies of many such projects in Theme 3 of this programme.

Nevertheless, change is not coming fast enough and there is grave concern that many of the indicator targets for the Millennium Development Goals will not be meet. Progress seems to be uneven across different Goals and across different regions of the world. The different colours in the 2004 Status Chart show the wide range of uneven rate of progress being made.

The areas in green show positive trends where different regions are 'on target' or where targets have been met already.

However, all the areas in pink and orange in the Status Chart indicate targets that are unlikely to be met in different parts of the world unless there are major improvements in government activity, business support and international development assistance.

THE EARTH CHARTER

The Earth Charter

We stand at a critical moment in Earth's history, a time when humanity must choose its future. As the world becomes increasingly interdependent and fragile, the future at once holds great peril and great promise. To move forward we must recognize that in the midst of a magnificent diversity of cultures and life forms we are one human family and one Earth community with a common destiny. We must join together to bring forth a sustainable global society founded on respect for nature, universal human rights, economic justice, and a culture of peace. Towards this end, it is imperative that we, the peoples of Earth, declare our responsibility to one another, to the greater community of life, and to future generations.

A People's Charter

The need for an Earth Charter was first raised in the Stockholm declaration's call for "a common outlook and for common principles to inspire and guide the peoples of the world." In its 1987 report, Our Common Future, the UN World Commission of Environment and Development issued a call for a new charter that would consolidate and extend relevant legal principles, creating "new norms … needed to maintain livelihoods and life on our shared planet" and "to guide state behaviour in the transition to sustainable development."

An attempt was made to take up the challenge of drafting the Earth Charter at the Rio Earth Summit, but the time was not right. In the wake of the Rio Earth Summit, a new Earth Charter Initiative began in 1994 under the leadership of the Earth Council and Green Cross International.

A global consultation process was instigated to help provide widespread input into the deveopment of the Earth Charter. Hundreds of groups and thousands of individulas became involved in this process. For example, between 1997 and 1999 over forty national Earth Charter committees were formed, and numerous Earth Charter conferences were held, all under the general coordination of the Earth Charter Commision.

The first draft of the Earth Charter (Benchmark Draft I) was largely based on a review of values and principles embedded within existing international laws, treaties and declarations. This document was released at the Rio+5 conference, and then circulated around the world for comment. Recommendations were integrated into a new version (Benchmark II) released in April 1999. The global reivew and consultation process continued throughout 1999, culminating in the launch of the Earth Charter at a meeting in March 2000 at UNESCO's Paris headquaters. The aim now is to both circulate the document as a People's Treaty and to have it taken to the United Nations General Assembly for endorsement.

A Declaration of Interdependence and Sustainable Development

A key feature of the Earth Charter campaign has been an investigation of local and national cultures in order to identify the common beliefs and values that underlie a global ethic for living sustainably.

The Earth Charter is a declaration of interdependence and universal responsibility as well as an urgent call to build a global partnership for sustainable development.

The focus of the Earth Charter is sustainable human development, which as we saw in Activity 3, includes the care and protection of the Earth. The Earth Charter recognizes that environmental, economic, social, cultural, ethical, and spiritual problems are interconnected.

The Earth Charter is a layered document with a Preamble, 16 guiding principles and 59 supporting principles that, together, outline an integrated vision for human rights and sustainable development.

Preamble

The Preamble to the Earth Charter provides an expanded sense of our responsibilities for sharing the Earth as part of a global community. These responsibilities embrace all people, future generations, and the larger community of life on Earth. This is a rationale for sustainable human development.

Guiding Principles

The first four guiding principles of the Earth Charter are:

1. To respect Earth and life in all its diversity;
2. To care for the community of life with understanding, compassion and love;
3. To build democratic societies that are just, sustainable, participatory and peaceful; and
4. To secure Earth's bounty and beauty for present and future generations.

These four principles illustrate that the concept of sustainable development in the Earth Charter embraces the view that the problems of poverty, environmental degradation, ethnic and religious conflict, and social injustice are all interdependent, and that policies that address one problem can impact and improve other issues.

The remaining guiding principles are organised into three groups:

Ecological Integrity

Protect and restore the integrity of Earth's ecological systems, with special concern for biological diversity and the natural processes that sustain life. Prevent harm as the best method of environmental protection and, when knowledge is limited, apply a precautionary approach. Adopt patterns of production, consumption, and reproduction that safeguard Earth's regenerative capacities, human rights, and community well-being. Advance the study of ecological sustainability and promote the open exchange and wide application of the knowledge acquired.

Social and Economic Justice

Eradicate poverty as an ethical, social, and environmental imperative. Ensure that economic activities and institutions at all levels promote human development in an equitable and sustainable manner. Affirm gender equality and equity as prerequisites to sustainable development and ensure universal access to education, health care, and economic opportunity.

Uphold the right of all, without discrimination, to a natural and social environment supportive of human dignity, bodily health, and spiritual well-being, with special attention to the rights of indigenous peoples and minorities.

Democracy, Nonviolence, and Peace

Strengthen democratic institutions at all levels, and provide transparency and accountability in governance, inclusive participation in decision making, and access to justice. Integrate into formal education and life-long learning the knowledge, values, and skills needed for a sustainable way of life. Treat all living beings with respect and consideration. Promote a culture of tolerance, nonviolence, and peace.

Supporting Principles

The supporting principles clarify the meaning of the main principles and provide an overview of the many issues that were raised by various groups in the course of the international consultation process to develop the Earth Charter. Taken together, they outline major strategies and provide an action plan for achieving sustainable development.

Using The Earth Charter

The Earth Charter Initiative recommends many ways for individuals and groups to become involved. The following recommended strategies can be used by classes, schools, teachers' unions and other professional associations of educators:

- Explore the Earth Charter website for more information.
- Study the Earth Charter and discuss it with friends, family and colleagues.
- Join a group to reflect on the Earth Charter and how its principles can be put into action and implemented in our schools, communities, and workplaces.
- Use the Earth Charter as an educational resource to promote a sustainable way of life by integrating it into curriculum as a framework for learning about sustainable development.

- Use the Earth Charter as a theme for workshops, conferences, forums and meetings.
- Begin a campaign for your school, university, or local government to endorse the Earth Charter.

SUSTAINABLE DEVELOPMENT: IMPLICATIONS FOR WORLD PEACE

This year the Tom Slick Conference will address the issue of "sustainable development and its relationship to the construction and maintenance of peace". It is our expectation that the exploration of this relationship will lead to some profound conclusions. Sustainability represents an approach to development which addresses the fundamental concerns of poverty, environment, equality, and democracy. With the end of the Cold War, the pursuit of lasting peace and an end to conflict has become, together with sustainable development, a global imperative. By examining the synergies of these two concepts which have come to dominate policy discussions in the 1990s, this international conference takes a significant step in the direction of a more complete understanding of both sustainable development and the peace process for the next century.

While the link between development and peace has been frequently examined, the results of its examination remain largely inconclusive. Although it might appear intuitive that meeting the basic needs of poor communities holds the promise of eliminating many of the types of situations which favor the outbreak of conflict, in many cases development can be shown to contribute to or benefit from the existence or possibility of armed conflicts. The concept of sustainable development modifies this relationship considerably. The 1992 Rio Declaration, presented at the United Nations Conference on Environment and Development, asserted in Principle 25 that "Peace, development and environmental protection are interdependent and indivisible". In other words, the idea of peace forms an integral part of the idea of sustainable development. In the next century, these two concepts are likely to become inseparable:

To understand the events of the next fifty years, then, one must understand environmental scarcity, cultural and racial clash, geographic destiny, and the transformation of war.

It is therefore necessary to understand clearly what is meant by sustainable development. The idea was popularized in 1987 by the United Nations Commission on Environment and Development through the Brundtland Report. That report, entitled *Our Common Future*, produced the most widely accepted definition of sustainable development, that is "development that meets the needs of the present without compromising the ability of future generations to meet their own needs." In 1992, the concept was formalized at the United Nations Conference on Environment and Development, and since then it has become part of the vocabulary of governmental, inter-governmental, and non-governmental institutions in practically all languages. The United Nations has an International Commission for Sustainable Development which meets each year, with representatives from all the countries of the world. The World Bank has a Vice-President for Environmentally Sustainable Development. The Government of the United States, like the governments of many other countries, has a National Commission for Sustainable Development. International business leaders created a Business Commission on Sustainable Development, with representation at the regional and global level. International development agencies actively promote development projects which pursue sustainability.

Such success in the dissemination of the idea and in its institutionalization has contributed to a certain erosion of the concept of sustainable development. Many mainstream economists, for example, together with some politicians and business leaders, have rejected sustainable development as the latest buzz-word whose "very appeal lies in its vagueness". For some, sustainable development is a utopian concept, for others demagoguery, while others continue to consider it to be inherently limited in its applicability to cases of extreme poverty and the corresponding difficulties confronted by a large part of the world's population.

This reaction to the proliferation of the term is understandable. In many cases the idea has been misappropriated and misused. Opponents may also mistrust the economic implications of sustainable development. After all, the idea of sustainability implies the introduction of a host of restrictions on the process of economic growth, based on the trade-offs necessary to address the question of inter and intra generational distribution. The introduction of authoritative sustainable policies may at times have a negative impact on short term profit margins, although with the balance of a longer useful life given to investments. On the whole, however, the environmental debate and the continuing puzzle of underdevelopment have had an impact on the way we think about growth and progress. Today there is a general recognition of a need to modify our approach to development in the present, even without taking into consideration the more difficult question of how future generations will cope with our legacy of spiraling population growth, expanding energy consumption, and the inevitable depletion of vital natural resources. Whether confronting the challenges of the present or the future, a new approach to development is a necessity which requires a real revolution in our behavior and mentality:

Unless our life-style is subjected to considerable reevaluation, including the adoption of far-reaching self-control regarding the satisfaction not of real wants but of self-gratifying desires, the emphasis on ecology could become yet another intensifier of the conflict between the rich and the poor.

Sustainable development represents an opportunity to construct a new approach, and the success of that effort has powerful implications for issues of peace and security.

In order to discuss in depth the relationship between sustainable development and peace, it will be necessary to progressively define the concept of sustainability throughout the course of the symposium. Through structured discussions, the Symposium will attempt to specify the aspects of sustainable development which have direct ramifications for the pursuit and maintenance of peace both in the context of today's international realities, and in light of the trends which will carry us into the Twenty-first Century.

Annotated Agenda

The Symposium will present a series of presentations and discussions, divided into specific themes addressing in a logical progression the relationship between sustainable development and the peace process. These themes are as follows:

Importance of the Topic

The first presentation will be made by the Key Note Speaker, the objective of which is to call attention to the importance of the theme of the conference, and raise some of the fundamental questions which will be elaborated in the following sessions.

"...the central fact remains that humanity's ability to define for itself a meaningful existence is increasingly threatened by the contradiction between subjective expectations and objective socioeconomic conditions. Inherent in the potential collision between these two broad trends is the danger that world politics- both in terms of international affairs and of internal societal conditions- could simply spiral out of control, generating massive political disorder and philosophical confusion."

"Large scale environmental degradation, exacerbated by rapid population growth, threatens to undermine political stability in many countries."

Peace and Sustainable Development

The next discussion will examine the relationship between peace and sustainable development, beginning with a review of some of the principal causes of conflict (that is, the lack of peace). Whether we consider isolated episodes, or whether the broader question of peace and war as part of a historical process is examined (this, by the way, was the theme developed brilliantly by the first Slick Professor at the LBJ School, Kenneth Boulding), many of the primary causes of conflict are closely related to the question of sustainable development, or better stated, unsustainable development.

Examples of current global trends which present formidable challenges to the achievement of both peace and sustainabity include: the problem of population growth above the carrying capacity of the known natural resource base and the predominant technology, mounting pressure on diminishing quantities of fresh water and topsoil, disputed jurisdiction over territorial areas containing strategic resources, the destabilizing impact of widespread poverty and increasing social inequality, and a rising flow of migrants fleeing war, famine, and other vestiges of political, social, and economic breakdown.

While exploring common challenges, the discussion will also address the question of how peace contributes to the sustainable development process. The panel considering the issue of Peace and Sustainable Development will have to consider the question from a variety of viewpoints.

The evidence on this issue is unequivocal. The lack of peace, that is, a situation of war or conflict, drains away resources that otherwise might be applied (although not necessarily) to promote the well-being of a nation's citizens. In addition, armed conflicts destroy natural resources, infrastructure, and human lives. The establishment of peace permits the recuperation of stable conditions for development and liberates resources for needed investments, although it does not ensure in and of itself that the resulting development will be sustainable.

"Certain aspects of the issues of peace and security bear directly upon the concept of sustainable development. Indeed, they are central to it ."

"Renewable resource scarcities of the next 50 years will probably occur with a speed, complexity and magnitude unprecedented in history...We have come to understand that scarcities of renewable resources often produce insidious and cumulative social effects, such as population displacement and economic disruption. These events can, in turn, lead to clashes between ethnic groups as well as to civil strife and insurgency."

"Also, some conflicts begin when people without shelter or other basic human rights seek better lives for themselves and their families by taking up weapons to fight against their own government or their neighbors ."

"Possible destruction of the environment during warfare is a threat to every human... The Science for Peace Institute at the University of Toronto estimates that 10 to 30 percent of all environmental degradation in the world is a direct result of the various militaries."

Defining Sustainable Development

At this point in the Conference, it becomes important to define and conceptualize the idea of sustainable development, already mentioned and debated in preceding discussions, and to show how it has been operationalized at various levels. At the outset, sustainable development was little more that an abstract concept. In recent years, however, there have been advances in raising the theoretical precision of the term. As a result, the usefulness of the concept in practical applications, case studies, and sustainable development planning has expanded dramatically. As a part of this process a need to disaggregate the various dimensions of the concept arose, particularly the need to specify the ideas of social, environmental, political, economic, and cultural sustainability.

There has also been a great deal of effort dedicated to the definition of general and specific indicators, as well as methodologies, for the construction of tendencial and desired 'scenarios' that reveal the possible and probable paths for development from the perspective of sustainability. These steps have been taken at various levels. The United Nations, for example, has been attempting to refine and promote the idea on a global scale and in individual nations through its International Commission on Sustainable Development and related organs and programs. More and more local, state, and national governments are doing the same.

"A proposal has to be economically and... ecologically sustainable...However, equally important is the social side, and here we mean equity, social mobility, social cohesion, participation, empowerment, cultural identity, and institutional development...It is, to my mind, an essential part of the definition of sustainability, because, let me remind you, the neglect of that side leads to institutions that are incapable of responding to the needs of society. We see the consequences of that in tragedies from Somalia to Rwanda and from Liberia to Bosnia."

"Democracy in the twenty-first century faces no task more pressing than to generate a nobler, more sustainable vision of the aim of life and society. Yes, it faces also the task of generating a cultural consensus which would make non-coercive conflict resolution possible. It has urgently to devise means of massive redistribution of resources, globally and within individual societies, to prevent cataclysmic conflict between the opulent and the impoverished. Yet most fundamentally, it needs to generate a vision of being human which would make a sustainable human presence on this earth possible."

Sustainable Development and Peace

The foregoing discussions identified how the situation or the process of peace contributes to the process of sustainable development, and how the idea of sustainable development is being defined, promoted, and put into action across the globe. This has prepared the way for an in depth look at the different ways in which sustainable development can aid in the process of peace construction.

This is not a trivial question, and consequently must be examined carefully. Sustainable development, if achieved, contributes decisively to the dissipation, if not the elimination, of several of the primary causes of conflict discussed at the beginning of the Conference. If a sustainable development strategy has been successful in terms of the reduction of poverty, the leveling of social inequalities, and the optimum allocation of scarce resources, then certainly many of the situations that exacerbate conflict between different groups, communities, and nationalities will be avoided. Improving the conditions for social justice in particular is fundamental to the promotion of peace in a variety of contexts throughout the world.

There are many ways in which sustainable development can lead to a situation of stability, security, and peace. Sustainable development, if comprehensive, represents a multi-disciplinary idea which acts not just economically, nor solely ecologically, not only politically, but on all of these fronts. Beyond this, sustainable development has implications for improvement of the institutional structure. The modification or reform of institutions for the purpose of resolving potentially contentious situations democratically lies at the heart of the idea of sustainability.

Take for example the question of fresh water resources. The dispute over water rights at one level or another represents one of the principal causes of real or potential conflict in many different parts of the world. There are some who say that the great war of the next century could arise from a struggle for the control of fresh water, a resource increasingly under pressure from demographic expansion and economic activities. Experiences of sustainable development in the area of management of supply and consumption of water have led to the creation of mechanisms which aid users to define the norms for allocation, use, and transfer of water rights in a democratic process. In practice, this democratic process of resource allocation replaces previously existing conflict. At the local level, the gun has been replaced by meetings between neighbors or committees who must learn to share a watershed. On a macro level, international accords and committees can serve as substitutes for wars.

"A dramatic turnaround story has occurred on the left bank of the Gal Oya irrigation project (in Sri Lanka)...mutual trust and reciprocity were nourished on a face-to-face basis prior to attempts to organize farmers into large groups...the level of conflict among farmers has also declined. Now with the assured water supply and the availability of a forum...to discuss and settle disputes..., the frequency and the seriousness of conflicts have been greatly reduced..The extent of that mutual respect was demonstrated in 1981 when communal violence broke out in the district, with some roving bands of Sinhalese youth burning Tamil shops in the marketplace: The reaction of the Sinhalese farmer-representatives was to go to the homes of the Tamil Irrigation Department officials in order to protect them from violence."

The Lessons of Experience

At this stage, case studies will provide illustrations of the topics brought up in the preceding discussions. These cases will depend on the invited speakers who will present them, but the possibilities are numerous.

An obvious example from history would be the situation of Europe after World War I as compared to the circumstances which followed World War II. In the first case, the post-war negotiation strategy of the victorious nations attempted to further cripple the vanquished by exacting a high price in the form of restitution. After the second world war, the victorious allies sought first and foremost the reconstruction and rehabilitation of Europe. While in 1918-19 the conditions for the renewal of bitter conflict were established, so in 1947 the foundations for the construction of a peaceful and sustainable Europe were laid.

Many other examples, positive and negative, can be discussed, from the experiences of Europe and the Americas, to those from the Middle East, from Asia, and from Africa. By way of further illustration, of growing importance is the question of rising pressure from the migration of poor populations fleeing from areas of low carrying capacity and economic underdevelopment. Such migrations throw host societies into disequilibrium as the hosts attempt to assimilate the displaced population. This situation often leads to conflict, as can be seen in Europe with the African immigrant community and in the United States with the migrations from Mexico and Central America. By raising the carrying capacity of the countries of origin, sustainable development strategies can offer an effective solution, reducing over time the causes of such migrations which represent today one of the principal potential causes of conflict between North and South.

"Almost 1 million Haitian `boat-people', one-sixth of the entire populace, have fled that island nation, an exodus fueled in large part by environmental degradation...El

Salvador, one of the most troubled nations of Central America, is also one of the most environmentally impoverished, with some of the worst erosion rates in the region. 'The fundamental causes of the present conflict are as much environmental as political, stemming from problems of resource distribution in an overcrowded land'."

"As political fragmentation and instability spread, national governments can no longer provide the physical and economic infrastructure for development. Countries in this category include Afghanistan, Haiti, Liberia, Sierra Leone, and Somalia ."

"A television newsman, overhearing the argument, asked her: `But if the Mexicans and Central Americans keep pressing in, won't this mean that eventually most of Texas will become Mexican?' and she said, looking defiantly into the camera: `If Hispanic mothers in Central America have many babies and anglo mothers in Texas have few, I suppose there will have to be an irresistible sweep of immigrants to the north. Yes, Texas will become Spanish .

Policy Implications

Finally, a panel of participants from diverse backgrounds will reflect on some of the major themes arising from previous discussions, and attempt to extract relevant conclusions for policy makers. What are, after all, the useful lessons for international institutions, for governments, for organized civil society, for citizens from the experiences mentioned? What can be done to promote sustainable development? How does one increase the scope of development programs in the interest of peace promotion? How should these lessons be applied in today's world, in light of future trends and forecasts (at least one generation)?

"As part of its prevention strategy, the Administration is vigorously promoting sustainable development, both at home and abroad "

"At the international level, our projections suggest that the population-driven environmental deterioration/political disintegration scenario...is not inevitable. This future can be averted if security is redefined, recognizing that food scarcity, not military aggression, is the principal threat to our future. This would lead to a massive reordering of priorities- giving top place to filling the family planning gap; to attacking the underlying causes of high fertility, such as illiteracy and poverty; to protecting soil and water resources; and to raising investment in agriculture."

"Adopting a central organizing principle- one agreed to voluntarily- means embarking on an all-out effort to use every policy and program, every law and institution, every treaty and alliance, every tactic and strategy, every plan and course of action- to use in short, every means to halt the destruction of the environment and to preserve and nurture our ecological system ."

SUSTAINABILITY AND DISASTER RECOVERY

Applying the principles of sustainability when making decisions can help communities avoid the pitfalls of adopting a course of action without realizing it will have detrimental impacts at another place or time. Ideally, all communities would routinely adopt a long-term view and incorporate sustainability ideals into all aspects of their comprehensive planning process—whether making development decisions, preparing for a disaster, implementing mitigation, or undertaking any other program.

In the absence of this ideal situation, however, a person concerned with avoiding losses due to hazards and disasters must look for opportunities to integrate sustainability with mitigation measures wherever possible. One fertile field for this integration is the disaster recovery period.

A disaster brings temporary changes to a community. People think about problems they normally do not consider—the risks they face from hazards, the quality of local housing, ways in which the community could be better planned and constructed, the local scenic and other natural resources, livability. At the same time, public officials have the media attention that enables them to garner support for innovative ideas. A disaster forces a community to make a seemingly endless series of decisions—some large, some small, some easy, and some quite difficult. Technical and expert advice becomes available from public and private sources. Financial assistance flows into the community, enabling it to tackle more ambitious projects than would normally be the case.

These changes can be viewed as opportunities to rebuild in a better way, instead of succumbing to the natural desire to put things back the way they were as soon as possible. They can provide a chance for a community to implement forward-looking activities that for one reason or another (usually financial or political) have not been undertaken, including improvements in lifestyle, safety, economic opportunity, or the environment. After a disaster, a community must take action to recover, so incorporating principles of sustainability into that process often does not involve much additional effort.

Hazards managers already work to build mitigation into many recovery activities. For example, they often use the Federal Emergency Management Agency's postdisaster programs and other initiatives that in many cases specifically call for mitigation. However, they could go still further, and ensure that the mitigation measures that are put in place promote—or at least do not undermine—sustainable communities.

An Overview of Holistic Recovery

How can a community take advantage of the opportunity that disaster recovery brings? As a foundation for this effort, a framework for sustainable—or "holistic"—recovery from disaster has been developed within which the principles of sustainability become decisionmaking criteria to be applied to each and every recovery decision—not just those that involve mitigation. On the next page is a sample matrix that can be a guide to decisionmaking for holistic recovery. The sustainability principles (and some ways of implementing them) are shown on the vertical axis. Across the top of the matrix are listed some of the problem situations that could confront a community in the aftermath of a disaster: utilities must be restored, infrastructure re-established, housing repaired, social services reinstituted, and commercial sectors rehabilitated. At the intersection of the problem and the principle there are opportunities for a recovery decision and action that would be more sustainable than a return to the status quo (marked with an X on the matrix). It should be noted that this matrix is just a sample of a hypothetical disaster in a hypothetical community. A similar matrix developed by a real community to help it in recovery would have a different list of disaster situations across the top, and a different set of boxes marked with X. The principles would be the same as in this sample, as would many of the options for applying them.

This holistic recovery framework can be used either in pre-disaster planning for recovery or during the recovery period itself to ensure that people consider viable, sustainable options as decisions are made. The range of possibilities, alternatives (including returning to the status quo), and impacts of the proposed recovery actions are considered in light of the sustainability principles as decisions are made about recovery, so that sustainable options are considered in each and every disaster recovery opportunity. During this process, a community can tailor a unique set of sustainable activities for its recovery that satisfies its own particular concerns, takes advantage of its strengths, and uses the tools and techniques that are most appropriate to its situation.

This process can result in some unusual combinations of problems and solutions. For example, a stricken community with a damaged freeway overpass might well decide to incorporate seismic-resistant features into the repaired structure. However, a community striving for holistic recovery would also consider demolishing or relocating the overpass to enhance livability in the surrounding neighborhood (principle number 5), or rebuilding it to improve access to, and thus economic vitality for, a nearby commercial area that was previously difficult to reach from the highway (principle number 2). This is just one of many possible outcomes of a systematic process of analyzing recovery in light of the six sustainability principles. The possibilities are endless, because each community has unique attributes, needs, and concerns, and each disaster superimposes a distinct set of impacts.

What is Holistic Recovery?

Definition: A holistic recovery from a disaster is one in which the stricken locality systematically considers each of the principles of sustainability in every decision it makes about reconstruction and redevelopment.

This can be more appealing to a community than simply trying to impose mitigation measures, even with financial and other incentives, because it gives the members of a community a way to examine their other day-to-day goals within a broader context. Mitigation doesn't drive the process — community goals, buttressed by sustainability ideals, do. But mitigation gets considered in every decision about economic development, infrastructure repair, housing needs, and environmental protection. By the same process, concerns about economic development, local environmental quality, social equity, future generations, and other aspects of a healthy community are considered in every decision about mitigation.

The Process

The best way to ensure community sustainability after a future disaster is to have a thorough plan for a holistic recovery. But even without such a plan, there are many things that can be done during recovery that will increase community sustainability, simply by using the holistic recovery framework as a guide and the disaster recovery process as the catalyst. A community must strive to fully coordinate available assistance and funding while seeking ways to accomplish other community goals and priorities. Holistic disaster recovery does not differ from "normal" disaster recovery — it is part of what should be normal disaster recovery. A good recovery engenders a sustainable community.

A community does not need a new or separate planning or recovery process to build sustainability. The sustainability perspective can be accommodated in different ways and to varying degrees within most standard procedures used by localities for comprehensive planning, mitigation planning, disaster recovery, or other efforts.

The Red River of the North

After the disastrous 1997 floods on the Red River of the North, thousands of households in the Greater Grand Forks area had damage serious enough to necessitate the replacement of their furnaces and/or hot water heaters. The recovery decision makers realized that this was a chance to effect a massive upgrade of the heating systems in the area. Rebates of $200 were offered to each homeowner and small business owner who replaced his or her damaged furnace or water heater with an energy-efficient unit. About 5,500 households and businesses (about half of those flooded) took advantage of the rebates. These new furnaces consume less fuel and give off fewer pollutants, improving quality of life in the Red River Basin.

A good, all-purpose planning process — the so-called 10-Step Planning Process — is one that is recommended for localities seeking funding, technical assistance, or recognition under such federal programs as the Community Rating System of the National Flood Insurance Program, several flood control programs of the U.S. Army Corps of Engineers, and the Hazard Mitigation Grant Program and the Flood Mitigation Assistance Program administered by the Federal Emergency Management Agency. It follows the basic procedures of gathering information, analyzing problems, setting goals, and finding ways to implement and fund agreed-upon activities.

A 10-Step Process for Local Holistic Recovery

1. Get organized. At this stage a community makes a commitment to sustainability by designating appropriate responsibility for the recovery, delegating it to an individual or entity — new or existing — and setting up measures for integrating sustainability into ongoing disaster recovery and other community processes, as necessary. One way to do this would be to appoint a "sustainability liaison" to the planning and decisionmaking body or the recovery team. The person in this role would be an advocate for considering the principles of sustainability at each step of the process as well as knowledgeable about and supportive of all those principles: environment, social equity, consideration of the future, economic development, quality of life, and disaster resilience.

2. Involve the public. Participatory processes are an essential aspect of sustainability involving the inclusion of all the stakeholders in recovery and in creating the vision of what the community should be like after the recovery is complete. A community that seeks sustainability must be committed to such involvement and, at this point, the community begins to design public participation into all phases of its recovery. There are many techniques from which to choose, from the traditional public hearings and town meetings to lectures, planning charettes, workshops, call-in radio shows, and community-based events like fairs and festivals. To fulfill the goal of social equity, communities should pay particular attention to reaching out to those people who may have been historically excluded from conventional "public notice" techniques because of language differences, cultural constraints, temporal or spatial barriers to attending meetings, or other factors. The opportunities for participation should be publicized through a variety of media, including flyers, posters, local newspapers, local television stations, and the Internet.

3. Coordinate with other agencies, departments, and groups. To mastermind a holistic recovery, a community must expand representation on the recovery team to include those who can contribute expertise regarding each of the principles of sustainability. They could be in-house staffers, local experts, representatives from state or federal agencies, or consultants. Depending on the situation, social services personnel, environmental specialists, engineers, economic development directors, parks or wildlife department personnel, the business community, or social services personnel all might be included. Formal and informal ties need to be developed with every conceivable private entity; non-profit group; neighborhood coalition; church; state, local, federal, and regional agency; and others. This will increase the diversity of ideas and potential solutions, provide a ready-made labor pool (which will be needed when implementation begins), and make problem-solving more imaginative. It also will strengthen local capacity within and across groups and areas of expertise.

4. Identify post-disaster problems. During this step, the recovery team begins to systematically consider ways in which it can build sustainability as it plans for and manages the recovery. The team can start by simply listing all the disaster-caused situations that need to be remedied in the course of recovery. (Some possibilities are listed across the top of the matrix.)

For each problem situation, information should be gathered to gain a full picture. This is a broad exercise that likely will include many sub-steps spread over a wide array of issues, for example:

- Obtaining expert analysis of local economic trends, costs of rebuilding, and opportunities for economic growth, before and after the disaster;

- Mapping an environmentally sensitive area;
- Assessing the community's present and future vulnerability to hazards and disasters;
- Pinpointing social inequity and its impacts within the community, before and after the disaster;
- Determining what quality of life concerns are important to local residents, before and after the disaster.

Obviously it is preferable to have this information in hand before a disaster, rather than having to gather it afterward, when the situation is confused, and time and resources are at a premium. This step will culminate in a list of problem situations, accompanied by supporting information.

Principles of Sustainability and Some Options for Applying Them

1. Maintain and enhance quality of life

Options:

- Make housing available/affordable/better
- Provide education opportunities
- Ensure mobility
- Provide health and other services
- Provide employment opportunities
- Provide for recreation
- Maintain safe/healthy environs
- Have opportunities for civic engagement

2. Enhance Economic vitality

Options:

- Support area redevelopment and revitalization
- Attract/retain businesses
- Attract/retain work force
- Rebuild for economic functionality
- Develop/redevelop recreational, historic, touris attractions

3. Ensure social and intergenerational equity

Options:

- Preserve/conserve natural, cultures, historical resources
- Adopt a longer-term focus for all planning
- Avoid/remedy disproportionate impacts on groups
- Consider future generations' quality of life
- Value diversity
- Preserve social connections in and among groups

4. **Enhance environmental quality**

    Options:

    - Preserve/conserve/restore natural resources
    - Protect open space
    - Manage stormwater
    - Prevent/remediate pollutio

5. **Incorporate disaster resilience/mitigation**

    Options:

    - Make buildings and infrastructure damage-resistant
    - Avoid development in hazardous areas
    - Manage stormwater
    - Protect natural areas
    - Promote and obtain hard and other insurance

6. **Use a participatory process**

    Options:

    - Incorporate with all of the other principles

5. Evaluate the problems and identify opportunities. The implications of sustainability become clear during this step. The recovery team evaluates each of the problems identified in Step 4 in light of the six principles of sustainability to see where there are opportunities during recovery to enhance community sustainability and move toward the community's vision of its future rather than returning to the status quo. The list of options in the box (and listed on the left side of the matrix) can be used to stimulate thinking about sustainable approaches a locality can use to address each postdisaster problem. One or more approaches should be designated as possibilities for each problem, focusing on those that are applicable to the community's situation, needs, and concerns. Note that this is not an exhaustive list and also that some options apply to more than one principle.

This step results in a list of possible ways to combine remedying a disaster-caused problem and addressing an "unsustainable" situation. Each idea represents a way to further one or more aspects of sustainability, without regard (at this point) to cost or feasibility. The list is simply a series of specific things that, ideally, the community would like to do. For example, suppose the community has experienced a flood that, among other impacts, has seriously damaged a neighborhood of low-income houses along a polluted stream. One item identified during this step might be: "Expand stormwater management system to better handle street drainage and reduce streambank erosion" (thereby repairing flood-damaged infrastructure, improving livability by reducing street flooding, minimizing future flood damage by enlarging the carrying capacity of the stormwater system, and improving environmental quality by preserving soil and riparian vegetation from erosion). Another item might be: "Incorporate seismic-resistant features and insulation into damaged housing during repair" (thereby improving livability by making the houses warmer and cooler according to the time of year and less expensive to heat or cool, improving disaster resilience by strengthening the housing against earthquakes, and protecting environmental quality by reducing energy consumption). The team tries to consolidate multiple sustainability principles into each possibility it lists.

6. Set goals. During this step the recovery team agrees on what realistically can be done. The team pares down the list of possibilities identified in Step 5 to those measures preferred by most of the stakeholders and most consonant with local needs and situations, public support, cost-effectiveness, availability of technical expertise, other community goals, local regulations, and other factors. A range of possibilities is developed and prioritized in case some cannot be implemented. These final choices become the recovery goals — positive statements of what the community intends to accomplish. By this point it will become clear that the goals established for a holistic recovery are broader and have more far-reaching implications than those for simply returning to the status quo.

This step will result in an agreed-upon set of actions that have reasonable applicability to the community.

7. Develop strategies for implementation. Working with the list of goals developed in Step 6, the recovery team reviews the tools, financial support, and expertise available to achieve each of them. For each goal, an implementation strategy is be developed that describes

- What is to be accomplished;
- The lead agency/entity and what it will provide or prepare;
- Partnerships that will enhance effectiveness;
- Ways to obtain technical expertise and advice;
- Official local action needed (passage or amendment of zoning or subdivision ordinances, adoption of building codes, etc.); and
- Funding methods.

This will produce a "package" associated with each community goal that outlines what is needed to achieve that goal. This step weeds out the possibilities that are not feasible for whatever reason and results in a set of strategies that realistically can be implemented.

8. Plan for action. During this step the recovery team drafts a complete plan for holistic recovery activities that fits into the recovery plan or becomes part of the community's comprehensive plan. Like other plans, it should include

- a budget;
- details for obtaining funding;
- a schedule for team meetings, public participation, data collection, report writing, on-the-ground action;
- a monitoring and review process; and
- provision for public review and comment.

This plan should be coordinated with existing comprehensive, development, capital improvement, drainage, transportation, housing, and recreation plans and programs. After public and agency/entity review, the plan should be revised and finalized.

9. Get agreement on the plan for action. Depending on the circumstances, the state, county, and/or local government may formally adopt or approve a holistic recovery plan or otherwise officially incorporate it into the recovery or comprehensive plan. During this stage, the local community should obtain agreement from federal and state agencies as appropriate. It might also enter into memoranda of understanding with other partners. The agreement of other stakeholders, especially historically excluded groups, should be obtained.

10. Implement, evaluate, and revise. This final step ensures that the community maximizes the opportunities that began as a disaster. Having the persons and entities responsible for implementation of various aspects of the recovery actually involved in the decision-making during all the earlier steps helps ensure that the goals and activities agreed upon are actually carried out.

As recovery proceeds, it will be clear that some goals and strategies need to be modified. A formal monitoring process helps identify what changes are needed. It also can help keep certain initiatives from simply being abandoned when an unforeseen obstacle is reached. Wherever possible, stakeholders should participate in reviews (at least annually) and help develop indicators of progress.

Some Tools for Community Sustainability	
• Local redevelopment authority • Economic incentives • Loans for businesses • Housing authority • Insurance • Capital improvements • Low interest subsidy loans • Revolving loan funds • Public investment • Redistricting • Subdivision regulations • Building codes • Special ordinances • Tax incentives • Transfer of development rights • Easements • Land purchase • Voluntary agreements • Planning • Habitat protection • Riparian buffers	• Filter strips and vegetative buffers • Soil conservation and management • Ecosystem restoration • Zoning and rezoning • Public education and awareness campaigns and events • Special protection of critical facilities, utilities, and networks • Preserve and create public spaces • Limit public investment in hazardous areas • Relocation out of hazardous areas • Preservation of natural floodplain, coastal, wetland, and other functions • Private-public partnerships and networks • Ombudspersons • Targeted workshops • Community festivals and other activities

A Long-Term Outlook

Sustainable practices (and the awareness of the principles of sustainability) introduced during recovery planning or actual recovery can be institutionalized within the community's decision-making, budgeting, and planning processes to ensure that they endure over time. Ideally, a community would develop indicators and a schedule for monitoring and tracking change and needed improvements. Such institutionalization would help build awareness of the many aspects of sustainability as local residents, public officials, city staff, and businesses come and go. The heightened awareness would in turn nurture an acceptance of sustainable practices as a local, public value and a way of life.

Using the holistic recovery framework, applying the sustainability principles, and employing a process like the 10-step procedure create additional benefits for a community. For one thing, they promote links, conceptual and operational, among different community interests and the groups that seek to further them. For example, how many times have people discovered — inadvertently — that those responsible for local parks and recreation actually are interested in the same sort of open space improvements that the wildlife advocates want? This process makes such serendipitous convergence more likely and helps solidify future collaboration, thus making it easier and more cost-effective for the community to accomplish its overall goals and carry out routine activities.

Another benefit to hazards managers is that drawing on the broad range of sustainability principles instead of just thinking about hazards in isolation makes it more likely that the hazard mitigation approaches that are adopted and carried out will actually minimize losses in the long run. It helps ensure that the mitigation measure(s) implemented will be valuable because they are paired with other community desires, and long-lasting, because they do not detract from other aspects of overall sustainability. Losses will not have to be borne, damage repaired, and victims compensated again and again in future disasters.

THE NEED FOR MORE INTEGRATED VIEWS ON MATERIALS AND PROCESS CONDITIONS

Integrated Computational Materials Engineering

Integrated Computational Materials Engineering (ICME) is an approach to design products, the materials that comprise them, and their associated materials processing methods by linking materials models at multiple length scales. Key words are "Integrated", involving integrating models at multiple length scales, and "Engineering", signifying industrial utility. The focus is on the materials, *i.e.* understanding how processes produce material structures, how those structures give rise to material properties, and how toselect materials for a given application. The key links are process-structures-properties-performance. The National Academies report describes the need for using multiscale materials modeling to capture the process-structures-properties-performance of a material.

Standardization in ICME

A fundamental requirement to meet the ambitious ICME objective of designing materials for specific products resp. components is an integrative and interdisciplinary computational description of the history of the component starting from the sound initial condition of a homogeneous, isotropic and stress free melt resp. gas phase and continuing *via* subsequent processing steps and eventually ending in the description of failure onset under operational load.

Integrated Computational Materials Engineering is an approach to design products, the materials that comprise them, and their associated materials processing methods by linking materials models at multiple length scales. ICME thus naturally requires the combination of a variety of models and software tools. It is thus a common objective to build up a scientific network of stakeholders concentrating on boosting ICME into industrial application by defining a common communication standard for ICME relevant tools.

Standardization of Information Exchange

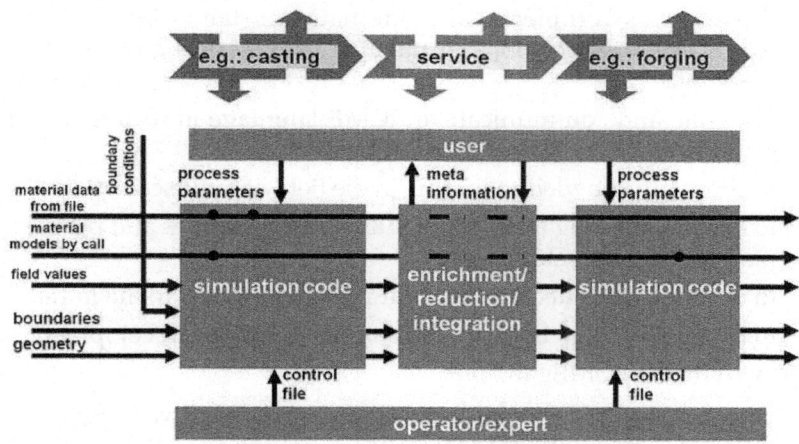

Efforts to generate a common language by standardizing and generalizing data formats for the exchange of simulation results represent a major mandatory step towards successful future applications of ICME. A future, structural framework for ICME comprising a variety of academic and/or commercial simulation tools operating on different scales and being modular interconnected by a common language in form of standardized data exchange will allow integrating different disciplines along the production chain, which by now have only scarcely interacted. This will substantially improve the understanding of individual processes by integrating the component history originating from preceding steps as the initial condition for the actual process. Eventually this will lead to optimized process and production scenarios and will allow effective tailoring of specific materials and component properties.

The ICMEg Project and its Mission

The ICMEg project aims to build up a scientific network of stakeholders concentrating on boosting ICME into industrial application by defining a common communication standard for ICME relevant tools. Eventually this will allow stakeholders from electronic, atomistic, mesoscopic and continuum communities to benefit from sharing knowledge and best practice and thus to promote a deeper understanding between the different communities of materials scientists, IT engineers and industrial users.

ICMEg will create an international network of simulation providers and users. It will promote a deeper understanding between the different communities (academia and industry) each of them by now using very different tools/methods and data formats. The harmonization and standardization of information exchange along the life-cycle of a component and across the different scales (electronic, atomistic, mesoscopic, continuum) are the key activity of ICMEg.

The mission of ICMEg is :

- to establish and to maintain a network of contacts to simulation software providers, governmental and international standardization authorities, ICME users, associations in the area of materials and processing, and academia
- to define and communicate an ICME language in form of an open and standardized communication protocol
- to stimulate knowledge sharing in the field of multiscale materials design
- to identify missing tools, models and functionalities and propose a roadmap for their development
- to discuss and to decide about future amendments to the initial standard
- to establish a legal body for a sustainable further development

The activities of ICMEg include :

- Organization of International Workshops on Software Solutions for Integrated Computational Materials Engineering
- Conducting market study and surwey on available simulation software for ICME
- Create and maintain forum for knowledge sharing in ICME
- Established legal body for future sustainable development: ICMEg e.V. has been founded on June 24, 2014 in Rolduc

Multiscale Modeling in Material Processing

Multiscale modeling aims to evaluate material properties or behavior on one level using information or models from different levels and properties of elementary processes. Usually, the following levels, addressing a phenomenon over a specific window of length and time, are recognized:

- Structural scale: Finite element, finite volume and finite difference partial differential equation are solvers used to simulate structural responses such as solid mechanics andtransport phenomena at large (meters) scales.

- process modeling/simulations: extrusion, rolling, sheet forming, stamping, casting, welding, etc.
- product modeling/simulations: performance, impact, fatigue, corrosion, etc.
- Macroscale: constitutive (rheology) equations are used at the continuum level in solid mechanics and transport phenomena at millimeter scales.
- Mesoscale: continuum level formulations are used with discrete quantities at multiple micrometre scale. "Meso" is an ambiguous term that means "intermediate" so it has been used as representing different intermediate scales. In this context it can represent modeling from crystal plasticity for metals, Eshelby solutions for any materials, homogenization methods, and unit cell methods.
- Microscale: modeling techniques that represent the micrometre scale such as dislocation dynamics codes for metals and phase field models for multiphase materials. Phase field models of phase transitions and microstructure formation and evolution on nanometer to millimeter scales.
- Nanoscale: semi-empirical atomistic methods are used such as Lennard-Jones, Brenner potentials, embedded atom method (EAM) potentials, and modified embedded atom potentials (MEAM) in molecular dynamics (MD), molecular statics (MS), Monte Carlo (MC), and kinetic Monte Carlo (KMC) formulations.
- Electronic scale: Schroedinger equations are used in computational framework as density functional theory (DFT) models of electron orbitals and bonding on angstrom to nanometer scales.

There are some codes that operate on different length scales such as:

- CALPHAD computational thermodynamics for prediction of equilibrium phase diagrams and even non-equilibrium phases.
- Phase field codes for simulation of microstructure evolution
- Databases of processing parameters, microstructure features, and properties from which one can draw correlations at various length scales
- GeoDict virtual material laboratory

Examples of Model integration

- Small scale models calculate material properties, or relationships between properties and parameters, *e.g.* yield strength vs. temperature, for use in continuum models
- CALPHAD computational thermodynamics software predicts free energy as a function of composition; a phase field model then uses this to predict structure formation and development, which one may then correlate with properties.
- An essential ingredient to model microstructure evolution by phase field models and other microstructre evolution codes are the initial and

boundary conditions. While boundary conditions may be taken *e.g.* from the simulation of the actual process, the initial conditions (*i.e.* the initial microstructure entering into the actual process step) involve the entire integrated process history starting from the homogeneous, isotropic and stress free melt. Thus - for a successful ICME - an efficient exchange of information along the entire process chain and across all relevant length scales is mandatory. The models to be combined for this purpose comprise both academic and/or commercial modelling tools and simulation software packages. To streamline the information flow within this heterogeneous variety of modelling tools, the concept of a modular, standardized simulation platform has recently been proposed. A first realisation of this concept is the AixViPMaP® - the Aachen Virtual Platform for Materials Processing.

- Process models calculate spatial distribution of structure features, *e.g.* fiber density and orientation in a composite material; small-scale models then calculate relationships between structure and properties, for use in a continuum models of overall part or system behavior

- Large scale models explicitly fully couple with small scale models, *e.g.* a fracture simulation might integrate a continuum solid mechanics model of macroscopic deformation with an FD model of atomic motions at the crack tip

- Suites of models (large-scale, small-scale, atomic-scale, process-structure, structure-properties, etc.) can be hierarchically integrated into a systems design framework to enable the computational design of entirely new materials. A commercial leader in the use of ICME in computational materials design is QuesTek Innovations LLC, a small business in Evanston, IL co-founded by Prof. Greg Olson of Northwestern University. QuesTek's high-performance Ferrium® steels were designed and developed using ICME methodologies.

- The Mississippi State University Internal State Variable (ISV) plasticity-damage model (DMG) developed by a team lead by Prof. Mark F. Horstemeyer (Founder ofPredictive Design Technologies) has been used to optimize the design of a Cadillac control arm, the Corvette engine cradle, and a powder metal steel engine bearing cap.

- ESI Group through its ProCast and SYSWeld are commercial finite element solutions used in production environments by major manufacturers in aerospace, automotive and government organizations to simulate local material phase changes of metals prior to manufacturing. PAMFORM is utilized for tracking material changes during composite forming manufacturing simulation.

THE CONCEPT OF SUSTAINABLE DEVELOPMENT

The satisfaction of human needs and aspirations in the major objective of development. The essential needs of vast numbers of people in developing countries for food, clothing, shelter, jobs - are not being met, and beyond their basic needs these people have legitimate aspirations for an improved quality of life. A world in which poverty and inequity are endemic will always be prone to ecological and other crises. Sustainable development requires meeting the basic needs of all and extending to all the opportunity to satisfy their aspirations for a better life.

Living standards that go beyond the basic minimum are sustainable only if consumption standards everywhere have regard for long-term sustainability. Yet many of us live beyond the world's ecological means, for instance in our patterns of energy use. Perceived needs are socially and culturally determined, and sustainable development requires the promotion of values that encourage consumption standards that are within the bounds of the ecological possible and to which all can reasonably aspire.

Meeting essential needs depends in part on achieving full growth potential, and sustainable development clearly requires economic growth in places where such needs are not being met. Elsewhere, it can be consistent with economic growth, provided the content of growth reflects the broad principles of sustainability and non-exploitation of others. But growth by itself is not enough. High levels of productive activity and widespread poverty can coexist, and can endanger the environment. Hence sustainable development requires that societies meet human needs both by increasing productive potential and by ensuring equitable opportunities for all.

An expansion in numbers can increase the pressure on resources and slow the rise in living standards in areas where deprivation is widespread. Though the issue is not merely one of population size but of the distribution of resources, sustainable development can only be pursued if demographic developments are in harmony with the changing productive potential of the ecosystem.

A society may in many ways compromise its ability to meet the essential needs of its people in the future - by overexploiting resources, for example. The direction of technological developments may solve some immediate problems but lead to even greater ones.

Settled agriculture, the diversion of watercourses, the extraction of minerals, the emission of heat and noxious gases into the atmosphere, commercial forests, and genetic manipulation are all examples or human intervention in natural systems during the course of development. Until recently, such interventions were small in scale and their impact limited. Today's interventions are more drastic in scale and impact, and more threatening to life-support systems both locally and globally. This need not happen. At a minimum, sustainable development must not endanger the natural systems that support life on Earth: the atmosphere, the waters, the soils, and the living beings.

Growth has no set limits in terms of population or resource use beyond which lies ecological disaster. Different limits hold for the use of energy, materials, water, and land. Many of these will manifest themselves in the form of rising costs and diminishing returns, rather than in the form of any sudden loss of a resource base. The accumulation of knowledge and the development of technology can enhance the carrying capacity of the resource base. But ultimate limits there are, and sustainability requires that long before these are reached, the world must ensure equitable access to the constrained resource and reorient technological efforts to relieve the presume.

Economic growth and development obviously involve changes in the physical ecosystem. Every ecosystem everywhere cannot be preserved intact. A forest may be depleted in one part of a watershed and extended elsewhere, which is not a bad thing if the exploitation has been planned and the effects on soil erosion rates, water regimes, and genetic losses have been taken into account. In general, renewable resources like forests and fish stocks need not be depleted provided the rate of use is within the limits of regeneration and natural growth. But most renewable resources are part of a complex and interlinked ecosystem, and maximum sustainable yield must be defined after taking into account system-wide effects of exploitation.

As for non-renewable resources, like fossil fuels and minerals, their use reduces the stock available for future generations. But this does not mean that such resources should not be used. In general the rate of depletion should take into account the criticality of that resource, the availability of technologies tor minimizing depletion, and the likelihood of substitutes being available. Thus land should not be degraded beyond reasonable recovery. With minerals and fossil fuels, the rate of depletion and the emphasis on recycling and economy of use should be calibrated to ensure that the resource does not run out before acceptable substitutes are available. Sustainable development requires that the rate of depletion of non renewable resources should foreclose as few future options as possible.

Development tends to simplify ecosystems and to reduce their diversity of species. And species, once extinct, are not renewable. The loss of plant and animal species can greatly limit the options of future generations; so sustainable development requires the conservation of plant and animal species.

So-called free goods like air and water are also resources. The raw materials and energy of production processes are only partly converted to useful products. The rest comes out as wastes. Sustainable development requires that the adverse impacts on the quality of air, water, and other natural elements are minimized so as to sustain the ecosystem's overall integrity.

In essence, sustainable development is a process of change in which the exploitation of resources, the direction of investments, the orientation of technological development; and institutional change are all in harmony and enhance both current and future potential to meet human needs and aspirations.

Equity and the Common Interest

Sustainable development has been described here in general terms. How are individuals in the real world to be persuaded or made to act in the common interest? The answer lies partly in education, institutional development, and law enforcement. But many problems of resource depletion and environmental stress arise from disparities in economic and political power. An industry may get away with unacceptable levels or air and water pollution because the people who bear the brunt of it are poor and unable to complain effectively. A forest may be destroyed by excessive felling because the people living there have no alternatives or because timber contractors generally have more influence then forest dwellers.

Ecological interactions do not respect the boundaries of individual ownership and political jurisdiction. Thus:

- In a watershed, the ways in which a farmer up the slope uses land directly affect run-off on farms downstream.
- the irrigation practices, pesticides, and fertilizers used on one farm affect the productivity of neighbouring ones, especially among small farms.
- The efficiency of a factory boiler determines its rate of emission of soot and noxious chemicals and affects all who live and work around it.
- The hot water discharged by a thermal power plant into a river or a local sea affects the catch of all who fish locally.

Traditional social systems recognized some aspects of this interdependence and enforced community control over agricultural practices and traditional rights relating to water, forests, and land. This enforcement of the 'common interest' did not necessarily impede growth and expansion though it may have limited the acceptance and diffusion of technical innovations.

Local interdependence has, if anything, increased because of the technology used in modern agriculture and manufacturing. Yet with this surge of technical progress, the growing 'enclosure' of common lands, the erosion of common rights in forests and other resources, and the spread of commerce and production for the market, the responsibilities for decision making are being taken away from both groups and individuals. This shift is still under way in many developing countries.

It is not that there is one set of villains and another of victims. All would be better off if each person took into account the effect oœ" his or her acts upon others. But each is unwilling to assume that others will behave in this socially desirable fashion, and hence all continue to pursue narrow self-interest. Communities or governments can compensate for this isolation through laws, education, taxes, subsidies, and other methods. Well-enforced laws and strict liability legislation can control harmful side effects. Most important, effective participation in decision-making processes by local communities can help them articulate and effectively enforce their common interest.

Interdependence is not simply a local phenomenon. Rapid growth in production has extended it to the international plane, with both physical and economic manifestations. There are growing global and regional pollution effects, such as in the more than 200 international river basins and the large number of shared seas.

The enforcement of common interest often suffers because areas of political jurisdiction and areas of impact do not coincide. Energy policies in one jurisdiction cause acid precipitation in another. The fishing policies of one state affect the fish catch of another. No supranational authority exists to resolve such issues, and the common interest can only be articulated through international cooperation.

In the same way, the ability of a government to control its national economy is reduced by growing international economic interactions. For example, foreign trade in commodities makes issues of carrying capacities and resource scarcities an international concern. If economic power and the benefits of trade were more equally distributed, common interests would be generally recognized. But the gains from trade are unequally distributed, and patterns of trade in, say, sugar affect not merely a local sugar-producing sector, but the economies and ecologies of the many developing countries that depend heavily on this product.

The search for common interest would be less difficult if all development and environment problems had solutions that would leave everyone better off. This is seldom the case, and there are usually winners and losers. Many problems arise from inequalities in access to resources. An inequitable landowner ship structure can lead to overexploitation of resources in the smallest holdings, with harmful effects on both environment and development. Internationally, monopolistic control over resources can drive those who do not share in them to excessive exploitation of marginal resources. The differing capacities of exploiters to commandeer 'free' goods - locally, nationally, and internationally - is another manifestation of unequal access to resources. 'Losers' in environment/development conflicts include those who suffer more than their fair share of the health, property, and ecosystem damage costs of pollution.

As a system approaches ecological limits, inequalities sharpen. Thus when a watershed deteriorates, poor farmers suffer more because they cannot afford the same anti-erosion measures as richer farmers. When urban air quality deteriorates, the poor, in their more vulnerable areas, suffer more health damage than the rich, who usually live in more pristine neighbourhoods. When mineral resources become depleted, late-comers to the industrialization process lose the benefits of low-cost supplies. Globally, wealthier nations are better placed financially and technologically to cope with the effects of possible climatic change.

Hence, our inability to promote the common interest in sustainable development is often a product of the relative neglect of economic and social justice within and amongst nations.

Strategic Imperatives

The world must quickly design strategies that will allow nations to move from their present, often destructive, processes of growth and development onto sustainable development paths. This will require policy changes in all countries, with respect both to their own development and to their impacts on other nations' development possibilities.

Critical objectives for environment and development policies that follow from the concept of sustainable development include:

- reviving growth;
- changing the quality of growth;
- meeting essential needs for jobs, food, energy, water, and sanitation;
- ensuring a sustainable level of population;
- conserving and enhancing the resource base:
- reorienting technology and managing risk; and
- merging environment and economics in decision making.

Reviving Growth

As indicated earlier, development that is sustainable has to address the problem of the large number of people who live in absolute poverty - that is, who are unable to satisfy even the most basic of their needs. Poverty reduces people's capacity to use resources in a sustainable manner; it intensifies pressure on the environment. Most such absolute poverty is in developing countries; in many, it has been aggravated by the economic stagnation of the 1980s. A necessary but not a sufficient condition for the elimination of absolute poverty is a relatively rapid rise in per capita incomes in the Third World. It is therefore essential that the stagnant or declining growth trends of this decade be reversed.

While attainable growth rates will vary, a certain minimum is needed to have any impact on absolute poverty. It seems unlikely that, taking developing countries as a whole, these objectives can be accomplished with per capita income growth of under 3 per cent. Given current population growth rates, this would require overall national income growth of around 5 per cent a year in the developing economies of Asia, 5.5 per cent in Latin America, and 6 per cent in Africa and West Asia.

Are these orders of magnitude attainable? The record in South and East Asia over the past quarter-century and especially over the last five years suggests that 5 per cent annual growth can be attained in most countries, including the two largest, India and China. In Latin America, average growth rates on the order of 5 per cent were achieved during the 1960s and 1970s, but fell well below that in the first half of this decade, mainly because of the debt crisis.

1. A revival of Latin American growth depends on the resolution of this crisis. In Africa, growth rates during the 1960s and 1970s were around 4-4.5 per cent, which at current rates of population growth would mean per capita income growth of only a little over 1 per cent.

2. Moreover, during the 1980s, growth nearly halted and in two-thirds of the countries per capita income declined.

3. Attaining a minimum level of growth in Africa requires the correction of short-term imbalances, and also the removal of deep-rooted constraints on the growth process.

Growth must be revived in developing countries because that is where the links between economic growth, the alleviation of poverty, and environmental conditions operate most directly. Yet developing countries are part of an interdependent world economy; their prospects also depend on the levels and patterns of growth in industrialized nations. The medium-term prospects for industrial countries are for growth of 3-4 per cent, the minimum that international financial institutions consider necessary if these countries are going to play a part in expanding the world economy. Such growth rates could be environmentally sustainable if industrialized nations can continue the recent shifts in the content of their growth towards less material- and energy-intensive activities and the improvement of their efficiency in using materials and energy.

Growth, Redistribution, and Poverty

1. The poverty line is that level of income below which an individual or household cannot afford on a regular basis the necessities of life. The percentage of the population below that line will depend on per capita national income and the manner in which it is distributed. How quickly can a developing country expect to eliminate absolute poverty? The answer will vary from country to country, but much can be learned from a typical case.

2. Consider a nation in which half the population lives below the poverty line and where the distribution of household incomes is as follows: the top one-fifth of households have 50 per cent of total income, the next fifth have 20 per cent, the next fifth have 14 per cent, the next fifth have 9 per cent, and the bottom fifth have just 7 per cent. This is a fair representation of the situation in many low-income developing countries.

3. In this case, if the income distribution remains unchanged, per capita national income would have to double before the poverty ratio drops from 50 to 10 per cent. If income is redistributed in favour of the poor, this reduction can occur sooner. Consider the case in which 25 per cent of the incremental income of the richest one-fifth of the population is redistributed equally to the others. The assumptions here about redistribution reflect three judgements. First, in most situations redistributive policies can only operate on increases in income. Second, in low-income developing countries the surplus that can be skimmed off for redistribution is available only from the wealthier groups. Third, redistributive policies cannot be so precisely targeted that they deliver benefits only to those who are below the poverty line, so some of the benefits will accrue to those who are just a little above it.

4. The number of years required to bring the poverty ratio down from 50 to 10 per cent ranges from:

    o   18-24 years if per capita income grows at 3 per cent,

    o   26-36 years if it grown at 2 per cent, and

    o   51-70 years if it grows only at 1 per cent.

In each case, the shorter time is associated with the redistribution of 25 per cent of the incremental income of the richest fifth of the population and the longer period with no redistribution.

5.   So with per capita national income growing only at 1 per cent a year, the time required to eliminate absolute poverty would stretch well into the next century. If, however, the aim is to ensure that the world is well on its way towards sustainable development by the beginning of the next century, it is necessary to aim at a minimum of 3 per cent per capita national income growth and to pursue vigorous redistributive policies.

As industrialized nations use less materials and energy, however, they will provide smaller markets for commodities and minerals from the developing nations. Yet if developing nations focus their efforts upon eliminating poverty and satisfying essential human needs, then domestic demand will increase for both agricultural products and manufactured goods and some services. Hence the very logic of sustainable development implies an internal stimulus to Third World growth.

Nonetheless, in large numbers of developing countries markets are very small; and for all developing countries high export growth, especially of non-traditional items, will also be necessary to finance imports, demand for which will be generated by rapid development. Thus a reorientation of international economic relations will be necessary for sustainable development.

Changing the Quality of Growth

Sustainable development involves more than growth. It requires a change in the content of growth, to make it less Material- and energy-intensive and more equitable in its impact. These changes are required in all countries as part of a package of measures to maintain the stock of ecological capital, to improve the distribution of income, and to reduce the degree of vulnerability to economic crises.

The process of economic development must be more soundly based upon the realities of the stock of capital that sustains it. This is rarely done in either developed or developing countries. For example, income from forestry operations is conventionally measured in terms of the value of timber and other products extracted, minus the costs of extraction. The costs of regenerating the forest are not taken into account, unless money is actually spent on such work. Thus figuring profits from logging rarely takes full account of the losses in future revenue incurred through degradation of the forest. Similar incomplete accounting occurs in the exploitation of other natural resources, especially in the case of resources that are not capitalized in enterprise or national accounts: air, water, and soil. In all countries, rich or poor, economic development must take full account in its measurements of growth of the improvement or deterioration in the stock of natural resources.

Income distribution is one aspect of the quality of growth, as described in the preceding section, and rapid growth combined with deteriorating income distribution may be worse than slower growth combined with redistribution in favour of the poor. For instance, in many developing countries the introduction of large-scale commercial agriculture may produce revenue rapidly, but may also dispossess a large number of small farmers and make income distribution more inequitable. In the long run, such a path may not be sustainable; it impoverishes many people and can increase pressures on the natural resource base through overcommercialized agriculture and through the marginalization of subsistence farmers. Relying more on smallholder cultivation may be slower at first, but more easily sustained over the long term.

Economic development is unsustainable if it increases vulnerability to crises. A drought may force farmers to slaughter animals needed for sustaining production in future years. A drop in prices may cause farmers or other producers to overexploit natural resources to maintain incomes. But vulnerability can be reduced by using technologies that lower production risks, by choosing institutional options that reduce market fluctuations, and by building up reserves, especially of food and foreign exchange. A development path that combines growth with reduced vulnerability is more sustainable than one that does not.

Yet it is not enough to broaden the range of economic variables taken into account. Sustainability requires views of human needs and well-being that incorporate such non-economic variables as education and health enjoyed for their own sake, clean air and water, and the protection of natural beauty. It must also work to remove disabilities from disadvantaged groups, many of whom live in ecologically vulnerable areas, such as many tribal groups in forests, desert nomads, groups in remote hill areas, and indigenous peoples of the Americas and Australasia.

Changing the quality of growth requires changing our approach to development efforts to take account of all of their effects. For instance, a hydropower project should not be seen merely as a way of producing more electricity; its effects upon the local environment and the livelihood of the local community must be included in any balance sheets. Thus the abandonment of a hydro project because it will disturb a rare ecological system could be a measure of progress, not a setback to development.

Nevertheless, in some cases, sustainability considerations will involve a rejection of activities that are financially attractive in the short run.

Economic and social development can and should be mutually reinforcing. Money spent on education and health can raise human productivity. Economic developments can accelerate social development by providing opportunities for underprivileged groups or by spreading education more rapidly.

Meeting Essential Human Needs

The satisfaction of human needs and aspirations is so obviously an objective of productive activity that it may appear redundant to assert its central role in the concept of sustainable development. All too often poverty is such that people cannot satisfy their needs for survival and well-being even if goods and services are available. At the same time, the demands of those not in poverty may have major environmental consequences.

The principal development challenge is to meet the needs and aspirations of an expanding developing world population. The most basic of all needs is for a livelihood: that is, employment. Between 1985 and 2000 the labour force in developing countries will increase by nearly 800 million, and new livelihood opportunities will have to be generated for 60 million persons every year. The pace and pattern of economic development have to generate sustainable work opportunities on this scale and at a level of productivity that would enable poor households to meet minimum consumption standards.

More food is required not merely to feed more people but to attack under-nourishment. For the developing world to eat, person for person, as well as the industrial world by the year 2000, annual increases of 5.0 per cent in calories and 5.8 per cent in proteins are needed in Africa; of 3.4 and 4.0 per cent, respectively, in Latin America; and of 3.5 and 4.5 per cent in Asia. Foodgrains and starchy roots are the primary sources of calories, while proteins are obtained primarily from products like milk, meat, fish, pulses, and oil-seeds.

Though the focus at present is necessarily on staple foods, the projections given above also highlight the need for a high rate of growth of protein availability. In Africa, the task is particularly challenging given the recent declining per capita food production and the current constraints on growth. In Asia and Latin America, the required growth rates in calorie and protein consumption seem to be more readily attainable. But increased food production should not be based on ecologically unsound production policies and compromise long-term prospects for food security.

Energy is another essential human need, one that cannot be universally met unless energy consumption patterns change. The most urgent problem is the requirements of poor Third World households, which depend mainly on fuelwood. By the turn of the century, 3 billion people may live in areas where wood is cut faster than it grows or where fuelwood is extremely scarce.

Corrective action would both reduce the drudgery of collecting wood over long distances and preserve the ecological base. The minimum requirements for cooking fuel in most developing countries appear to be on the order of 250 kilogrammes of coal equivalent per capita per year. This is a fraction of the household energy consumption in industrial countries.

The linked basic needs of housing, water supply, sanitation, and health care are also environmentally important. Deficiencies in these areas are often visible manifestations of environmental stress. In the Third World, the failure to meet these key needs is one of the major causes of many communicable diseases such as malaria, gastro-intestinal infestations, cholera, and typhoid. Population growth and the drift into cities threaten to make these problems worse. Planners must find ways of relying more on supporting community initiatives and self-help efforts and on effectively using low-cost technologies.

ENSURING A SUSTAINABLE LEVEL OF POPULATION

The sustainability of development is intimately linked to the dynamics of population growth. The issue, however, is not simply one of global population size. A child born in a country where levels of material and energy use are high places a greater burden on the Earth's resources than a child born in a poorer country. A similar argument applies within countries. Nonetheless, sustainable development can be pursued more easily when population size is stabilized at a level consistent with the productive capacity of the ecosystem.

In industrial countries, the overall rate of population growth is under 1 per cent, and several countries have reached or are approaching zero population growth. The total population of the industrialized world could increase from its current 1.2 billion to about 1.4 billion in the year 2025.

The greater part of global population increase will take place in developing countries, where the 1985 population of 3.7 billion may increase to 6.8 billion by 2025. The Third World does not have the option of migration to 'new' lands, and the time available for adjustment is much less than industrial countries had. Hence the challenge now is to quickly lower population growth rates, especially in regions such as Africa, where these rates are increasing.

Birth rates declined in industrial countries largely because of economic and social development. Rising levels of income and urbanization and the changing role of women all played important roles. Similar processes are now at work in developing countries. These should be recognized and encouraged. Population policies should be integrated with other economic and social development programmes female education, health care, and the expansion of the livelihood base of the poor. But time is short, and developing countries will also have to promote direct measures to reduce fertility, to avoid going radically beyond the productive potential to support their populations. In fact, increased access to family planning services is itself a form of social development that allows couples, and women in particular, the right to self-determination.

Population growth in developing countries will remain unevenly distributed between rural and urban areas. UK projections suggest that by the first decade of the next century, the absolute size of rural populations in most developing countries will start declining. Nearly 90 per cent of the increase in the developing world will take place in urban areas, the population of which in expected to rise from 1.15 billion in 1985 to 3.25 million in 2025. The increase will be particularly marked in Africa and, to a lesser extent, in Asia.

Developing-country cities are growing much faster than the capacity of authorities to cope. Shortages of housing, water, sanitation, and mass transit are widespread. A growing proportion of city-dwellers live in slums and shanty towns, many of them exposed to air and water pollution and to industrial and natural hazards. Further deterioration is likely, given that most urban growth will take place in the largest cities. Thus more manageable cities may be the principal gain from slower rates of population growth.

Urbanization is itself part of the development process. The challenge is to manage the process so as to avoid a severe deterioration in the quality of life. Thus the development of smaller urban centres needs to be encouraged to reduce pressures in large cities. Solving the impending urban crisis will require the promotion of self-help housing and urban services by and for the poor, and a more positive approach to the role of the informal sector, supported by sufficient funds for water supply, sanitation, and other services.

Conserving and Enhancing the Resource Base

If needs are to be met on a sustainable basis the Earth's natural resource base must be conserved and enhanced. Major changes in policies will be needed to cope with the industrial world's current high levels of consumption, the increases in consumption needed to meet minimum standards in developing countries, and expected population growth. However, the case for the conservation of nature should not rest only with development goals. It is part of our moral obligation to other living beings and future generations.

Pressure on resources increases when people lack alternatives. Development policies must widen people's options for earning a sustainable livelihood, particularly for resource-poor households and in areas under ecological stress. In a hilly area, for instance, economic self-interest and ecology can be combined by helping farmers shift from grain to tree crops by providing them with advice, equipment, and marketing assistance. Programmes to protect the incomes of farmers, fishermen, and foresters against short-term price declines may decrease their need to overexploit resources.

The conservation of agricultural resources is an urgent task because in many parts of the world cultivation has already been extended to marginal lands, and fishery and forestry resources have been overexploited. These resources must be conserved and enhanced to meet the needs of growing populations. Land use in agriculture and forestry must be based on a scientific assessment of land capacity, and the annual depletion of topsoil, fish stock, or forest resources must not exceed the rate of regeneration.

The pressures on agricultural land from crop and livestock production can be partly relieved by increasing productivity. But short-sighted, short-term improvements in productivity can create different forms of ecological stress, such as the loss of genetic diversity in standing crops, salinization and alkalization of irrigated lands, nitrate pollution of ground-water, and pesticide residues in food. Ecologically more benign alternatives are available. Future increases in productivity, in both developed and developing countries, should be based on the better controlled application of water and agrochemicals, as well as on more extensive use of organic manures and non-chemical means of pest control. These alternatives can be promoted only by an agricultural policy based on ecological realities.

In the case of fisheries and tropical forestry, we rely largely on the exploitation of the naturally available stocks. The sustainable yield from these stocks may well fall short of demand. Hence it will be necessary to turn to methods that produce more fish, fuelwood, and forest products under controlled conditions. Substitutes for fuelwood can be promoted.

The ultimate limits to global development are perhaps determined by the availability of energy resources and by the biosphere's capacity to absorb the by-products of energy use. These energy limits may be approached far sooner than the limits imposed by other material resources. First, there are the supply problems: the depletion of oil reserves, the high cost and environmental impact of coal mining, and the hazards of nuclear technology. Second, there are emission problems, most notably acid pollution and carbon dioxide build up leading to global warming.

Some of these problems can be met by increased use of renewable energy sources. But the exploitation of renewable sources such as fuelwood and hydro-power also entails ecological problems. Hence sustainability requires a clear focus on conserving and efficiently using energy.

Industrialized countries must recognize that their energy consumption is polluting the biosphere and eating into scarce fossil fuel supplies. Recent improvements in energy efficiency and a shift towards less energy-intensive sectors have helped limit consumption. But the process must be accelerated to reduce per capita consumption and encourage a shift to non polluting sources and technologies. The simple duplication in the developing world of industrial countries' energy use patterns is neither feasible nor desirable. Changing these patterns for the better will call for new policies in urban development, industry location, housing design, transportation systems, and the choice of agricultural and industrial technologies.

Non-fuel mineral resources appear to pose fewer supply problems. Studies done before 1960 that assumed an exponentially growing demand did not envisage a problem until well into the next century. Since then, world consumption of most metals has remained nearly constant, which suggests that the exhaustion of non-fuel minerals is even more distant. The history of technological developments also suggests that industry can adjust to scarcity through greater efficiency in use, recycling, and substitution. More immediate needs include modifying the pattern of world trade in minerals to allow exporters a higher share in the value added from mineral use, and improving the access of developing countries to mineral supplies, as their demands increase.

The prevention and reduction of air and water pollution will remain a critical task of resource conservation. Air and water quality come under pressure from such activities as fertilizer and pesticide use, urban sewage, fossil fuel burning, the use of certain chemicals, and various other industrial activities. Each of these is expected to increase the pollution load on the biosphere substantially, particularly in developing countries. Cleaning up after the event is an expensive solution. Hence all countries need to anticipate and prevent these pollution problems, by, for instance, enforcing emission standards that reflect likely long-term effects, promoting low-waste technologies, and anticipating the impact of new products, technologies, and wastes.

Reorienting Technology and Managing Risk

The fulfilment of all these tasks will require the reorientation of technology the key link between humans and nature. First, the capacity for technological innovation needs to be greatly enhanced in developing countries so that they can respond more effectively to the challenges of sustainable development. Second, the orientation of technology development must be changed to pay greater attention to environmental factors.

The technologies of industrial countries are not always suited or easily adaptable to the socio-economic and environmental conditions of developing countries. To compound the problem, the bulk of world research and development addresses few of the pressing issues facing these countries, such as arid-land agriculture or the control of tropical diseases. Not enough is being done to adapt recent innovations in materials technology, energy conservation, information technology, and biotechnology to the needs of developing countries. These gaps must be covered by enhancing research, design, development, and extension capabilities in the Third World.

In all countries, the processes of generating alternative technologies, upgrading traditional ones, and selecting and adapting imported technologies should be informed by environmental resource concerns. Most technological research by commercial organizations is devoted to product and process innovations that have market value. Technologies are needed that produce 'social goods', such as improved air quality or increased product life, or that resolve problems normally outside the cost calculus of individual enterprises, such as the external costs of pollution or waste disposal.

The role of public policy is to ensure, through incentives and disincentives, that commercial organizations find it worthwhile to take fuller account of environmental factors in the technologies they develop. Publicly funded research institutions also need such direction, and the objectives of sustainable development and environmental protection must be built into the mandates of the institutions that work in environmentally sensitive areas.

The development of environmentally appropriate technologies is closely related to questions of risk management. Such systems as nuclear reactors, electric and other utility distribution networks, communication systems, and mass transportation are vulnerable if stressed beyond a certain point. The fact that they are connected through networks tends to make them immune to small disturbances but more vulnerable to unexpected disruptions that exceed a finite threshold. Applying sophisticated analyses of vulnerabilities and past failures to technology design, manufacturing standards, and contingency plans in operations can make the consequences of a failure or accident much less catastrophic.

The best vulnerability and risk analysis has not been applied consistently across technologies or systems. A major purpose of large system design should be to make the consequences of failure or sabotage less serious. There is thus a need for new techniques and technologies - as well as legal and institutional mechanisms - for safety design and control, accident prevention, contingency planning, damage mitigation, and provision of relief.

Environmental risks arising from technological and developmental decisions impinge on individuals and areas that have little or no influence on those decisions. Their interests must be taken into account. National and international institutional mechanisms are needed to assess potential impacts of new technologies before they are widely used, in order to ensure that their production, use, and disposal do not overstress environmental resources. Similar arrangements are required for major interventions in natural systems, such as river diversion or forest clearance. In addition, liability for damages from unintended consequences must be strengthened and enforced.

Merging Environment and Economics in Decision Making

The common theme throughout this strategy for sustainable development is the need to integrate economic and ecological considerations in decision making. They are, after all, integrated in the workings of the real world. This will require a change in attitudes and objectives and in institutional arrangements at every level.

Economic and ecological concerns are not necessarily in opposition. For example, policies that conserve the quality of agricultural land and protect forests improve the long-term prospects for agricultural development. An increase in the efficiency of energy and material use serves ecological purposes but can also reduce costs. But the compatibility of environmental and economic objectives is often lost in the pursuit of individual or group gains, with little regard for the impacts on others, with a blind faith in science's ability to find solutions, and in ignorance of the distant consequences of today's decisions. Institutional rigidities add to this myopia.

One important rigidity is the tendency to deal with one industry or sector in isolation, failing to recognize the importance of intersectoral linkages. Modern agriculture uses substantial amounts of commercially produced energy and large quantities of industrial products. At the same time, the more traditional connection - in which agriculture is a source of raw materials for industry - is being diluted by the widening use of synthetics. The energy-industry connection is also changing, with a strong tendency towards a decline in the energy intensity of industrial production in industrial countries. In the Third World, however, the gradual shift of the industrial base towards the basic material producing sectors is leading to an increase in the energy intensity of industrial production.

These inter sectoral connections create patterns of economic and ecological interdependence rarely reflected in the ways in which policy is made. Sectoral organizations tend to pursue sectoral objectives and to treat their impacts on other sectors as side effects, taken into account only if compelled to do so. Hence impacts on forests rarely worry those involved in guiding public policy or business activities in the fields of energy, industrial development, crop husbandry, or foreign trade. Many of the environment and development problems that confront us have their roots in this sectoral fragmentation of responsibility. Sustainable development requires that such fragmentation be overcome.

Sustainability requires the enforcement of wider responsibilities for the impacts of decisions. This requires changes in the legal and institutional frameworks that will enforce the common interest. Some necessary changes in the legal framework start from the proposition that an environment adequate for health and well-being is essential for all human beings including future generations. Such a view places the right to use public and private resources in its proper social context and provides a goal for more specific measures.

The law alone cannot enforce the common interest. It principally needs community knowledge and support, which entails greater public participation in the decisions that affect the environment. This is best secured by decentralizing the management of resources upon which local communities defend, and giving these communities an effective say over the use f these resources. It will also require promoting citizens' initiatives, empowering people's organizations, and strengthening local democracy.

Some large-scale projects, however, require participation on a different basis. Public inquiries and hearings on the development and environment impacts can help greatly in drawing attention to different points of view. Free access to relevant information and the availability of alternative sources of technical expertise can provide an informed basis for public discussion. When the environmental impact of a proposed project is particularly high, public scrutiny of the case should be mandatory and, wherever feasible, the decision should be subject to prior public approval, perhaps by referendum.

Changes are also required in the attitudes and procedures of both public and private-sector enterprises. Moreover, environmental regulation must move beyond the usual menu of safety regulations, zoning laws, and pollution control enactments; environmental objectives must be built into taxation, prior approval procedures for investment and technology choice, foreign trade incentives, and all components of development policy.

The integration of economic and ecological factors into the law and into decision making systems within countries has to be matched at the international level. The growth in fuel and material use dictates that direct physical linkages between ecosystems of different countries will increase. Economic interactions through trade, finance, investment, and travel will also grow and heighten economic and ecological interdependence. Hence in the future, even more so than now, sustainable development requires the unification of economics and ecology in international relations.

Conclusion

In its broadest sense, the strategy for sustainable development aims to promote harmony among human brings and between humanity and nature. In the specific context of the development and environment crises of the 1980s, which current national and international political and economic institutions have not and perhaps cannot overcome, the pursuit of sustainable development requires:

- a political system that secures effective citizen participation in decision making.
- an economic system that is able to generate surpluses and technical knowledge on a self-reliant and sustained basis
- a social system that provides for solutions for the tensions arising from disharmonious development.
- a production system that respects the obligation to preserve the ecological base for development,
- a technological system that can search continuously for new solutions,
- an international system that fosters sustainable patterns of trade and finance, and
- an administrative system that is flexible and has the capacity for self-correction.

These requirements are more in the nature of goals that should underlie national and international action on development. What matters is the sincerity with which these goals are pursued and the effectiveness with which departures from them are corrected

PROCESS DESIGN AND SUSTAINABILITY IN THE PRODUCTION OF BIOETHANOL FROM LIGNOCELLULOSIC MATERIALS

Bioethanol has been used as a biofuel in large scale since the implementation of the Brazilian alcohol program. However, its production from sugar cane in Brazil and corn grains in the United States has been subject of debate regarding its effects on food prices and land use change (LUC). Production of bioethanol from lignocellulosic materials (LC), also called second-generation bioethanol (bioethanol 2G), is proposed as an alternative without such adverse effects. Bioethanol 2G has been an active field of research in the last decades and nowadays there are several companies scaling up their process in what seems to be the beginning of a learning process for having commercial scale production processes no later than 2020.

Compared to the accessibility of sucrose in sugar cane and starch in grains, cellulose and hemicellulose in LC are hardly available for saccharification and fermentation due to the presence of lignin, and a stage of "pre-treatment" is required to facilitate its conversion to fermentable sugars. A number of methods have been described, proposed and tested as a pre-treatment stage where the biomass structure is significantly altered to improve the accessibility of cellulolytic enzymes to the cellulose matrix. These enzymes convert cellulose into a mixture of cellobiose and glucose that fermenting microorganisms (yeasts and bacteria) will transform into bioethanol. Distillation is used for bioethanol recovery from the fermented, which is concentrated up to the azeotropic point (ca. 95% w/w bioethanol) and then dehydrated. An overview of the process structure and the many alternatives developed. The number of feasible process alternatives makes the decision of the most appropriate sequence of operations a difficult problem, since the stages are interconnected, and changes in process conditions change the behaviour of the entire process and the economic and environmental indicators as well.

Many authors have emphasized the importance of Life Cycle Assessment (LCA) as a tool for measuring sustainability indicators in the production of bioethanol 2G, comparing its environmental impacts with those of corn-based and sugar based-bioethanol, also called first generation bioethanol, and also with gasoline. The LCA is a process of compilation and evaluation of the inputs, such as feedstock and energy, the outputs, like sub-products and pollutant emissions, and the potential environmental impacts of a production system throughout its life cycle. The results obtained allow the comparison of processes, products and services currently available or under development. LCA can identify potential environmental impacts at an early stage of a process design, and provides the opportunity for making decisions to improve its sustainability before process implementation. The environmental performance of proposed processes for the production of lignocellulosic bioethanol are still under discussion due to the state of development of the technology and the differences in focus and assumptions of the LCA practitioners.

Process Design

Process synthesis and design are engineering activities oriented to the identification of novel process configurations or flow sheets able to produce the target product. These activities are part of a much broader field known as Process System Engineering, characterized for analyzing the productive systems as a whole, studying how the components and their interactions contribute to the overall 'behavior' of the system. Quoting Grossmann and Westerberg (2000): "Process Systems Engineering is concerned with the improvement of decision-making processes for the creation and operation of the chemical supply chain. It deals with the discovery, design, manufacture, and distribution of chemical products in the context of many conflicting goals". From the above definition, it is clear that process design deals not only with the selection of equipment and operations, but also with social, economic and environmental issues that the production process can affect. Conceptual process synthesis and design emerged from the concept of unit operation process, allowing decomposition of a complex interconnected network of equipment and activities into well-defined blocks with specific objectives and common features. This concept was introduced by A.D. Little in 1915. He stated that any chemical process can be represented as a series of "unit operations" such as solid separation, reaction or heat exchange. This set the cornerstone for the heuristic process system synthesis approach, which builds upon the knowledge and expertise of the designer to achieve suboptimal designs, limiting time and resources consumed when the synthesis problem is too big or too complex. A heuristic method proposed by Rudd (1968) is based on decomposition, whereby a design problem for which no previous technology existed is broken down into a sequence of sub-design problems until the level of available technology is reached.

The optimization based approach for process synthesis relies on the formulation of a superstructure of process flow sheets among which the optimal process is selected *via* mathematical programming. This approach often leads to a class of problem known as *Mixed Integer Non-Linear Programming problem* (MINLP) composed of integer variables (binary variables in most cases), representing the selection of one of the candidate processes in the superstructure, and continuous variables to account for compositions, temperatures and other decision variables. Since many of the unit operations commonly used in chemical engineering are described by non-convex equations, the synthesis problem involves the solution of a non-convex MINLP hampering the detection of the global optimum. Despite difficulties in MINLP methods, this and other optimization-based process design methodologies are expected to contribute to the development of biorefineries and bioethanol production processes.

Knowledge based and optimization based process synthesis have been applied for the design of bioethanol production processes. Since, formulation of well posed optimization problems leading to the synthesis of bioethanol processes requires mathematical models that adequately represent the behaviour of the unit operations under analysis, this method has not been extensively used compared to the knowledge-based process design; however, its application has been largely dominated by heuristic methods, with the exception of a recently published work in knowledge-based hierarchical decomposition.

To help a more structured revision of the relevant advances in this area, a generic bioethanol production process will be divided into stages and the process alternatives within it will be analyzed. Each stage is composed of several unit operations, and material and energy exchange occur between stages not only in an acyclic fashion, but also with possible recycling and energy integration between them. The process stages are defined by their objectives:

(i)   pre-treatment, aimed to improve enzymatic accessibility to cellulose while recovering monosaccharides from hemicelluloses;

(ii) saccharification, aimed to convert cellulose and hemicellulose to glucose and other monosaccharides;

(iii) fermentation, for bioethanol production from the released monosaccharides;

(iv) bioethanol recovery, aimed to get a biofuel able to be mixed or readily used, and

(v)   waste water treatment, aimed to comply with regulations and to recover as much energy as possible. Due to excellent reviews in enzymatic hydrolysis and fermentation, only new and integrated strategies for these stages will be reviewed in this work.

Waste water treatment is a surprisingly overlooked subject in the literature., despite being a key issue in bioethanol production since it contributes with nearly 20% of the total installed equipments cost and represents up to 15% of the minimum bioethanol selling price (MESP), this is the product selling price required to pay for the operational and capital expenditures having no further income. Regarding experimental studies, the performance of anaerobic treatment on stillage produced by conventional and second generation bioethanol production processes were compared and no relevant differences were found on biogas yield nor in COD removal, but lower biogas productivity was found for waste water treatment of second generation stillage, Concerning process design and evaluation, several possible configurations of biological waste water treatment suitable to replace the energy consuming evaporation of stillage in a simulated second generation bioethanol facility were compared in a techno-economical study. Finally, Harris Group and NREL designed a waste water treatment process based in anaerobic digestion, activated sludge systems and reverse osmosis that forms the basis of the NREL cellulosic bioethanol model.

Knowledge-based Process Design Pre-treatment

Pre-treatment processes have been in the core of bioethanol production research for the last decades. As an effective pre-treatment for hemicellulose removal, dilute sulfuric acid hydrolysis has been extensively used. The most common acid catalyst is sulphuric acid; a comparison between phosphoric and sulphuric acid concluded that hemicellulose removal and subsequent enzymatic hydrolysis after dilute sulphuric acid pre-treatment outperforms phosphoric acid. A key variable in this pre-treatment is acid consumption, since after pre-treatment the remaining acid has to be neutralized in order to proceed with saccharification and fermentation. Neutralization is often made with lime, producing insoluble sulphate salts that precipitate and have to be separated and disposed. Efforts to reduce sulphuric acid load lead to identify two key process variables in pre-treatment: acid concentration in reaction liquor and acid dosage per unit mass of LC (corn stover). The authors found that, when acid is dosed on reaction liquor basis, glucose yield after enzymatic hydrolysis is highly dependent on the solid loading. However, when using constant acid dosage based on the mass of corn stover (g acid/ g dry LC), saccharification yield is relatively constant between 15 and 30% w/w solids. Hence, it was found that acid loading based on total solids (g acid/g dry biomass) rather than the acid concentration (g acid/g pre-treatment liquid) governs the pre-treatment efficiency. Additionally, using highly active commercial enzyme preparations (CellicCTEC2) allows maintaining high conversions even when acid dosage was reduced by 50% and much milder pre-treatment conditions were used: 170°C for 15-20 min compared with 190°C for 1-5 min. Auto-hydrolysis pre-treatment, also known as hot liquid water or hydrothermal pre-treatment, has several conceptual advantages over dilute acid hydrolysis since no corrosive resistant alloys are needed and solid waste (gypsum) generation is minimized. The process involves heating an aqueous suspension of LC, typically at temperatures ranging from 200 to 230°C. Operated under suitable conditions, hemicellulose is converted into soluble saccharides, whereas the treated solids show an increased susceptibility to cellulolytic enzymes. Recognizing the tradeoffs between pre-treatment and saccharification and fermentation processes, Romaní et al. (2012) use a response surface methodology (RSM) with the following independent variables: pre-treatment severity, solid fraction in simultaneous saccharification and fermentation and the enzyme to substrate ratio. As a result of the RSM optimization, 91% cellulose to bioethanol yield was achieved with 20% w/w solid loading and 16 filter paper units (FPU)/g glucan. A study on rice straw auto-hydrolysis used a semi-continuous fixed bed reactor achieving near complete saccharification. The authors claimed several advantages compared to other reactor configurations: high solid-to-water-ratios (1:10), prevention of degradation product formation and energy savings since no biomass comminuting is necessary.

Among the pre-treatments aimed to remove lignin as a way to enhance enzyme accessibility to recalcitrant substrates, organic solvent based processes have been extensively studied. The process involves contacting the lignocellulosic material with a mixture of water, mineral acids at low concentration –not always used- and an organic solvent at high temperatures. The use of solvent based pre-treatments for improving enzymatic hydrolysis of LC has been recently reviewed. As a strategy for improving hemicellulose recovery, Romaní et al. 2011 developed a pretreatment process based on sequential stages of auto-hydrolysis (to promote the solubilization of hemicellulose) and Organosolv pulping (to dissolve lignin, leaving solids enriched in cellulose). Other pre-treatments aimed to reduce the lignin content of the lignocellulosic material are Ammonia Fiber Expansion and lime treatment (AFEX). This process relies in the use of liquefied ammonia at elevated pressures instead of water. It is an alkaline method with delignification capability, where the material is treated for 10 to 60 min at a temperature close to 100°C. The process is more effective in low lignin content LC such as switchgrass or corn stover, being less efficient in softwoods and hardwoods, where harsher conditions are required. The main drawback of this method is that has to be recovered and compressed for a new pre-treatment cycle. Lime pretreatment is typically performed at 100°C, for approximately 1 hr using calcium hydroxide. However, if lignin content is above 18% w/w oxidative conditions or longer times (up to 8 weeks at 55-65°C) are necessary.

Process Integration in the Saccharification and Fermentation Stages

Saccharification and fermentation stages, as reacting systems, represent the core of any bioethanol production process. Process design in these areas has been hindered by the many tradeoffs between pre-treatment, saccharification and fermentation stages. Regarding pre-treatment, a high severity one enhances cellulose saccharification yield but often destroys part of the hemicellulose fraction and produce inhibitory compounds that impair the fermentation stage. Concerning the tradeoffs between saccharification and fermentation, cellulolytic enzymes have an optimal operation temperature close to 50°C, while most fermenting microorganisms have an optimal growing temperature between 30 and 35°C. Additionally, glucose and xylose produced during saccharification act as inhibitory compounds for cellulase and β-glucosidase, thus reducing the saccharification rate. As a way to overcome this inhibition, saccharification and fermentation could be performed in a single reactor, thus coupling the reactions of sugars release and consumption, so reducing sugars concentration and thus relieving end-product inhibition when compared to SHF. However, a compromise temperature must be used, decreasing the enzymatic saccharification rate and affecting the glucose to ethanol yield of the fermenting microorganism. In SHF and SSF, glucose is the main sugar reacted into ethanol; however pentoses like xylose and arabinose represent a relevant carbohydrate fraction that needs to be used to improve the LC to ethanol yield. Simultaneous saccharification and co-fermentation uses naturally fermenting -or engineered- microorganisms to ferment five carbon sugars.

The concept SSF was early suggested in a patent issued to Gauss *et al.* (1976), where they claim that a higher bioethanol yield and a lower end-product inhibition were observed compared with sequential hydrolysis and fermentation. A complete review on SSF can be found in Olofsson *et al.* (2008); a somewhat more general but remarkably clear presentation on the subject can be found in Sun and Cheng (2002), while in the work reported by Cardona and Sánchez (2007)the process was treated from a reaction-reaction process intensification viewpoint. In this review, new modifications to SSF and SHF processes are discussed. Also, relevant efforts to integrate enzyme production, saccharification and fermentation in the so called consolidated bioprocess are reviewed. The reaction mixture in SHF and SSF at high solid loadings shows a transition from a solid-like viscous material to a liquid suspension as reaction progresses. To promote its rapid liquefaction, a pre-hydrolysis step prior to SSF using optimal saccharification temperature has been proposed. However, lower yields using pre-hydrolysis followed by SSF has been found compared to pure SSF. A further modification involves an intermittent feeding of solids during the pre-hydrolysis stage to achieve up to 20% of solids in the reactor.

Kang et al (2012) studied a temperature shift SSF process (TS-SSF) as a way to tackle the temperature difference trade-off between the saccharification and fermentation processes. The process consists in a high temperature phase at 45°C, where AFEX pre-treated wheat straw (16% w/v solid loading) was contacted with cellulase, β-glucosidase and *K. marxianus* (CHY1612), a thermo-tolerant yeast. After a time ranging from 6 to 24 hrs, the fermenter was left to cool down to 35°C and the time course of SSF was monitored during 72 hrs.

Another approach to deal with the difference in the optimal temperatures of saccharification and fermentation is the so-called non-isothermal saccharification and fermentation (NSSF). In this configuration, saccharification and fermentation are performed simultaneously at their optimal temperatures, but in separate reactors. In a fixed bed saccharification reactor, pre-treated lignocellulose is retained and the withdrawn liquor is pumped to a heat exchanger, cooled to fermentation temperature and sent to a fermented coupled with a cell retention system. Using this configuration, a reduction of the overall enzyme requirement by 30-40% was obtained.

A batch cascade of simultaneous saccharification for xylose and glucose fermentation has been proposed as a way to improve the overall process yield by using the hemicellulose fraction. The system is composed of two sequential SSF phases operating on pentose and hexose. In the first phase, pentose conversion to ethanol is achieved using xylanase, endo-glucanase, and recombinant *E. coli* (KO11) with minimal glucose consumption. In the second phase, hexose conversion is performed using cellulase, β-glucosidase, and *S. cerevisiae* (D5A). Using four reactors and an equivalent solid loading of 6% w/v distributed in two second phase reactors; an ethanol yield of 60% was achieved at 27 g/L of ethanol. A previous work by the same authors used a straightforward method consisting in only two stages. A final bioethanol titer of 22.3 g/L was achieved, equivalent to 84% of the theoretical yield. Unfortunately, results from both studies are not comparable since only one load of solids was used to achieve an initial solid concentration of 6% w/v in <u>Li et al. (2010)</u>. CBP is believed to be an advanced solution to reduce process complexity and production cost. *Clostridium phytofermentans* was tested on AFEX-treated corn stover. At optimal conditions with 0.5% (w/w) glucan, *C. phytofermentans* hydrolyzed 76% of glucan and 88.6% of xylan in 10 days without any external enzyme dosage. Decomposition products from AFEX pre-treatment helped to increase bioethanol yield somewhat during CBP, although the explanation for this behaviour is still unknown.

An even more integrated one-step conversion method was published using the white-rot fungus *Phlebia* sp. MG-60. The fungus was able to delignify untreated oak wood chips in aerobic solid-state fermentation. After 56 days of aerobic incubation, 40.7% of the initial lignin was degraded with minimal glucan degradation. After this biological pre-treatment step, culture conditions were shifted to semi-aerobic triggering the saccharification ability of *Phlebia* sp. MG-60. After 20 days of simultaneous saccharification and fermentation, 43.9% of the theoretical yield based on pre-treated wood chips was achieved, without any external enzyme addition. It is worth mentioning that since no meaningful glucan or hemicellulose degradation occurred, this yield also corresponds to the yield based on the initial carbohydrate content of wood chips. It remains unclear if cellulase and β-glucosidase supplementation would be able to improve the reported yields. No improvement would suggest that the accessibility of enzymes or the fermentation capabilities are the process bottleneck. Regarding the fermentation capability of *Phlebia* sp. MG-60, bioethanol yields of 0.44 g/g glucose, 0.41 g/g mannose, 0.40 g/g galactose, 0.41 g/g fructose and 0.33 g/g xylose were found, but the fungus was unable to ferment arabinose. Time needed for fermentation of the above mentioned sugars, with initial concentrations ranging between 20 and 23 g/L, was 120 hrs, suggesting that fermentation capability is not the process limiting step.

Pilot and Demonstration Plants

Advances in bioethanol production have allowed the construction and operation of pilot and demonstration plants, mainly in Europe and in the United

States. The Lignol process, developed in Canada, uses the Organosolv process as pre-treatment of the LC to produce bioethanol, high quality and pure lignin and, potentially, furfural, acetic acid and xylose. Iogen Corporation, also based in Canada, uses an acid catalyzed steam explosion pre-treatment, followed by SHF without prior solid separation or washing. Enzymes are produced by Iogens's proprietary technology. After SHF, lignin and unreacted cellulose is separated for energy production. Steam explosion is also used by Abengoa Bioenergy, followed by SHF in their facilities in Babilafuente, Spain. FibreEtOH, developed in Finland by a consortium integrated by pulp and paper, enzyme production and process design industries, an ethanol distributor and VTT Technical Research Centre of Finland, aims to process a mixture of solid recovered fibers (SRF) and Kraft pulp fibers. SRF are treated by acid catalyzed steam explosion and then mixed with pulp fibers. The mixture is fed to a continuous liquefaction system (cL), with a residence time of 6 hrs, and the outlet product is sent to a batch SSF reactor. Several projects are based in dilute acid processes. POET, a US based ethanol producer in partnership with Royal DSM, uses a single stage dilute acid pre-treatment followed by saccharification and fermentation, solids are not separated prior to fermentation and to promote aggregation of the solid particles, before entering to a distillation column, a thermal treatment is used. Earlier, Inbicon (a subsidiary company of DONG Energy, the mayor energy producer in Denmark) worked with a two-step pre-treatment based in dilute acid; now it uses a one-step auto-hydrolysis (AH), followed by two stage SSF. In the first step LC is liquefied in a horizontal reactor designed for handling high concentrations of solids, and then transferred to a conventional fermenter. The product broth is stripped under vacuum and the removed ethanol distillated and rectified. The solids in the column bottoms are separated, and part of the liquid is recycled to the pre-hydrolysis reactor. The rest is combined with pre-treatment liquors, concentrated in evaporators and commercialized as molasses rich in 5-carbon sugars. The most relevant feature of the process is that it is entirely continuous.

Finally, Mascoma (US based company linked to Dartmouth College's Thayer School of Engineering) has claimed to produce bioethanol *via* CBP, using proprietary engineered microorganisms, *e.g.* modified thermophilic bacteria as *Thermoanaerobacterium saccharolyticum*.

Optimization-based Process Design

Integration of first and second generation bioethanol production process is considered to be an intermediate step towards full scale lignocellulosic bioethanol production. Recognizing that in sugarcane based industries bagasse is already used in electricity generation, a simulation and optimization approach was used to optimize plant operation. Only two decision variables were considered: the fraction of feedstock entering the boiler, and the fraction of the bagasse surplus (not burned) used for second generation bioethanol production, being the rest sold. The process was simulated using the software EMSO, and the optimization variables were handled by a particle swan optimization algorithm. The authors showed that the production of 2G bioethanol increases thermal demands in at least 25%, which narrows the decision feasible range on how much bagasse, can be diverted to production of second generation fuel. Furthermore, electric power surplus is diminished in at least 31%.

On the other hand, a flow sheet for the production of bioethanol from switch grass *via* biochemical route was proposed by optimizing a superstructure embedding a number of alternatives. Two technologies were considered in the pre-treatment stage: dilute acid and ammonia fiber explosion. Cellulose to bioethanol conversion was achieved by using SHF and a number of distillation and dehydration alternatives were considered: rectification, adsorption in corn grits, molecular sieves, and pervaporation. The problem was formulated as a mixed-integer nonlinear programming (MINLP). Since, only two pre-treatments were included, the superstructure is optimized by decomposing the MINLP for each of the pre-treatments. After this decomposition, each problem is a nonlinear programming problem, and hence can be solved more easily. The optimal flow sheet consists of dilute acid hydrolysis and molecular sieves. With this configuration, a production cost of 21 cUSD/L was calculated.

Bioethanol is one of the many products within the concept of biorefinery. Selection of the optimal products and processes for producing them is an immense combinatorial problem. Recently, a shortcut method for the synthesis and screening of integrated biorefineries was presented. The method is based on a structural representation based on chemical species and conversion operators that relate them. An optimization approach is developed to screen and determine optimum network configurations for various technology pathways using simple data. Although the method was used for synthetic gasoline process selection, it can be readily adapted for bioethanol production and, more importantly, for a mix of products including bioethanol. A similar process synthesis approach was presented in the same year involving a two-step algorithm. In the first step, a set of possible intermediates was identified given a LC composition and flow rate. In the second step, given a desired product (ethanol among others), necessary species and pathways leading to it are identified and a matching is done of intermediates identified on both steps.

Sustainability of the Production of Bioethanol

The sustainability assessment of the production of 2G bioethanol must integrate the economic, social and environmental criteria, providing relevant information in the decisions of process design and, at the same time, increasing the inherent complexity of the process synthesis. This is mainly due to the high number of interacting parameters or indicators, in which a wide range of stakeholders: investors and insurers, employees and contractors, suppliers, customers and local communities are involved.

The sustainability indicator system has been developed focused on the impact of process operations, mainly environmental impacts and to a lesser extent social impacts. In the last decade the awareness about sustainability has increased and the concept of "life cycle thinking" has been implemented. It includes the determination of environmental, social and economic impacts of a product over its entire life cycle and in the latest years has been progressively more relevant and has driven companies to adopt cleaner technologies and better management techniques.

Azapagic *et al.* (2006) developed a methodology called Process Design for Sustainability (PDfS), establishing indicators for each criteria in each stage of the process under development, and the most relevant sustainability indicators were identified to obtain a preliminary assessment for each stage that allowed making decisions on the design of the stage. Microeconomic indicators, such as net present value, cash flow analysis, return rate on capital invested and others were taken into account to assess the economical sustainability. The environmental sustainability was determined using LCA, while in the social criteria, indicators like occupational health, safety or public acceptability were used. Nevertheless, Azapagic *et al.* (2006) excluded an integrated sustainability assessment, but they related the theoretical principles of sustainability to the design practice.

Economic Criteria in Bioethanol Production

Economic criteria for process design of bioethanol production are the commonly used in traditional design approaches, which can be grouped in Microeconomic (Capital costs, Operating costs, Profitability and Investments) and Macroeconomic (Value added, Taxes paid and Other Investments) aspect. However, due to the inclusion of environmental criteria in the design stage, the economic performance can be improved by obtaining "green" incentives, *i.e.*: carbon credits.

Social Criteria in Bioethanol Production

The transformation of conventional farms into farms dedicated to the production of raw materials for bioethanol production has changed the income, employment, and quality of life of farming communities in rural areas, particularly in developing countries. In order to determine the influence of bioethanol production on the social issues, the main indicators suggested are: provision of employment, health and safety (of employees/contractors, customers and citizens), nuisance (by odour, noise and visual impact) and public acceptability of process and product. In addition, bioethanol production is indirect competition with agriculture for land use (land use change), affecting food security. In the case of bioethanol, these indicators allow designing frameworks of social criteria for bioethanol feedstock production for the international acceptance.

Environmental Criteria in Bioethanol Production

LCA has allowed quantifying environmental sustainability criteria for process assessment in bioethanol production. Current studies on LCA are focused on comparing a technology scheme for bioethanol production to fossil fuels or other alternative processes for bioethanol production.

Some of the most relevant sustainability criteria for environmental assessment are directly related to the source of raw material, energy consumption, emissions (into air, water and land) and environmental impacts (global warming, ozone depletion, acidification, human toxicity, eco-toxicity, summer smog and, eutrophication). LCA studies have shown environmental and social benefits associated with lignocellulosic bioethanol and its use in light-duty vehicles compared to gasoline and corn bioethanol. To date, most of the studies have focused on modeling the upstream agricultural processes in specific geographic locations as well as the end uses, and have considered only a limited set of feedstocks in addition to a single pre-treatment technology.

The production and use of biofuels has the potential to reduce greenhouse gas (GHG) emissions, particularly with the development of bioethanol from agricultural wastes or lignocellulosic crops such as switch grass. Alizadeh *et al.* (2005) proposed that in order to validate the lignocellulosic bioethanol production a reduction of 20% in GHG emissions must be achieved.

Technologies and Sustainability

Different technologies for converting lignocellulosic feedstocks to bioethanol are under development at laboratory, pilot, and demonstration scales as was described above. As these developing technologies may differ in their feedstock requirements as well as in process energy and chemical inputs, they exhibit a range of life cycle energy and environmental performance, aspects that have not been well acknowledged in the literature. Optimum process conditions for maximizing bioethanol production with minimum environmental impacts depend on the type of feedstock and conversion technology. There are several technologies proposed for lignocellulosic bioethanol production, and the different routes and system boundaries considered are the main reasons for the different results obtained in the LCA reports.

Most of the LCA studies on bioethanol production have been focused on assessing the farming systems with general assumptions on the process. Very few have addressed the environmental issues related to the process, probably due to the process uncertainties and the lack of a commercial scale plant. The industrial scale up of this process appears to be delayed by technological issues or by the lack of information from demonstration facilities. This lack of industrial scale information contributes to the data uncertainty within the life cycle inventories required in these studies.

Enzyme Production

According to Zhi-Fu *et al.* (2003) the cultivation of raw material and the production of the enzyme are the main contributors to eutrophication. In enzyme production main impacts come from feed production, since grains are usually needed for the fermentation broth, and the process have air and water emissions that have to be considered. In the impact category of acidification, the contribution comes from: the emissions involved in the generation of the energy required by the process, the agricultural activities for feedstock production, and the fermentation. The contribution from feedstock cultivation is mainly due to the emissions from the nutrients use in farming, field operations and phosphate manufacturing. MacLean and Spatari (2009) reported a range emissions from 1000 to 10,000 g CO_2 eq./kg enzyme depending upon the implemented technology. On the other hand, Kumar and Murthy (2012) determined the emissions from enzyme production in the range from 278.3 to 340 kg CO_2 eq./10,000 MJ of produced bioethanol.

Kadam (2000) analyzed the effect of enzymatic hydrolysis on the environmental impact of bioethanol production. He also compared the alternative of using acid hydrolysis for cellulose saccharification, showing that it was better than enzyme saccharification with respect to GHG reduction, natural resource depletion, acidification potential and eutrophication potential. Due to the higher proportion of biomass sent to the boiler for electricity production, it leads to lower SOx, NOx, and fossil CO_2 emissions. MacLean and Spatari (2009) showed that the environmental impact of chemical and enzyme inputs contributed only in 3% of the fossil energy use and in 3% of the GHG emissions for corn bioethanol, but in the case of lignocellulosic bioethanol the enzyme is responsible for 30 to 40% of the fossil energy use and 30 to 35% of GHG emissions of the total life cycle.

Pre-treatment

Kumar and Murthy (2012) compared four different pre-treatment methods for bioethanol production (dilute acid, dilute alkali, hot water, and steam explosion). Authors found that technologies used for ethanol production had major impact on total fossil energy use and GHG emissions, having the steam explosion pre-treatment the lowest contribution in these impact categories. In addition, they showed that it was possible to reduce the fossil energy demand between 57% and 112% to produce 10,000 MJ of ethanol compared to gasoline; however, about 0.35 ha of land is required to produce this energy. For all evaluated pre-treatments, co-product energy from the cogeneration system using lignin residues and biogas produced from waste water treatment produced enough steam to exceed the process requirements; however, excess electricity was obtained only in the case of steam explosion Therefore, fossil energy use during ethanol production in that case was a negative value. Fossil energy reductions higher than 100% are due to the low process energy used (thermal and electricity) and high co-product energy produced during the ethanol production process. The authors assumed that this steam and electricity will replace the energy and emissions associated with the required steam and electricity production from fossil energy sources. In this sense, a negative value of fossil energy demand was due to the energy replaced by excess electricity, which was otherwise produced from fossil fuels. On the other hand,Wang et al. (2012) evaluated different bioethanol production scenarios using waste paper from different sources without pre-treatments and two cases considering acid and lime pre-treatment from office paper and newspaper, respectively. They found that an oxidative lime pre-treatment reduced GHG emissions and the overall environmental impact when compared to bioethanol production with non pre-treated newspaper, whereas a dilute acid pre-treatment increased GHG emissions when compared to the non pre-treated office paper.

Energy cogeneration and energy ratio

Zhi-fu *et al.* (2003) determined that E-10 fuel will produce less GHG than traditional gasoline, but if fossil-based electricity is used in bioethanol production it has a negative effect on the impact category of GHG. González-García *et al.* (2012c)have shown that most of the emissions from the bioethanol conversion process are due to emissions from the energy cogeneration system. In this sense, cogeneration takes relevance in the environmental impact of the process.

Kumar and Murthy (2012) compared the effect of including cogeneration in different pre-treatments technologies and found that fossil energy used was maximum when using dilute alkali pre-treatment because of the decrease in the available lignin for cogeneration, while net fossil energy use was found negative for bioethanol production in LCA using steam explosion as pre-treatment. They showed that the energy supply by cogeneration exceeded the needs of energy for the process.

THE USE OF HYBRID PROCESSES AND NEW OPERATING MODES: THE KEY TO MANY PROBLEMS

Hybrid Kernel

A hybrid kernel is an operating system kernel architecture that attempts to combine aspects and benefits of microkernel and monolithic kernel architectures used in computeroperating systems. The traditional kernel categories are monolithic kernels and microkernels (with nanokernels and exokernels seen as more extreme versions of microkernels). The "hybrid" category is controversial, due to the similarity of hybrid kernels and ordinary monolithic kernels; the term has been dismissed by Linus Torvalds as simple marketing.

The idea behind a hybrid kernel is to have a kernel structure similar to that of a microkernel, but to implement that structure in the manner of a monolithic kernel. In contrast to a microkernel, all (or nearly all) operating system services in a hybrid kernel are still in kernel space. So there are none of the reliability benefits of having services in user space, as with a microkernel. However, just as with an ordinary monolithic kernel, there is none of the performance overhead for message passing and context switching between kernel and user mode that normally comes with a microkernel.

EXAMPLES

NT Kernel

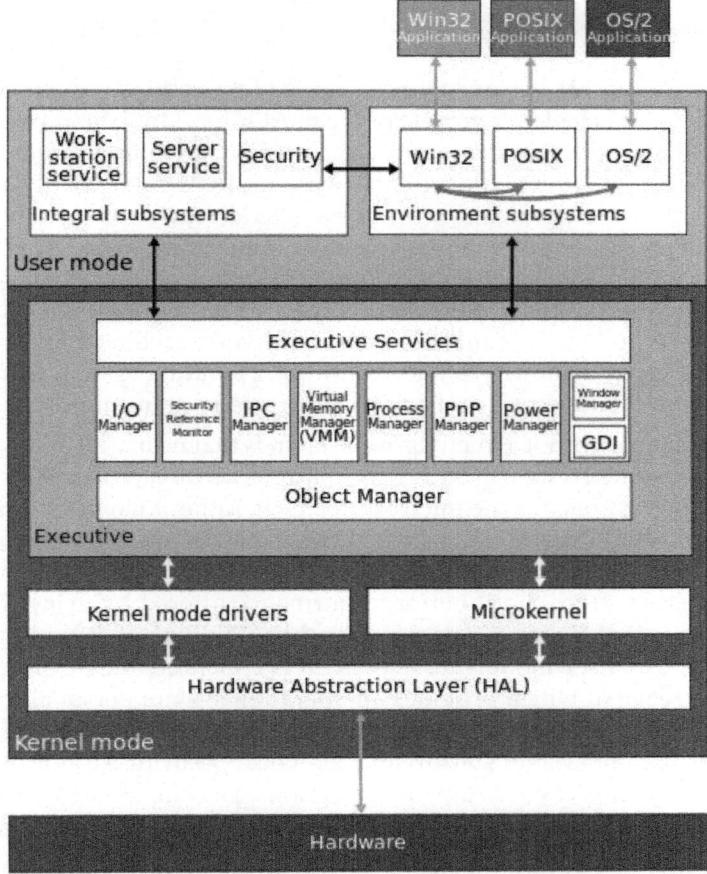

The Windows NT operating system family's architecture consists of two layers (user mode andkernel mode), with many different modules within both of these layers.

One prominent example of a hybrid kernel is the Microsoft NT kernel that powers all operating systems in the Windows NTfamily, up to and including Windows 10 and Windows Server 2012, and powers Windows Phone 8, Windows Phone 8.1, andXbox One. NT-based Windows is classified as a hybrid kernel (or a macrokernel) rather than a monolithic kernel because the emulation subsystems run in user-mode server processes, rather than in kernel mode as on a monolithic kernel, and further because of the large number of design goals which resemble design goals of Mach (in particular the separation of OS personalities from a general kernel design). Conversely, the reason NT is not a microkernel system is because most of the system components run in the same address space as the kernel, as would be the case with a monolithic design (in a traditional monolithic design, there would not be a microkernel per se, but the kernel would implement broadly similar functionality to NT's microkernel and kernel-mode subsystems).

Description

The Windows NT design included many of the same objectives as Mach, the archetypal microkernel system, one of the most important being its structure as a collection of modules that communicate *via* well-known interfaces, with a small microkernel limited to core functions such as first-level interrupt handling, thread scheduling and synchronization primitives. This allows for the possibility of using either direct procedure calls or interprocess communication (IPC) to communicate between modules, and hence for the potential location of modules in different address spaces (for example in either kernel space or server processes). Other design goals shared with Mach included support for diverse architectures, a kernel with abstractions general enough to allow multiple operating system personalities to be implemented on top of it and an object-oriented organisation.

The reason NT is not a microkernel system is that nearly all of the subsystems providing system services, including the entire Executive, run in kernel mode, in the same address space as the microkernel itself, rather than in user-mode server processes, as would be the case with a microkernel design. This is an attribute NT shares with early versions of Mach, as well as all commercial systems based on Mach, and stems from the superior performance offered by using direct procedure calls in a single memory space, rather than IPC, for communication amongst subsystems. The user-mode subsystems on NT include one or more emulation subsystems, each of which provides an operating system personality to applications, the Session Manager Subsystem (smss.exe), which starts the emulation subsystems during system startup and the Local Security Authority Subsystem Service, which enforces security on the system. The subsystems are not written to a particular OS personality, but rather to the native NT API (or Native API).

The primary operating system personality on Windows is the Windows API, which is always present. The emulation subsystem which implements the Windows personality is called the Client/Server Runtime Subsystem. On versions of NT prior to 4.0, this subsystem process also contained the window manager, graphics device interface and graphics device drivers. For performance reasons, however, in version 4.0 and later, these modules (which are often implemented in user mode even on monolithic systems, especially those designed without internal graphics support) run as a kernel-mode subsystem.

As of 2007, one other operating system personality, UNIX, is offered as an optionally installed system component on certain versions of Windows Vista and Windows Server 2003 R2. The associated subsystem process is the Subsystem for UNIX-Based Applications, which was formerly part of a Windows add-on called Windows Services for UNIX. An OS/2 subsystem was supported in older versions of Windows NT, as was a very limited POSIX subsystem. The POSIX subsystem was supplanted by the UNIX subsystem, hence the identical executable name.

Applications that run on NT are written to one of the OS personalities (usually the Windows API), and not to the native NT API for which documentation is not publicly available (with the exception of routines used in device driver development). An OS personality is implemented *via* a set of user-mode DLLs, which are mapped into application processes' address spaces as required, together with an emulation subsystem server process. Applications access system services by calling into the OS personality DLLs mapped into their address spaces, which in turn call into the NT run-time library (ntdll.dll), also mapped into the process address space. The NT run-time library services these requests by trapping into kernel mode to either call kernel-mode Executive routines or make Local Procedure Calls (LPCs) to the appropriate user-mode subsystem server processes, which in turn use the NT API to communicate with application processes, the kernel-mode subsystems and each other.

XNU Kernel

XNU is the kernel that Apple Inc. acquired and developed for use in the OS X and iOS operating systems and released as free and open source software as part of the Darwin operating system. *XNU is an acronym for X is Not Unix.*

Originally developed by NeXT for the NeXTSTEP operating system, XNU was a hybrid kernel combining version 2.5 of the Mach kernel developed at Carnegie Mellon Universitywith components from 4.3BSD and an object-oriented API for writing drivers called Driver Kit.

After Apple acquired NeXT, the Mach component was upgraded to 3.0, the BSD components were upgraded with code from the FreeBSD project and the Driver Kit was replaced with a C++ API for writing drivers called I/O Kit.

Description

Like some other modern kernels, XNU is a hybrid, containing features of both monolithic and microkernels, attempting to make the best use of both technologies, such as themessage passing capability of microkernels enabling greater modularity and larger portions of the OS to benefit from protected memory, as well as retaining the speed of monolithic kernels for certain critical tasks.

Currently, XNU runs on ARM, IA-32, and x86-64 based processors, both single processor and SMP models.

Chapter 4

THE SYNTHESIS OF PROPENE OXIDE: A SUCCESSFUL EXAMPLE OF SUSTAINABLE INDUSTRIAL CHEMISTRY

Propene oxide, which is one of the major commodity chemicals used in chemical industry, desperately requires a new process for its production, because of the disadvantages that are encountered with the currently available processes. The existing processes used for the production of propene oxide — the chlorohydrin and hydroperoxide processes — and their advantages and disadvantages. Furthermore, the new processes and catalysts under development for the propene oxide production are discussed, as well as the challenges that are still limiting the applications of some of those prospects. The most important new developments for the production of propene oxide are: the hydrogen peroxide combination process, the ethene oxide alike silver catalysts, the molten salt systems, and the gold–titania catalyst systems.

INTRODUCTION CURRENT INDUSTRIAL PROPENE OXIDE PRODUCTION

Propene oxide (PO) is one of the largest propene derivatives in production, ranking second behind polypropene; it is used primarily as a chemical intermediate. Worldwide, approximately 8 million tons are produced annually, the major suppliers are Dow (1.9 M-ton yr^{-1}), Lyondell (2.1-ton yr^{-1}) and Shell/BASF (0.9M-ton yr^{-1}); other key producers are Sumitomo, Repsol and Huntsman.

One of the fastest growing markets is China; in fact, nowadays there are more than 30 Chinese PO producers. From 1990 to 2000, PO consumption in China grew by more than 30% per year, and, with the rapid development in the fields of polyurethane, propylene glycol and surfactants, it is expected to increase even further over the next few years. By 2005, the output capacity of China reached almost 600 000 ton yr^{-1}, and by 2010, it is expected to surpass 1.1 M-tonyr-1. For instance, in 2006, 250 000 ton of PO per year were put on stream in the CNOOC-

Shell petrochemical complex in Nanhai. Sinopec Zhenhai Ref Chem and Lyondell Chemical have received approval for a joint PO/SM plant in Ningbo, China; the plant will be able to produce 270 000 ton of PO per year and start-up is expected in 2010. Outside China, one of the largest plants in the world, which use PO/SM technology, the Maasvlakte facility, The Netherlands, is a joint venture between Lyondell and Bayer and has an annual capacity of almost 300 000 ton of PO per year. The Dow facility in Stade, Germany, produces almost 630 000 tons ofPO per year with CHPO technology.

Approximately 65% of the PO produced is used for the synthesis of polyether polyols (in a reaction with polyhydric alcohols), one of the main components used in the manufacture of polyurethanes, propene glycol (20%), glycol ethers (5%) and butanediol, amongst others.

Polyurethane products and formulated systems are used in rigid foams, flexible foams, adhesives, sealants, coatings and elastomers, as well as in many other applications. Propene glycols are used in a wide variety of end-use and industrial applications, from unsaturated polyester resins, cosmetics and householddetergents, to paints and automotive brake fluids.

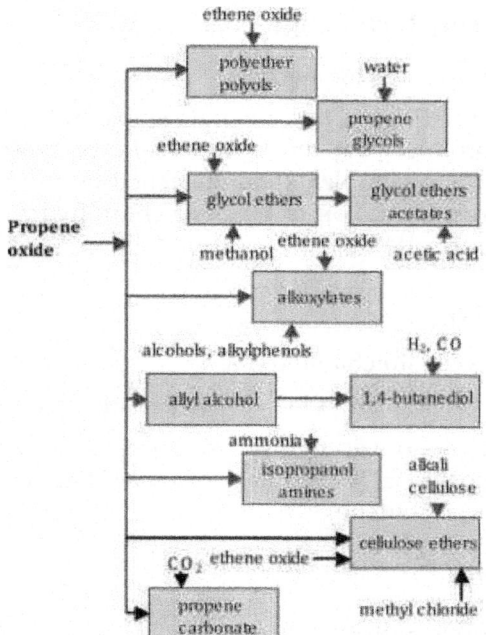

Fig. : Overview of the uses of propene oxide (PO).

Propene glycol ethers are commonly used as solvents and coupling agents in paints and in the production of coatings, inks, resins and cleaners. PO is also used to manufacture butanediol, propoxylated or specialty organic compounds, flame-retardants, modified carbohydrates (starches), synthetic lubricants, oil-field drilling chemicals and textile surfactants.

The overall annual growth rate for polyether polyols, propene glycol and propene glycol ethers is more than 3%, which corresponds to a demand of approximately 200 000-300 000 tons of PO. This demand is primarily driven by end-use applications, including unsaturated polyester resins, food, cosmetics, anti-freeze and aircraft de-icing fluids. The improved biodegradability and biocompatibility of propene glycol and polyalkylene glycols, with respect to the analogous products obtained from ethene oxide, allow them to be used in food product additives, pharmaceuticals and cosmetics. Demand for polyether polyols is expected to grow at a rate of 5-6% per year, driven by its considerable requirement in the automotive, construction and home furnishing industries. The global demand for PO is therefore forecast to grow at 4-5% per year, with the greatest growth observed in China and India. European demand is expected to grow at a slower rate of 3-3.5% per year. In the USA, demand grew at an average rate of 3.5% per year between 2001 and 2006.

Currently, the industrial production of PO mainly comes from the oxidation of propene with other chemicals. The main technologies employed in its production are outlined below.

CHPO (Chlorohydrin) Technology

The overall process stoichiometry is:

Synthesis is carried out in two separate steps. In the first reactor, propene reacts with Cl2 to produce propene chlorohydrin *via* intermediate formation of the propene chloronium complex, then quenched by water. In the epoxidation reactor, the dehydrochlorination of propene chlorohydrin occurs using a base (usually calcium hydroxide).

PO/TBA Technology

The overall process stoichiometry is:

Propene + isobutane + O_2 → PO +1 — butyl alcohol (TBA)

Synthesis is carried out in two separate steps. The oxidant for propene is t-butyl hydroperoxide (t-BuOOH). The coproduct TBA can be used as such, or it can be dehydrated to isobutene.

PO/SM Technology

The overall process stoichiometry is:

Propene + ethylbenzene + O_2 → PO + styrene (SM) + H_2O

Synthesis is carried out in three separate steps. The oxidant for propene is ethylbenzene hydroperoxide (EBHP).

While in the 1970s more than 95% of the PO world production capacity was achieved by means of CHPO technology, in the 1980s the TBA and SM processes gained increasing market shares. In 2005, CHPO technology accounted for approximately 45% of the capacity (Dow, Asahi Glass), SM technology for 35% (Lyondell, Shell/BASF, Repsol) and TBA technology for around 20% (Lyondell, Huntsman).

In the hydroperoxidation routes (PO/SM and PO/TBA), the conversion of the oxidant is greater than 95%, with a selectivity for PO greater than 95%; the main C_3 by-product is acetone. In the PO/TBA process, one disadvantage is that a portion of the t-BuOOH, once formed, immediately decomposes to TBA. In the PO/SM process, styrene may yield heavy by-products through oligomerization, causing a loss in catalyst activity.

The hydroperoxidation method was first developed by Halcon Corp and Atlantic Richfield Oil Corp (now Lyondell) in the 1970s, and was then also implemented commercially by Shell. The catalysts for this reaction are either homogeneous Mo6 + complexes (Halcon/ARCO) or heterogeneous silica-supported Ti4 + (Shell).

The main drawback of current technologies is that they generate additional coproducts. In the CHPO process, for each ton of PO, 1.4 tons of chlorine and 1.0 ton of calcium hydroxide (or sodium hydroxide) are needed, and approximately 2.0 tons of $CaCl_2$ are obtained as a coproduct. In addition, there is a large excess of water; therefore, a large volume of wastewater (brine solution) containing the Ca is formed, with about 40 tons being produced per ton of PO. The main by-product is 1,2-dichloropropane (up to 10% yield), which is either sold as a solvent or incinerated together with the other organic chlorinated by-products to produce HCl from the off-gas. The minor by-products are dichloropropanols, which are produced from allyl chloride, itself being produced by a reaction between propene and Cl2.

The recycling of the solution produced in the dehydrochlorination step to the electrolysis cell of the chloro-alkali facility, and the hydrodechlorination of the chlorinated propane to propane, have been considered as possible options to make the process environmentally more sustainable. However, these processes have not yet been implemented at the industrial level. No new plants using the CHPO technology are expected to be built - because of the large investment required a new plant would not be economically viable. However, existing plants that have been fully depreciated, are technologically up-to-date and are integrated with Cl2 production are still competitive and will keep operating.

In the hydroperoxidation route, either a-methylbenzyl alcohol or t-butanol are the reaction coproducts; both molecules are dehydrated to yield either a styrene monomer (2.5 ton per ton of PO) or isobutene (2.1 ton per ton of PO), respectively. t-Butanol may also be the direct feed for MTBE production. Balancing the markets for PO and the coproducts has proven difficult with these routes, thus leading to a volatile economic performance over time. Furthermore, the hydroperoxidation routes require relatively large capital investments. Existing hydroperoxidation plants continue to operate and are incrementally improved; however, future investments in these technologies will decline.

PO only Routes Several Approaches for Sustainable Alternatives

Alternative routes that do not produce sizeable quantities of coproducts and that do not use chlorine-based chemistry have already been, or will be, implemented at the commercial level. In April 2003, Sumitomo Chemical commercialized the first "PO-only" plant in Japan, which produces PO by oxidation of propene with cumyl hydroperoxide (the latter being obtained by hydroperoxidation of cumene) without a significant formation of coproducts. Nowadays, the plant located at the Chiba factory, a joint venture between Nihon Oxirane Co and Lyondell, produces around 200 000 tons of PO/year. A second plant was started in May 2009 in Saudi Arabia, a joint project with Saudi Arabian Oil Co.

This process is a variant of the PO/SM process that uses cumene instead of ethylbenzene and recycles the coproduct cumyl alcohol *via* dehydration to a-methyl-styrene and hydrogenation back to cumene (the latter two steps can be combined into a single hydrogenolysis step). Cumene hydroperoxidationtechnology is well-known for its use in phenol and acetone production.

Much attention has been directed recently towards processing approaches based on the epoxidation of propene using hydrogen peroxide (HP) as the oxidant. In principle, this technology can employ commercial HP, HP produced in an integrated facility or even that generated by the direct combination ofhydrogen and oxygen. Recently developed technologies include:

- liquid-phase epoxidation with conventionally produced HP (HPPO: hydrogen- peroxide-propene-oxide);
- liquid-phase epoxidation with in situ generated HP (in situ HPPO).

The estimate for the future PO production market share indicates that by 2012 roughly 18% of all PO production will be based on HPPO technology, corresponding to a total capacity of almost 1.5 million-tons per year, 15% will be based on PO/TBA technology, 29% on PO/SM technology, 34% on CHPO technology and 4% on the Sumitomo cumene/PO process. Current PO production capacities are mainly located in Europe, North America and Asia; there are a few small-capacity plants in the Middle East. However, this region is of interest with regards to the natural gas reserves and the new routes of natural gas transformation, such as the methanol-to-olefins, methanol-to-propene and propane dehydrogenation routes. In fact, nowadays propene is produced mainly in steam-cracking units of naphtha and in FCC plants; 97% of the propene is of oil-based origin, while only 3% is gas-based. In the next few years, this ratio is expected to shift in favor of natural gas.

New approaches under investigation include the direct oxidation of propene with molecular oxygen, eventually in the presence of hydrogen:

- direct oxidation of propene with oxygen (DOPO: direct-oxidation-propene-oxide);
- hydro-oxidation of propene with oxygen and hydrogen (HOPO: hydrogen-oxygen- propene-oxide).

Several companies are working on the direct oxidation of propene; for instance Lyondell is operating a pilot plant in Newtown Square, PA, and intends to commercialize the technology by 2010. Shell Chemical is also working on a direct route to PO production, based on variations of the gold and silver catalysts it uses to make ethene oxide.

Evidently, PO synthesis is a compendium of industrial chemistry. In fact, the different approaches and technologies used nowadays for PO synthesis are emblematic not only of the limitations and drawbacks that have burdened the chemical industry, but also of how the discovery of new catalysts and catalytic technologies may lead to the development of more economical and moresustainable processes. Furthermore, it is an example of how the oxidation of an organic substrate can be achieved through quite different approaches, with various oxidants (organic hydroperoxides, HP, molecular oxygen) and catalyst types, either in the liquid or in the gasphase. However, in all the recently developed technologies, the discovery and use of a new heterogeneous catalyst has been the turning point for the successful commercial implementation of the corresponding process.

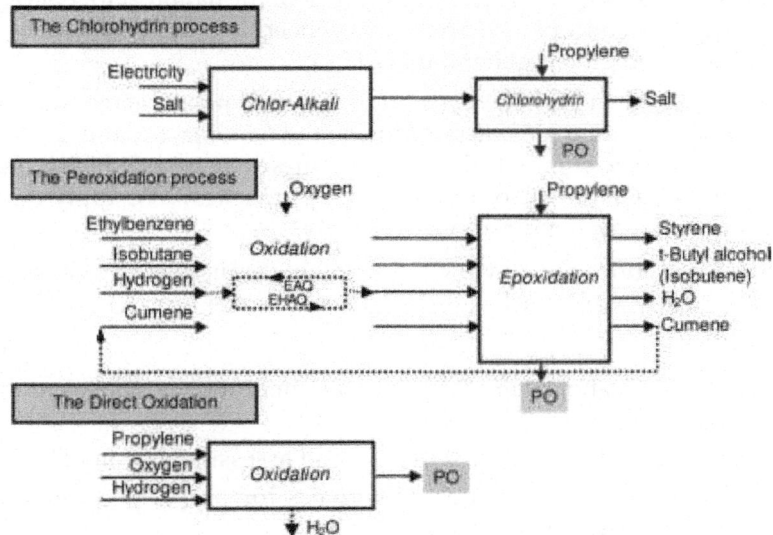

Fig.: Summary of the various technologies for PO production.

INTEGRATION OF NANOREACTOR AND MULTISITE CATALYSIS FOR A SUSTAINABLE CHEMICAL PRODUCTION

From μm- to nm-reactors: Actual microreactors have channels of μm size. We will develop a new concept based on the use of nm size channels (using Al_2O_3, TiO_2 or SiO_2 nanomembranes - diameters of the ordered and straight channels varies from 50 to 300 nm). The nanoreactor concept is based on the use of two different gas phase reactants (entering from one side each of the two opposite sides of the nanomembranes), and a thin fluid layer between the two stacked membranes. This design is important to develop intrinsically safe reactors. We apply this concept for the direct synthesis of H_2O_2 from H_2 and O_2. The use of nanomembranes (with respect to micron-sized channels in actual microreactors) shows various advantages:

(i) large increase of degree of intensification (nm versus μm section of the channels);

(ii) better control for consecutive reactions and thus higher selectivity (shorter length of the channels, μm versus mm length);

(iii) vectorial design in multistep reactions;

(iv) better integration in multiphase reactions.

Vectorial pathway for multisite catalytic reactions: A limit in cascade (or domino) reactions is that there is the inability to control the sequence of reactions of transformation of a reactant in a multisite catalyst. The concept of vectorial pathway for multisite catalytic reactions is based on the idea of an ordered sequence of catalytic sites along the axial direction of the channels of a membrane, to control the transformation. The development of the nanoreactor concept allows implementing this concept. We apply this concept for the synthesis of propene oxide (PO) with integrated generation of H_2O_2.

Dynamic nanoreactor: The concept of dynamic nanoreactor is based on the transient generation of toxic reactants inside the nanoreactor and the immediate conversion, in order to eliminate the storage of these reactants (which is minimised, but not eliminated in on-site or on-demand approaches). We apply this concept for the synthesis of diphenylcarbonate (DPC).

Under a scientific point of view, the following main activities have been made in the active tasks/ work packages (WPs):

WP1:

i) Successful in-house manufacture of membrane materials and its characterisation.

ii) Development of catalyst synthesis and assessment protocols.

iii) Identification of carbon characteristics for preparing carbon-based or carbon-coated membranes for line 2.

iv) Start of a quantum mechanical approach to define optimal catalyst cluster sizes and chemistries for the key catalytic reactions.

WP2:

i) Analysis of the nanoreactor configuration and materials to be use (in particular the membrane) in terms of their characteristics, suitability for applications, properties and limits, reactivity towards the reactants and products, *etc.*

ii) Preliminary experiments for the construction and testing of nanoreactor prototype according to the requirements for the two lines investigated in the project.

iii) Start the engineering analysis of nanoreactors concepts developed in this WP.

iv) Identification of alternative nanoreactor configurations.

WP3:

i)   Analysis of the state-of-the-art, to find a priori the preferable reaction conditions, and identify safety and engineering constrains and issues.

ii) Review of relevant literature and patents related to engineering and scale-up of the nanoreactor.

iii) Preliminary tests of H_2O_2 and PO synthesis, and set-up of relative apparatus and reactors.

WP4:

i)   Definition of overall research strategy, and research goals.

ii) Set up of experimental equipment.

iii) Investigations on mechanism of phosgene formation over carbon catalysts and catalyst deactivation.

iv) Structure activity relationships.

v)   Investigations on Cl_2 activation over carbon materials using pulse reactor.

vi) Phosgene free DPC synthesis.

vii) Engineering and safety aspects.

WP5:

i)   INCAS websites.

ii) Definition of procedures for dissemination.

iii) Exploitation committee.

WP6:

i)   Project management.

ii) Project coordination.

Potential impact

The project vision for safer and sustainable industrial chemistry is centred on the following aspects:

(i)   realise an efficient modular-design of industrial chemical synthesis, also for large-scale processes, is a decisive factor to accelerate the introduction of novel processes, and for competiveness;

(ii) a further progress with respect to actual microreactors is necessary, by integrating nanoreactor, membrane and advanced catalytic concepts; and

(iii) this novel design is a key element to improve safety. It is more effective (time, cost) to accelerate the transition to a safer and sustainable industrial chemistry developing novel reactors than to redesign the process using alternative reactants requiring large investments costs.

From Micro- to Nano-reactors

Actual microreactors have channels of micrometric size. We will develop a new concept based on the use of nanometric size channels (using alumina, titania or silica nanomembranes - diameters of the ordered and straight channels varies from 50 to 300 nm). It is possible a design using multistacked nanomembranes where each layer contains a different catalyst or even each membrane may have an asymmetric catalyst composition along the axial channel direction. The nanomembrane is based on an oxide and metal or other oxide nanoparticles can be deposited, or homogeneous complexes anchored. The nanoreactor concept is based on the use of two different gas phase reactants (entering from one side each of the two opposite sides of the nanomembranes), and a thin fluid layer between the two stacked membranes. Due to capillary forces, the nanoholes can be partially or completely wet, but at the same time the pressure differential across the membrane avoids back-diffusion of the liquid solvent.

This design is important to improve safety and develop intrinsically safe reactors. An example is for the direct synthesis of H_2O_2 from H_2 and O_2 which may be feed from the two sides of the reactor (feeds 1 and 3 in the scheme), while H_2O_2 forms on the liquid phase (methanol or methanol/ CO_2). There is a separate feed for the two gas reactants, which eliminates the problems of explosivity mixtures and without the limitations given from low gas solubility, the high wall to volume ratio provides additional safety. Similar concepts may be applied for the safer design of other reactions, and for performing in a single reactor complex multisteps reactions. The use of nanomembranes (with respect to micron-sized channels in actual microreactors) shows various advantages:

(i) large increase of degree of intensification (nm versus μm section of the channels);

(ii) better control for consecutive reactions and thus higher selectivity (shorter length of the channels, μm versus mm length);

(iii) vectorial design in multistep reactions;

(iv) better integration in multiphase reactions.

Vectorial Pathway for Multisite Catalytic Reactions

A limit in cascade (or domino) reactions is that there is no possibility to control the sequence of reactions of transformation of a reactant in a multisite catalyst. The concept of vectorial pathway for multisite catalytic reactions is based on the idea of an ordered sequence of catalytic sites along the axial direction of the channels of a membrane, to control the transformation.

In a micrometric-sized channel, the diameter of the channel is too large (with respect to the diffusional path) to make effective this concept. It is necessary a nanometric size of the channels to expect a controlled (forced) hopping of the transforming molecule from one catalytic site to the other, and thus an effective control of the selectivity. The development of the nanoreactor concept shown above allows implementing this concept. Worth to note is the possibility to integrate homo- and hetero-geneous catalysis.

Dynamic Nanoreactor

The concept of dynamic nanoreactor is based on the transient generation of toxic reactants inside the nanoreactor and the immediate conversion, in order to completely eliminate the storage of these reactants (which is minimised, but not eliminated in on-site or on-demand approaches). We apply this concept for the synthesis of diphenylcarbonate (DPC), using the dynamic nanoreactor concept, CO and Cl_2 (the latter eventually deriving from a separate stage of HCl catalytic oxidation) is feed to a nanoporous oxide membrane which side at contact with the gas-phase is functionalised with nanoporous carbon (either by coating with a porous carbon layer, or by growing carbon nano-tubes/ -fibres) to catalyse the phosgene synthesis. The other membrane side acts as catalyst for the solvent-free conversion of phenol to diphenylcarbonate (DPC) in liquid phase. In order to control the temperature (the phosgene synthesis is highly exothermic), a cross-flow micro-heat exchanger (coated with an oxide layer to avoid side reactions and further enhance the catalyst amount) is inserted in the nanoreactor. The enhanced turbulence in the microfluidic thin liquid layer favours the control of the temperature, a critical problem in conventional reactors. The oxide layer could be modified to enhance the catalytic synthesis of DPC.

Scalability of the approach for industrial applications is a critical aspect that will be considered in the project. Although for the initial, lab-scale studies planar nanomembranes would be preferable to proof-the-concept, cylindrical nanoreactors would be probably more suitable for scale-up. Therefore, the project is organised in two phases, a first one (approximately the first two years) which aims is to proof-the-concept, while a second one (the other two years) focused at investigating this issue of scalability of the concepts. A main review and assessment point between these two phases allows to eventually refocusing the activities on the more promising research lines.

PROPYLENE OXIDE

Propylene oxide is an organic compound with the molecular formula CH_3CH-CH_2O. This colourless volatile liquid is produced on a large scale industrially, its major application being its use for the production of polyether polyols for use in making polyurethaneplastics. It is a chiral epoxide, although it is commonly used as a racemic mixture.

This compound is sometimes called 1,2-propylene oxide to distinguish it from its isomer 1,3-propylene oxide, better known asoxetane.

Production

Industrial production of propylene oxide starts from propylene. Two general approaches are employed, one involving hydrochlorination and the other involving oxidation. In 2005, about half of the world production was through chlorohydrintechnology and one half *via* oxidation routes. The latter approach is growing in importance.

Hydrochlorination Route

The traditional route proceeds *via* the conversion of propylene to chloropropanols:

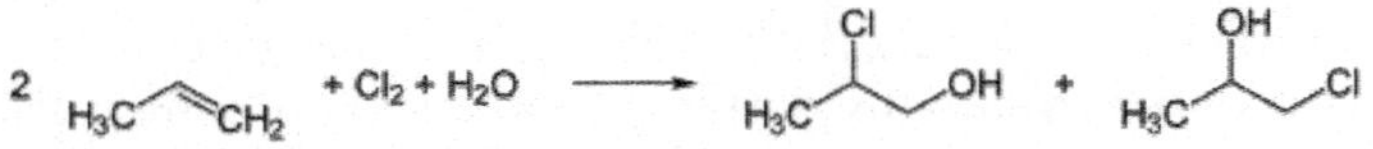

The reaction produces a mixture of 1-chloro-2-propanol and 2-chloro-1-propanol, which is then dehydrochlorinated. For example:

Lime is often used as a chlorine absorber.

Co-oxidation of Propylene

The other general route to propylene oxide involves co-oxidation of the organic chemicals isobutane or ethylbenzene. In the present of catalyst, air oxidation occurs as follows:

$$CH_3CH{=}CH_2 + Ph\text{-}CH_2CH_3 + O_2 \rightarrow CH_3CHCH_2O + Ph\text{-}CH{=}CH_2 + H_2O$$

The coproducts of these reactions, *t*-butyl alcohol or styrene, are useful feedstock for other products. For example, *t*-butyl alcohol reacts with methanol to give MTBE, an additive for gasoline. Before the current US ban of MTBE, propylene/isobutane was one of the most important production process.

Oxidation of Propylene

In April 2003, Sumitomo Chemical commercialized the first PO-only plant in Japan, which produces propylene oxide from oxidation of cumene without significant production of other products. This method is a variant of the POSM process (co-oxidation) that uses cumene hydroperoxide instead of ethylbenzene hydroperoxide and recycles the coproduct (alpha-hydroxycumene) *via* dehydrationand hydrogenation back to cumene.

In March 2009, BASF and Dow Chemical started up their new HPPO plant in Antwerp. In the HPPO-Process, propylene is oxidized with hydrogen peroxide:

$$CH_3CH{=}CH_2 + H_2O_2 \rightarrow CH_3CHCH_2O + H_2O$$

In this process no side products other than water are generated.

Uses

Between 60 and 70% of all propylene oxide is converted to polyether polyols for the production of polyurethane plastics. About 20% of propylene oxide is hydrolyzed into propylene glycol, *via* a process which is accelerated by acid or base catalysis. Other major products are polypropylene glycol, propylene glycol ethers, and propylene carbonate.

Historic and Niche Uses

Propylene oxide was once used as a racing fuel, but that usage is now prohibited under the US NHRA rules for safety reasons. It has also been used in glow fuel for model aircraft and surface vehicles, typically combined in small percentages of around 2% as an additive to the typical methanol, nitromethane, and oil mix. It is also used in thermobaric weapons, and microbial fumigation.

Fumigant

The United States Food and Drug Administration has approved the use of propylene oxide to pasteurize raw almonds beginning on September 1, 2007 in response to two incidents of contamination by *Salmonella* in commercial orchards, one incident occurring in Canada, and one incident in the United States. Pistachio nuts can also be subjected to propylene oxide to control *Salmonella*.

Microscopy

Propylene oxide is commonly used in the preparation of biological samples for electron microscopy, to remove residual ethanol previously used for dehydration. In a typical procedure, the sample is first immersed in a mixture of equal volumes of ethanol and propylene oxide for 5 minutes, and then four times in pure oxide, 10 minutes each.

Safety

Propylene oxide is a probable human carcinogen, and listed as an IARC Group 2B carcinogen.

CHLORINE-FREE SYNTHESIS FOR GREEN CHEMISTRY

In the last 20 years, chemists have put enormous effort into designing chemicals with various applications ranging from medicines and cosmetics to materials and molecular machines. However, for the most part, their work demonstrated a quite surprising lack of interest in taking hazards into consideration in the design process. The goal was often to design substances that were robust and could last as long as possible. This philosophy has resulted in a legacy of persistent toxic and bio-accumulative substances and lingering toxic waste sites. Nowadays, it is known that it is more desirable to avoid substances that persist indefinitely in the environment or in a landfill, and to replace them with substances designed to degrade after use. Polymeric materials, for instance, should have no negative effect on the environment during their production, utilization, or disposal. Therefore, the design of safer chemicals cannot be limited to hazards associated with the manufacture and use of the chemical, but also to its disposal (*i.e.*, its full life cycle).

Among halogens, chlorine is by far the most abundant chemical in nature and also the easiest to produce and use. This explains its predominant and seemingly irreplaceable role in the chemical industry. Five hundred companies at 650 sites around the world have the capacity to produce over 58×10^6 tonnes of chlorine and 62×10^6 tonnes of its co-product, caustic soda, per year.

For example, the European chloro-alkali industry had a production in 2009 of 9.1×10^6 tonnes at about 80 plants, mostly (about 95 percent) *via* electrolysis-based techniques (chlor-alkali industry); the sector directly employs about 40 000 people in 20 European Union countries. Germany is Europe's largest chlorine producer, accounting in 2009 for 43.5 percent of European production. Owing to their peculiar characteristics, halogens are widely used by all sectors of the chemical industry to produce solvents, catalysts, building blocks, additives, and drugs. Chlorine is a major building block in today's chemistry. More than 90 percent of pharmaceuticals contain or are manufactured using chlorine, which is also used in the production of 86 percent of crop protection chemicals. Furthermore, halogens are contained in several commodities that we all use daily as plastics (*e.g.*, chlorine is contained in polyvinyl chloride, PVC, one of the most widely used plastic materials), solvents for dry cleaning and metal degreasing, textiles, agrochemicals and pharmaceuticals, insecticides, dyestuffs, household cleaning products, and disinfectants. Chlorine is used extensively in organic and inorganic chemistry as an oxidizing agent (*i.e.*, water disinfectant) and as a leaving group in substitution and elimination reactions.

Chlorine compounds find use as intermediates in the production of a number of important commercial products that do not contain chlorine. Foremost examples are polycarbonates, polyurethanes, silicones, polytetrafluoroethylene, carboxymethyl cellulose, and propylene oxide.

Through a chain of chemical derivatives and relatively easily made compounds and intermediates, such molecules have used the intrinsic energy available through the use of chlorine primarily produced *via* electrolysis.

Environmental and Health Concerns

The widespread use of halocarbons was often driven by observations that most of them were more stable than other substances. Their stability tended to encourage beliefs that they were mostly harmless, but starting from the mid-1920s it was discovered that they can cause chloracne or fatal liver diseases in workers in the chemical industries. By the 1950s, toxicity and health hazards related to halocarbons were widely reported. Concerns about the environmental and health impact of halocarbons were first raised early in the 1960s in studies about DDT and other halocarbon pesticides. Today, they are widely recognized as persistent pollutants and doubts were recently (2006) raised even about very stable molecules such as perfluorooctanoic acid (PFOA) and Teflon, just to mention a few.

Today, European and international legislation for environmental protection is becoming stricter and recognizes that there is a growing need for replacement of halogenated compounds at a productive and end-user level. Besides EU directives, which tackle global environmental concerns (*e.g.*, sustainable development) through prescription about chemical production in the frame of a multi-sector approach (political, scientific, economical, and social), some regulations are specific to ban or restrict the production of some chemical compounds or byproducts at an industrial level.

The EU has recently established REACH (Registration, Evaluation, Authorization and Restriction of Chemicals), a single integrated system for the registration, evaluation, and authorization of chemicals, together with the European Chemicals Agency (ECHA). REACH requires firms that manufacture and import chemicals to evaluate the risks resulting from the use of those chemicals and to take the necessary steps to manage any identified risk. Industry has the burden of proving that chemicals placed on the market are safe. By the end of 2008, ECHA received approximately 2.75 million requests for preliminary authorization concerning about 150 000 compounds collected in an ECHA online database. Among them, over 20 000 substances are chlorinated and more than 4000 are brominated (this excludes those where "chlorine is used in the making", which would increase these numbers considerably). The process for authorization will be long (possibly ending in 2018), but a significant fraction of halogenated substances are at risk of rejection, thus forcing their replacement at a productive level.

Furthermore, some international conventions ban or restrict production of specific chemicals. At least two need to be mentioned. The Stockholm Convention on Persistent Organic Pollutants (most of which contain chlorine; byproducts of both domestic and industrial activities) and the Montreal Protocol on Substances that Deplete the Ozone Layer which control the phase-out of ozone depleting substance (ODS) production and use. The application of both is managed by the United Nations Environmental Program, and all their prescriptions will come into force within 2020. Finally, we must cite the Kyoto Protocol about greenhouse gas (GHG) emissions, which was adopted almost worldwide and then mostly disregarded. International agreements about a follow-up to the Kyoto Protocol are currently in progress.

Today, the European and international legislation and EU directives for environmental protection are becoming stricter and recognize that there is a growing need for replacement of particular halogenated compounds at a productive and end-user level. In 2007, the Intergovernmental Panel on Climate Change stated that some halocarbons were a direct cause of global warming, owing to their nature as GHGs. Some of them are ODSs. Halogen-based materials may show an indirect toxicity or eco-toxicity, releasing halogen atoms/molecules and/or harmful compounds (*e.g.*, in case of accidental fires, a case of particular interest for the electric distribution industry).

Energy

Since the industrial revolution, the halogen chlorine remains "an iconic molecule" for industrial chemical production. Even though its production by the electrolysis of sodium chloride is very energy-intensive, it is still used because it allows the manufacture of chlorinated derivatives in a very easy way, owing to its high energy and reactivity; for example, easily obtained from chlorine are $AlCl_3$, $SnCl_4$, $TiCl_4$, $SiCl_4$, $ZnCl_2$, PCl_3, PCl_5, $POCl_3$, $COCl_2$, *etc.*, which in turn are pillar intermediates in the production of numerous everyday goods. This kind of chloride chemistry is widely utilized because the energy is transferred to these intermediates, making further syntheses easy; a good number of fundamental reactions of the industrial production are based on the synthesis of chloride compounds obtained by reaction with $SOCl_2$, $COCl_2$, or $AlCl_3$ as a catalyst.

Besides their (eco-) toxicity, a major concern with chlorine derivatives is the large amount of energy necessary for their production, and this is why chlorinated molecules have both a direct (as GHGs) and indirect (CO_2 production) impact on climate change at a global level.

Estimates of the global warming potential (GWP) resulting from chlorine use by the chlorine industry in Europe is 0.29 percent of the total GWP, while the estimate from the primary energy consumed is 0.45 percent of the total GWP.

The chlorine industry is extremely energy-intensive: its CO_2 emissions is comparable to that of the iron and steel industry and higher than for cement (1.5 kg CO_2/kg for the chlorine industry vs. 1.7 for iron and steel and, 0.95 for cement) and that for a world production of the order of 40×10^6 tonnes (covering Europe and China in 2008) vs. 1.2×10^9 tones of iron and steel, and 2.3×10^9 tonnes of cement.

Beyond Chlorine Chemistry

Can we pursue an intrinsically safer, cleaner, and more energy-efficient alternative to chlorine chemistry? Many of society's greatest challenges and fortunes depend on the development of the chloro-alkali industry; but is chlorine-based chemistry sustainable?

Chlorine-based chemistry very often does not obey the principles of atom economy and waste minimization introduced, respectively, by B. Trost and R. Sheldon; in fact, halogen anions are byproducts of many organic reactions and represent a waste to be disposed of. The environmental and health constraints (toxicity and eco-toxicity, ozone layer depletion) and the growing need for energy (energy efficiency, climate change) force us to take advantage of available knowledge to develop a new chemical strategy. By the motivation given, it seems appropriate to refer to chlorine-free chemistry as "beyond-chlorine chemistry."

The beyond-chlorine strategy has two approaches, bottom-up and top-down, as follows:

- The bottom-up approach involves investigating halogen-free reactions and processes on a lab scale in academies and then scaling them up for production by industry (the most appropriate term would be green chemistry).

- The top-down approach involves industry collaboration with academic partners to design halogen-free alternatives for industrial products and processes (the most appropriate term would be sustainable chemistry).

Beyond-chloro does not mean, of course, that we should avoid chloride anion, for example in foods or as disinfectant. The substitution of chlorine compounds and of compounds where "chlorine is used in the making" means that we avoid electrolysis as the primary energetic source; however, this makes chemistry "without chlorine" considerably more difficult and illustrates why it has not been adopted before.

The rationale behind this Special Topic issue is to seek useful and industrially relevant examples for alternatives to chlorine in synthesis, so as to facilitate the development of industrially relevant and implementable breakthrough technologies. The 29 papers included in the issue are partitioned into the following parts:

1. Chlorine-Free Reagents And Reaction Selectivity

2. Chlorine-Free Catalysts

3. Carbonate Chemistry

4. Chlorine-Free Solvents

5. Benign Chloro-Free Methodologies

6. Metrics On Chlorine-Free Syntheses

MULTIFUNCTIONAL MATERIALS AND THEIR IMPACT ON SUSTAINABILITY

Multifunctional materials are the materials that perform multiple functions in a system due to their specific properties. Multifunctional materials can be both naturally existing and specially engineered.

For example, some traditional materials that provide, for instance, high mechanical strength can be modified at the nanoscale to attain other properties such as energy absorption, self-healing, *etc.* The applications of such new "smart" materials include energy, medicine, nanoelectronics, aerospace, defense, semiconductor, and other industries.

Numerous examples of multifunctional materials can be found in the nature. Bio-materials routinely contain sensing, healing, actuation, and other functions built into the primary structures of an organism. For example, the human skin consists of many layers of cells, each of which contains oil and perspiration glands, sensory receptors, hair follicles, blood vessels, and other components with functions other than providing the basic structure and protection for the internal organs. Through biological evolution, these structures were seamlessly integrated into the body to serve their functions.

The ability for materials to respond to their environment in a useful manner has broad technological impact. Such "smart" systems are being developed in which material properties (such as optical, electrical, or mechanical characteristics) respond to external stimuli. Materials of this kind have tremendous potential to impact new system performance by reducing size, weight, cost, power consumption, and complexity while improving efficiency, safety, and versatility. The multifunctionality of materials often occurs at scales from nano through macro and on various temporal and compositional levels.

Innovative advanced materials make a direct and positive impact on economic growth, the environment, and quality of life. They allow for improved processes and products and create several avenues to increasing sustainability.

Note the following areas of impact:

1. Reduction of environmental effects
2. Increased efficiency of processes
3. Light weight products
4. Lowered power consumption
5. Reducing system size
6. Reducing system weight
7. Reducing system cost
8. Reducing complexity
9. Increasing safety
10. Increasing fuel flexibility
11. Increasing versatility

Most of these impacts may result in higher efficiency of the system and cost savings.

Examples of Advanced Materials Studies

The following are several examples of sustainable solutions through improved materials chemistry or using alterative innovative materials.

Power-generating Structural Composites

"Researchers at ITN Energy Systems and SRI International have integrated a power-generating function into fiber-reinforced composites. Individual fibers are coated with cathodic, electrolytic, and anodic layers to create a battery. The use of the surface area of fibers as opposed to that of a foil in a thin film battery allows greater energy outputs, measured on the order of 50 Wh/kg in a carbon fiber-reinforced epoxy laminate. These batteries may be deposited on various substrates, including glass, carbon, and metallic fibers."

Thermostructural Materials for Gas Turbines

Gas turbines are a core technology in aero-propulsion and industrial power generation. Technological progress in this area depends on advances in thermostructural materials. The requirements to reduce emissions, increase fuel flexibility, and resist environmental attack call for development of new material systems with multifunctional properties. University of California Santa Barbara researchers employ a holistic approach that embraces and integrates all critical aspects of materials technology, including alloys, coatings, and composites, processing, and simulations to create the thermostructural materials that combine mechanical strength and exceptional thermal stability. Materials issues relevant to the high-pressure turbine include higher temperature single crystal alloys that act in concert with coatings, advanced bond coat alloys for environmental protection with improved thermo-chemical and thermo-mechanical compatibility with the load-bearing alloy, and thermal barrier oxides with new compositions that enhance temperature capabilities. Ceramic matrix composites (CMCs) and associated environmental barrier coatings are also incorporated in next generation engines, especially for combustors.

Nanoparticle Assembly Using DNA Strands

"Scientists at the U.S. Department of Energy's Brookhaven National Laboratory have developed a general approach for combining different types of nanoparticles to produce large-scale composite materials with special properties. The approach takes advantage of the attractive pairing of complementary strands of synthetic DNA — based on the molecule that carries the genetic code in its sequence of matched bases known by the letters A, T, G, and C. After coating the nanoparticles with a chemically standardized "construction platform" and adding extender molecules to which DNA can easily bind, the scientists attach complementary lab-designed DNA strands to the two different kinds of nanoparticles they want to link up. The natural pairing of the matching strands then "self-assembles" the particles into a three-dimensional array consisting of billions of particles. Varying the length of the DNA linkers, their surface density on particles, and other factors gives scientists the ability to control and optimize different types of newly formed materials and their properties."

Organic Batteries Provide Better Recyclability

A typical battery consists of two electrodes - anode and cathode, electrolyte layer, separator, and current collectors. Most of traditional battery technologies use metals or metal oxides as electrode-active materials, and metals are not renewable resources. This study describes the use of organic materials as electrodes. The advantage of such organic-based batteries over Li-ion batteries in terms of sustainability is improved recyclability, safety, adaptability to wet fabrication process, and extraction of starting material from less limited resources. One recently developed type of organic battery is based on organic radical polymers - "aliphatic or nonconjugated redox polymers with organic robust radical pendant groups as the redox site". The organic batteries have lower energy density compared to Li-ion technology, but this limitation is expected to be overcome in the near future.

PROPYLENE OXIDE PRODUCTION TECHNOLOGY ROADMAP PROGRESS

Propylene oxide is an important basic organic chemical raw materials in the production of propylene derivatives after polypropylene and acrylonitrile. Because of propylene oxide has a great ring oxygen tension, chemically active, mainly for (an important raw material for the production of polyurethane) production of polyether polyols, propylene glycol, but also the production of non-ionic surfactants, oil demulsifier, pesticide emulsifier many important chemical raw material.

2008 of PO production capacity of 780.8wt/ a, predicted in 2013, the world production capacity PO 990wt/ a, the demand for the 915wt,2008-2013 the average annual growth rate of 6.2% annual demand, higher productivity growth rate of 1.3% over the same period. In the long run, the global market outlook remains optimistic PO. The main driver of growth in demand from the polyurethane industry demand for polyether polyols growth.

TRADITIONAL PRODUCTION PROCESS

Chlorohydrin

Propylene, chlorine, alcohol and water to produce chlorine, lime and then processed to generate propylene oxide. Process is divided into chlorohydrination, saponification, refined three steps. Reaction conditions: pressure,40-90C, conversion: 97% acrylic, ~ 100% chlorine, ethylene oxide yield: > 96%

Byproducts: hydrochloric acid, dichloro propane dichloride, diisopropyl ether.

Epichlorohydrin production process short, mature technology, working load flexibility, without catalyst, good selectivity and high yield. However, multi-product waste, especially with CaCl2 and produce large amounts of chlorinated organic waste, chlorine and acrylic mix may explode, another serious equipment corrosion.

Indirect Oxidation (Halcon Act)

Propylene oxide with an organic peroxide to obtain propylene oxide. The organic peroxide may be hydrogen peroxide or an organic percarboxylic acid compound. These peroxides peroxygen portion itself is transferred to the oxygen-selective olefin epoxide is generated, its conversion to alcohol or ketone organic percarboxylic acid is converted. There are isobutane and ethylbenzene Law Act.

Isobutane law: a two-step reaction. The first step: The reaction of isobutane with oxygen to give tert -butyl hydroperoxide (TBHP) and tert-butanol (TBA) reaction conditions : 137C, 3.15Mpa; Second step: the reaction of propylene oxide and TBHP (PO). Reaction conditions : 121C, 3.61Mpa.

Isobutane method of propylene oxide yield> 95%, and the generation TBA, MTBE can be further reacted with methanol, or can be dehydrated to the olefin hydrogenation recycle isobutane, cogeneration ratio : PO/ TBA = 1/ (2.43-2.67).

Ethylbenzene law: a two-step reaction. The first step: the reaction of ethylbenzene with oxygen to ethylbenzene hydroperoxide (EBHP) and benzyl alcohol (MBA), reaction conditions : 141C, 0.35Mpa; Second step: the reaction of propylene with EBHP propylene oxide (PO), the reaction conditions : 115C, 2.24Mpa.

Method of propylene oxide ethylbenzene yield> 95%, can be further generation of MBA styrene (SM) acetophenone is reacted with cogeneration ratio : PO/ SM = 1/ (2.25-2.29).

Advantages indirect oxidation is to produce large-scale, high yield, pollution-free; disadvantage is complex process, long process, equipment and materials for the high quality requirements, harsh operating conditions, a large co-product markets need to be considered.

The new production line:

1. Cumene. This method also belongs to an indirect oxidation method. In 2003, Japan's Sumitomo Chemical Company in Chiba, Japan built the world 's first industrial unit, placed in 2009 by a second suit Sumitomo and Saudi joint production.

The law cumene as raw material, in two steps. First step: in the air oxidation of cumene cumene hydroperoxide (CMHP), the reaction conditions :90-130C,0.1-1.0 Mpa; Second step: the reaction of propylene with PO and CMHP dimethylbenzyl alcohol (CMA). CMA dehydration, hydrogenation of cumene, cycle. The reaction conditions :25-200C, 0.1-10Mpa TS-1 as catalyst byproduct is only water.

2. Hydrogen peroxide method : This method is used to produce propylene oxide over the hydrogen peroxide and water, and a direct oxidation of propylene. In 2009, BASF and Dow joint development, production in Antwerp, Belgium world 's first industrial unit, the production scale of 30kt/ a.

Advantages of hydrogen peroxide law byproduct is water pollution, mild reaction conditions : pressure, 40 ~ 60 , catalyst activity, selectivity product is extremely high. Propylene conversion of substantially all the hydrogen peroxide, the PO yield > 95%. The disadvantage is that the cost of hydrogen peroxide is too high, relatively expensive catalyst TS-1, PO in the presence of large amounts of water will be hydrolyzed to produce the alcohol. Because this method requires hydrogen peroxide as raw materials, so from economic considerations, are generally produced with a combination of hydrogen peroxide.

According to the production process of hydrogen peroxide, and had made directly using synthetic materials for the hydrogen peroxide route step synthesis of propylene oxide, there Anthraquinone synthesis and H_2-O_2.

CARCINOGENIC EFFECTS OF EXPOSURE TO PROPYLENE OXIDE

The purpose of this bulletin is to disseminate recent information on the potential carcinogenicity of propylene oxide. Recent studies of the chronic effects of this chemical in animals have produced evidence that cancer is associated with exposure to propylene oxide. This bulletin describes the results of those animal studies, presents the known human health effects of propylene oxide, and suggests guidelines for minimizing occupational exposures.

Physical and Chemical Properties

Propylene oxide at room temperature is a volatile, colorless, highly flammable liquid with a sweet, ether-like odor. The odor threshold for propylene oxide vapor is reported to be 200 parts of propylene oxide per million parts of air (200 ppm) in humans.

Production, Use, and Potential for Occupational Exposure

Propylene oxide can be produced by the chlorohydrin process (involving the reaction of propylene with chlorine) or by the hydroperoxide process. U.S. production of propylene oxide in 1980 was approximately 1,767 billion pounds. Most propylene oxide is used as an intermediate in the production of polyether polyols for polyurethane foams, and in the production of propylene glycol for unsaturated polyester resins. Minor quantities are used for sterilizing medical equipment and for fumigating foodstuffs.

The National Institute for Occupational Safety and Health (niosh) estimates that approximately 209,000 U.S. workers are occupationally exposed to propylene oxide. An industrial hygiene study conducted at a propylene-oxide-producing plant in the United States found time-weighted average (TWA) exposures to propylene oxide ranging from 0.2 to 2.0 ppm. Peak air concentrations ranged from 10 to 3,800 ppm, with the highest exposures occurring during maintenance operations.

Other data on occupational exposures to propylene oxide have been reported from a large chemical manufacturing facility that produced more than 200 chemical products, including derivatives of propylene oxide. Propylene oxide was detected in only one of seven personal samples collected at one worksite. That sample contained 1.5 ppm and was obtained for an operator in an area where flexible polyols were produced.

In a similar study, occupational exposure to propylene oxide was evaluated at three production areas of another large chemical manufacturing facility that produced derivatives of propylene oxide. Worker exposures were reported to range from 0.2 to 2.5 ppm in the polymer polyol and oxide adduct production areas. With an analytical limit of 0.01 mg per sample, propylene oxide was not detected in any samples collected in the flexible polyol production area.

Exposure Limits

The Occupational Safety and Health Administration (OSHA) has recently established an 8-hr TWA of 20 ppm for propylene oxide to protect workers against the risk of irritation and central nervous systemdepression. However, during the OSHA rulemaking process, niosh disagreed with the proposed permissible exposure limit (PEL), recommending that propylene oxide be designated as a potential occupational carcinogen.

The American Conference of Governmental Industrial Hygienists (ACGIH) threshold limit value (TLV®) for propylene oxide is 20 ppm (50 mg/m^3) as an 8-hr TWA. The ACGIH TLV is based on the acute toxicity of propylene oxide and its "lesser toxicity in relation to ethylene oxide".

Table : Chemical and Physical Properties of Propylene Oxide*·

Item	Description
CAS$^=$ registry number	75-56-9
RTECS$^=$ accession number	TZ2975000
Synonyms	Epoxypropane 1,2-epoxypropane Methyl ethylene oxide Methyloxirane Propene oxide Propylene epoxide 1,2-propylene oxide
Molecular formula	C_3H_6O
Structural formula	$CH_2\text{-}CH\text{-}CH_3$
Molecular weight	58.08
Flash point	-30°C (-22°F)
Color	Colorless
Odor	Ether-like, sweet, alcoholic
Boiling point	34.2°C (93.6°F) at 760 mm Hg
Freezing point	-112°C (-169.6°F)
Vapor pressure	445 mm Hg at 20°C (68°F)
Vapor density	(Air = 1) 2.0
Specific gravity	0.826 at 25°C
Flammability limits	2.1% to 38.5% by volume in air
Odor threshold	200 ppm
Solubility	59% by wt in water at 25°C; miscible with acetone, benzene, carbon tetra-chloride, ether, and methanol

Results of Animal Studies

Acute Toxicity

Acute toxicity has been reported in several animal species exposed to propylene oxide by various routes of administration. Results of these studies are summarized in Table 2. The gavage-administered dose of propylene oxide that was lethal for 50% of the animals tested (LD_{50}) was 1,140 mg/kg of body weight in rats and 690 mg/kg in guinea pigs. The percutaneous LD_{50} in rabbits was 1.5 ml/kg. The lethal concentrations for 50% of the animals (LC_{50}) for a 4-hr inhalation exposure were 4,126 mg/m^3 in mice and 9,486 mg/m^3 in rats.

Mutagenic Effects

Propylene oxide has been found to be a direct-acting mutagen (*i.e.*, it does not require metabolic activation) in *Salmonella* and *Escherichia coli* assays. A summary of the positive mutagenic responses is presented in Table 3. Reverse mutations (base-pair substitutions) have been demonstrated in *Salmonella* and *E. coli*. Propylene oxide has not caused frameshift mutations in *Salmonella typhimurium* strains TA1537, TA1536, and TA98.

Propylene oxide has been shown to be mutagenic in *Bacillus subtilis*, yeast, and *Drosophila melanogaster*. Exposure to propylene oxide vapor caused increased sex-linked recessive lethal mutations in two germ-cell stages of *D. melanogaster*.

Propylene oxide induced DNA damage (single-strand breaks) in rat hepatocytes and chromosomal aberrations (chromatid gaps, chromosomal gaps, breaks, and fragments) in rat liver cells and human lymphocytes. Dominant, lethal mutations were not induced in mice or rats. Mouse sperm-head morphology examinations following propylene oxide exposure did not reveal an increase in abnormal forms.

Lynch *et al.* exposed groups of 12 cynomologus monkeys to 0 (filtered air), 100, of 300 ppm of propylene oxide for 7 hr/day, 5 days/week over a 2-year period. Blood was collected during the final month of exposure and used to culture lymphocytes to assay sister chromatid exchanges (SCEs) and chromosomal aberrations. The incidence of SCEs; or chromosomal aberrations was not significantly altered compared with the control group.

Table : Acute Toxicity of Propylene Oxide for Various Animal Species and Routes of Administration.

Species	Route	Acute toxicity (LD_{50}/LC_{50})	References
Rat	Oral	1,140 mg/kg	Smyth *et al.* 1941
Guinea pig	Oral	690 mg/kg	Smyth *et al.* 1941
Rat	Inhalation	9,486 mg/m^3	Jacobson *et al.* 1956
Mouse	Inhalation	4,126 mg/m^3	Jacobson *et al.* 1956
Rabbit	Dermal	1.5 ml/kg	Weil *et al.* 1963

Table : Summary of Positive Mutagenic Responses to Propylene Oxide.

Mutation	Organism	References
Reverse mutation (base-pair substitutions)	Escherichia coli	McMahon et al. 1979 Bootman et al. 1979 Hemminki et al. 1980 Dean et al. 1985
	Salmonella typhimurium TA1535, TA100	Wade et al. 1978 Bootman et al. 1979 McMahon et al. 1979
	Bacillus subtilis	Phillips et al. 1980 Bootman et al. 1979
Forward mutation	Schizosaccharomyces pombe	Migliore et al. 1982
	Bacillus subtilis phage ø 105	Garro and Phillips 1980
Sex-linked recessive lethal mutation	Drosophila melanogaster	Hardin et al. 1983
DNA damage (single-strand breaks)	Rat hepatocytes	Sina et al. 1983
Chromosome damage	Human lymphocytes Epithelial rat liver	Bootman et al. 1979 Dean and Hodson-Walker 1979

Carcinogenic and Other Chronic Effects

Inhalation

The National Toxicology Program (NTP) has completed a bioassay to determine the carcinogenicity of propylene oxide in F344/N rats and B6C3F$_1$ mice. Groups of 50 animals of each sex and species were exposed to propylene oxide (greater than 99.9% pure) by inhalation at concentrations of 0 (chamber control), 200, or 400 ppm for 6 hr/day, 5 days/week for a period of 103 weeks. The survival rate of the exposed rats was comparable with that of the controls; however, the survival rate for the mice was lower than that of the controls. Rats and mice exposed at 400 ppm had lower terminal body weights than controls.

In both rats and mice, the primary tissue affected by inhalation of propylene oxide was the respiratory epithelium of the nasal turbinates. Suppurative inflammation, epithelial hyperplasia, and squamous metaplasia occurred in male and female rats in a concentration-related manner. Papillary adenoma of the nasal turbinate epithelium and underlying submucosal glands were observed in 3/50 female rats and 2/50 male rats exposed at 400 ppm. These tumors were not observed in the control group or in animals exposed at 200 ppm. The NTP Peer Review Panel concluded that under the conditions of these studies there was "some evidence of carcinogenicity" for F344/N rats, as indicated by the increased incidence of papillary adenomas of the nasal turbinates in the male and female rats exposed at the higher concentration (400 ppm).

Inflammation of the respiratory epithelium was observed in a concentration-related pattern among male and female mice exposed to propylene oxide. One squamous-cell carcinoma and one papilloma were found in the nasal cavities of male mice exposed at 400 ppm, and two adenocarcinomas were seen in the nasal cavities of female mice exposed at 400 ppm. In addition, nasal cavity hemangiomas were observed in 5/50 male and 3/50 female mice exposed at 400 ppm, and hemangiosarcomas were observed in 5/50 males and 2/50 females. The increased incidences of hemangiomas and hemangiosarcomas in males were statistically significant (p = 0.028, Fisher exact tests). Because of the rarity of these vascular tumors in the B6C3F$_1$ strain, the NTP Peer Review Panel concluded that under the conditions of these studies, there was "clear evidence of carcinogenicity" for male and female B6C3F$_1$ mice.

In a study by Lynch *et al.*, groups of 80 male Fischer 344 rats were exposed to propylene oxide (98% pure) at concentrations of 0 (filtered air), 100, and 300 ppm (0, 237, and 711 mg/m^3) for 7 hr/day, 5 days/week over a period of 104 weeks. The survival rates for the two exposed groups were lower than those for the controls. Rats exposed to propylene oxide at 100 and 300 ppm had an increased incidence of inflammatory lesions of the respiratory system and a dose-dependent increase of complex epithelial hyperplasia in the nasal cavity. Two rats exposed at 300 ppm developed nasal cavity adenomas. Statistically significant (P<0.05) increases occurred in the incidences of adrenal phaeochromocytomas in the propylene-oxide-exposed rats (25/78 at 100 ppm and 22/80 at 300 ppm compared with 8/78 at 0 ppm). Increases also occurred in the incidences of peritoneal mesotheliomas (8/78 at 100 ppm and 9/80 at 300 ppm compared with 3/78 for the controls). All rat groups were affected by an outbreak of *Mycoplasma pulmonis* infection, which occurred about 16 months into the study. Both the infection and the propylene oxide exposure affected the survival of rats in this study. Although the proliferative lesions of the nasal mucosa appeared to be treatment-related, the authors could not ascertain how their development was influenced by the intercurrent inflammatory disease.

Reuzel and Kuper [1983] exposed groups of 100 Wistar rats of each sex to propylene oxide at concentrations of 0 (filtered air), 30, 100, or 300 ppm (71, 238. or 713 mg/m^3) for 6 hr/day, 5 days/week over a period of 28 months. Ten rats of each sex from each exposure group were killed for examination after 12, 18, or 24 months of exposure. Body weight reduction, increased mortality, hyperplasia of the mucosa, and degeneration of the olfactory epithelium were found in the male and female animals exposed at 300 ppm. Female rats exposed at 300 ppm had a higher incidence (P<0.05, chi-square test) of myocardial degeneration than did the controls (10/69 at 300 ppm compared with 3/69 for the controls). Females exposed at 30 or 100 ppm showed an increased number of fibroadenomas of the mammary glands, but the number of fibroadenoma-bearing animals did not increase compared with controls. However, for females exposed at 300 ppm, a statistically significant increase was reported in the number of rats with two or more fibroadenomas and in the number of fibroadenoma-bearing animals.

Compared with the controls, female rats exposed to propylene oxide had a statistically significant increase in the incidence of malignant mammary tumors. However, this incidence did not exceed the historical control incidences (0% to

15%) derived from six different long-term studies by the same laboratory on the same strain of rats. The number of females with benign mammary tumors and the mean number of fibroadenomas per fibroadenoma-bearing animal were both higher than the historical control data. No significant increase occurred in the incidence of any specific type of tumor other than mammary tumors in exposed Wistar rats compared with controls. However, the total incidence of all tumors other than mammary tumors was significantly increased ($p<0.01$) in the females exposed at 300 ppm compared with controls. In the male mid female rats exposed at 300 ppm, the total number of animals bearing malignant tumors other than mammary tumors was significantly higher than in controls ($p<0.05$ in males and $p<0.01$ in females, chi-square test).

The authors questioned whether the observed effects on the mammary glands and the increase in total tumor incidence were due to the direct action of propylene oxide or to some indirect mechanism. They concluded that propylene oxide enhances the development of malignant neoplasms through an indirect, non-specific mechanism.

Table : Incidence of Mammary Tumors in Female Wistar Rats Exposed to Propylene Oxide by Inhalation for 28 Month.

Item	Propylene oxide concentration (ppm)			
	0	30	100	300
Number of rats with benign mammary tumors	32/69	30/71	39/69	47/70
Mean number of fibroadenomas per fibroadenoma-bearing animal	1.3	2.1	2.2	2.4
Number of rats with malignant mammary tumors	3/69	6/71	5/69	8/70†

Oral Administration

Dunkelberg [1982] administered 15 or 60 mg of propylene oxide (99% pure) per kg of body weight in salad oil by gavage to groups of 50 female Sprague-Dawley rats 2 times/week for 150 weeks. Control groups consisted of 50 untreated female rats and 50 female rats dosed with salad oil. Survival rates for rats treated with propylene oxide were not statistically different from those of the controls. Treated animals developed hyperplasia, hyperkeratosis, or papillomas of the forestomach (incidence rates were 7/50 at the 15-mg/kg dose and 17/50 at the 60-mg/kg dose). Squamous-cell carcinomas of the forestomach developed in a dose-dependent manner in rats treated with propylene oxide (incidence rates were 0/50 for controls, 2/50 at the 15-mg/kg dose, and 19/50 at the 60-mg/kg dose). The first of these tumors was observed during the 79th week in the high-dose group. One additional animal in the 60-mg/kg group had an adenocarcinorna.

Subcutaneous Administration

Dunkelberg [1981] injected groups of 100 female NMRI mice subcutaneously with 0.1, 0.3, 1.0, or 2.5 mg of 99% pure propylene oxide (in 0.1 ml tricaprylin) once/week for 95 weeks. Control groups consisted of 200 untreated female mice and 200 female mice injected with 0.1 ml tricaprylin (*i.e.*, the vehicle control group). A dose-related increase in cancers (mostly fibrosarcomas) occurred at the injection site. The incidences of sarcomas (fibrosarcomas and pleomorphic sarcomas) at propylene oxide injection sites were 3/100 at 0.1 mg, 2/100 at 0.3 mg, 12/100 at 1.0 mg, and 15/100 at 2.5 mg. Four fibrosarcomas were observed in the vehicle control group, and no sarcomas were observed in the untreated control group.

Human Health Effects

The human health effects of propylene oxide exposure include corneal burns, contact dermatitis, and a reduced capacity to repair DNA lesions. McLaughlin reported that humans exposed to propylene oxide vapor received corneal burns. Contact dermatitis was involved in two case reports--one concerning an electron microscopy technician and the other concerning two laboratory assistants. All three individuals had positive responses to propylene oxide in standard allergy patch tests.

Twenty-three workers aged 25 to 59 were exposed to propylene oxide in a factory producing alkylated starch. Lymphocytes from these workers were examined for a reduced capacity for unscheduled DNA synthesis following the *in vitro* induction of DNA damage to their lymphocytes. Unscheduled DNA synthesis is a step in the enzymatic repair of DNA damage. Estimates of airborne exposure were obtained using both personal and area sampling. Eight-hour TWA exposure concentrations of propylene oxide were calculated for five of the most highly exposed workers over 5 workdays (8-hr shifts). These concentrations ranged from 0.6 to 12 ppm. The control group consisted of 12 workers aged 21 to 46 who were not exposed to propylene oxide. Under the conditions of this experiment, unscheduled DNA synthesis was significantly inhibited ($p<0.001$, t-test) in the group exposed to propylene oxide.

The research data presented in this Current Intelligence Bulletin (CIB) have focused primarily on the carcinogenic effects of propylene oxide in exposed animals. The animal studies described provide sufficient evidence to conclude that propylene oxide is carcinogenic in laboratory animals.

Several systems exist for classifying a substance as a carcinogen. Such classification systems have been developed by NTP, the International Agency for Research on Cancer (IARC), and OSHA, in its "Identification, Classification, and Regulation of Potential Occupational Carcinogens", also known as "The OSHA Cancer Policy." niosh considers the OSHA classification system the most appropriate for use in identifying occupational carcinogens.

Exposure to Propylene oxide has been shown to produce cancer and benign tumors in both rats and mice. niosh therefore recommends that propylene oxide be considered a potential occupational carcinogen in conformance with the OSHA Cancer Policy. The excess cancer risk for workers exposed to propylene oxide has not yet been established, but the probability of developing cancer should be decreased by minimizing exposure. As a matter of prudent public health policy, employers should assess the conditions under which workers may be exposed to propylene oxide and take reasonable precautions (such as appropriate engineering and work practice controls) to reduce exposures to the lowest feasible concentrations.

Guidelines for Minimizing Worker Exposure

The following guidelines for reducing worker exposure to propylene oxide are general and should be adapted to specific work situations as required.

Exposure Monitoring

NIOSH recommends that each employer who manufactures, transports, packages, stores, or uses propylene oxide in any capacity determine whether a potential exists for any worker to be exposed to the chemical. In work areas where exposures may occur, an initial survey should be done to determine the extent of worker exposure. TWA exposures should be determined by collecting samples over a full shift. When the potential for exposure is periodic, short-term sampling may be needed to replace or supplement full-shift sampling. Personal sampling (*i.e.*, sampling conducted in (the worker's breathing zone) is preferred over area sampling. If personal sampling is not feasible, area sampling can be substituted only if the results can be used to approximate the workers' exposure. Sampling should be used to:

(1) identify the sources of emissions so that effective engineering or work practice controls can be instituted, and

(2) ensure that controls already in place are operational and effective.

The limit of quantization is 0.1 mg propylene oxide per sample. which corresponds to 13 ppm (31 mg/m^3) for a 15-min sampling period, or 8 ppm (19 mg/m^3) for the maximum recommended sample volume (5 liters).

The niosh *Occupational Exposure Sampling Strategy Manual* provides guidance for developing effective strategies to monitor worker exposures to toxic chemicals. The manual contains information on determining the need for exposure monitoring, the number of samples to be collected, and the appropriate sampling times.

Controlling Worker Exposure

Maintaining equipment and educating workers are both vital components of a good program for controlling occupational exposures. Workers should be informed of any materials that may contain or be contaminated with propylene oxide, the nature of the potential hazard, and methods for reducing exposure. Every attempt should be made to minimize exposure to propylene oxide by using the following work practices and controls: product substitution, closed systems and ventilation, worker isolation, personal protective equipment (such as chemical protective clothing and equipment, and respiratory protective devices), and decontamination and waste disposal. These measures as well as medical monitoring procedures are discussed here briefly.

Product Substitution

When feasible, employers should substitute a less hazardous material for propylene oxide. However, extreme care must be used when selecting substitutes. Possible adverse health effects from exposure to the substitute should be evaluated before selection.

Closed Systems and Ventilation

Engineering controls should be the principal method for reducing propylene oxide exposure in the workplace. Achieving and maintaining reduced concentrations of airborne propylene oxide depend on adequate engineering controls such as closed-system operations and ventilation systems that are properly constructed and maintained.

Closed-system operations provide the most effective means for minimizing worker exposures to propylene oxide. Closed systems should be used for producing, storing, transferring, packaging, and processing propylene oxide. For quality control laboratories or laboratories where production samples are prepared for analyses, exhaust ventilation systems should be designed to capture and contain vapors.

One company has developed techniques that allow for the maintenance of a closed system while process and tank car samples of propylene oxide are obtained and analyzed for quality control. These automated techniques have virtually eliminated the need for manual sampling and have reportedly been successful in reducing exposures.

Worker Isolation

The area in which propylene oxide is produced or used should be restricted to workers who are essential to the process or operation. If feasible, these workers should be isolated from direct contactwith propylene oxide by use of automated equipment operated from a closed control booth or room. This room should be maintained at greater air pressure than that surrounding the process equipment so that air flows out rather than in. When workers must enter the general work area to perform process checks, adjustments, maintenance, assembly line tasks, and related operations, they should take special precautions such as the use of personal protective equipment.

PERSONAL PROTECTIVE EQUIPMENT

Chemical Protective Clothing and Equipment

To minimize skin contact and absorption, workers using propylene oxide should wear appropriate chemical protective clothing (CPC) such as gloves and aprons. CPC made from butyl rubber and Teflon® should provide adequate protection for at least 1 hr. Note, however, that the quality of gloves may vary significantly among glove producers. Product-specific chemical permeation data should therefore be obtained from the glove manufacturer. Splashproof goggles or face shields should be worn if there is any possibility that liquid propylene oxide will contact the eyes. Safety showers and eye wash stations should be located close to operations that involve propylene oxide.

Respiratory Protective Devices.

The use of respirators is the least desirable method of controlling worker exposures and should not be used as the primary control method during routine operations. However, niosh recognizes that respirators may be required to provide protection in certain situations such as implementation of engineering controls, certain short-duration maintenance procedures, and emergencies. niosh maintains that only the most protective respirators should be used for situations involving carcinogens. These respirators include:

- any self-contained breathing apparatus with a full facepiece operated in a pressure-demand or other positive-pressure mode, and
- any supplied-air respirator with a full facepiece operated in a pressure-demand or other positive-pressure mode in combination with an auxiliary self-contained breathing apparatus operated in a pressure-demand or other positive-pressure mode.

Any respiratory protection program must, at a minimum, meet the requirements of 29 CFR 1910-134. Respirators should be approved by niosh and the Mine Safety and Health Administration. A complete respiratory protection program should include regular training and medical evaluation of personnel, fit testing, periodic environmental monitoring, and maintenance, inspection, and cleaning of equipment. The program should be evaluated regularly.

Decontamination and Waste Disposal

If propylene oxide contacts the skin, promptly wash the contaminated area with soap or a mild detergent and water.

The following steps should be taken in the event of a propylene oxide spill:

1. Remove all ignition sources.
2. Ventilate the area of a spill or leak
3. Absorb small quantities on paper towels and permit the vapor to evaporate completely in a safe place (such as under a fume hood). Allow sufficient time for the vapor to clear completely from the hood ductwork; then burn the paper towels in a suitable location away from the combustible materials.

4. Collect large quantities and dissolve in alcohol of greater molecular weight than butyl alcohol. Atomize the solution and burn it in a suitable combustion chamber.

5. Do not allow propylene oxide to enter a confined space such as a sewer because it may explode.

Medical Monitoring

A medical monitoring program should be established for prevention or early detection of the acute, chronic, or carcinogenic effects of propylene oxide. Medical and work histories should be taken for each worker before job placement and updated periodically. The worker's physician should be given information on the adverse health effects of propylene oxide exposure and an estimate of the worker's potential for exposure. This information should include results of workplace sampling and a description of any protective devices or equipment the worker is required to use. The examining physician should direct particular attention to the skin and to the nasal and respiratory tracts, as these sites are the most likely to be affected by propylene oxide. The occurrence of disease or other work-related health effects requires immediate evaluation of primary preventive measures (*e.g.*, industrial hygiene monitoring, engineering controls. and personal protective equipment). Medical personnel should ensure that workers are informed of the health effects associated with propylene oxide exposure.

CURRENT INDUSTRIAL PROPENE OXIDE PRODUCTION

Propylene Oxide by the BASF-Dow HPPO Process

In recent years, the propylene oxide (PO) industry has been quite active in researching new process technologies to manufacture PO without co-producing large amounts of styrene or t-butyl alcohol, or generating chloride-containing waste streams. Many companies have investigated PO technologies using hydrogen peroxide (HPPO process), and effective catalysts have been developed to improve the selectivity to PO. In late 2002, BASF and Dow Chemical joined forces in the development of an HPPO technology process. A 300,000 t/yr plant using this technology was completed at Antwerp, Belgium, in March 2009. In Thailand, Dow and Siam Cement group (SCG) have broken ground on a 390,000 t/yr PO facility near Map Ta Phut, using the BASF-Dow HPPO technology. The plant is expected to come online in 2011.

This review presents a conceptual design and preliminary economics for a plant producing 200,000 t/yr of PO from propylene using the HPPO process. The plant is integrated with a unit that generates hydrogen peroxide by direct reaction of hydrogen and oxygen. We also compare the economics of the HPPO technology with those of the conventional PO/SM, PO/TBA and chlorohydrin processes.

Our analysis indicates that recent patented improvements in the product recovery configuration have resulted in a significant reduction in the overall steam consumption of the HPPO process. The improved version of the technology can be cost-competitive with the PO/SM and PO/TBA processes when market prices for the styrene and TBA co-products are relatively low. One of the main benefits of the HPPO technology, which could justify new capacity additions in developing regions, is the fact that it eliminates the need for additional infrastructure or markets for co-products. However, to be competitive, the HPPO process depends on the availability of a low cost source of hydrogen peroxide.

Chapter 5

DEVELOPMENT OF A PROTOTYPE MINIATURE SILICON MICROGYROSCOPE

Dunzhu Xia*, Shuling Chen and Shourong Wang

School of Instrument Science and Engineering, Southeast University, Nanjing City, Jiangsu Province, 210096, China; E-Mails: chenshuling318@126.com (S.C.); srwang@seu.edu.cn (S.W.)

*   Author to whom correspondence should be addressed; E-Mail: xiadz_1999@163.com; Tel.: +86-025-83793552; Fax: +86-025-83793644

ABSTRACT

A miniature vacuum-packaged silicon microgyroscope (SMG) with symmetrical and decoupled structure was designed to prevent unintended coupling between drive and sense modes. To ensure high resonant stability and strong disturbance resisting capacity, a self-oscillating closed-loop circuit including an automatic gain control (AGC) loop based on electrostatic force feedback is adopted in drive mode, while, dual-channel decomposition and reconstruction closed loops are applied in sense mode. Moreover, the temperature effect on its zero bias was characterized experimentally and a practical compensation method is given. The testing results demonstrate that the useful signal and quadrature signal will not interact with each other because their phases are decoupled. Under a scale factor condition of 9.6 mV/°/s, in full measurement range of ± 300 deg/s, the zero bias stability reaches 15°/h with worse-case nonlinearity of 400 ppm, and the temperature variation trend of the SMG bias is thus largely eliminated, so that the maximum bias value is reduced to one tenth of the original after compensation from -40 °C to 80 °C.

Keywords

Silicon microgyroscope (SMG); self-oscillating; dual-channel closed-loop; scale factor; zero bias stability; temperature compensation; miniature prototype

1. INTRODUCTION

The silicon microgyroscope (SMG) is an important kind of sensor found in inertial instruments that are widely used in aerospace measurement, balance control, and vehicle navigation systems. One of the major features for successful commercialization of microgyroscopes is their vacuum packaging under reduced pressure levels, which is essential in determining reliability, working performance, required drive voltage and the total cost of the sensor module. The performance of SMGs under vacuum has been studied intensively, and some gyroscopes are found to have a better performance in a vacuum chamber [1,2]. Some mature vacuum packaging technologies such as silicon-silicon bonding and silicon-glass wafer bonding are successfully used to package microgyroscopes [3]. In this paper, a special and timely effective packaging approach using laser welding has been realized in order to guarantee the gyroscope will function under high mechanical sensitivity conditions *i.e.*, with a high quality factor (Q factor), in a vacuum environment and for a long time.

Apart from the preparation of a suitable vacuum working environment, the meticulous design of structure and circuitry become the two main key technologies. Considering that the previous literature primarily used varying-distance detection which could cause unexpected nonlinearity errors, a fully decoupled structure with varying-area style is proposed. In most of the literature drive mode circuitry designs of AGC and PLL (Phase Lock Loop) modules have been independently or simultaneously used to attain a certain performance [4-6]. On this basis, a simplified drive circuit with a combination of AGC and modulation technologies was investigated to achieve a high SNR (Signal Noise Ratio) and good performance. As is known, open-loop detection is favorable for design simplification and to accomplish the integration and miniaturization of the measurement-control circuit, which can thus basically meet the specification requirements of low-precision SMGs, and has been successfully applied in many reports [2,3]. Unfortunately, the linearity of open-loop detection tends to be affected by the non-linear characteristics of detection displacement and its corresponding capacitance, which leads to a limited measurable angular range. Especially, the gyroscope's use is complicated, usually affected by many kinds of disturbances such as outer interferences, over shocks, vibrations, temperature and so on. Though some gyroscopes in the literature [7,8] use single channel closed-loops or double-channel closed-loops without modulation of high frequency carrier signal, it is very difficult for a separate charge amplifier to measure a minor varying signal due to some large existing unquenchable parasite signals.

Some ASIC gyroscopes using sigma-delta digital closed-loop has been realized in the lab [9-11], but this architecture is more complex to realize in separate circuitry. In [12], though digital control loops are successfully perfected for basic performance, two carrier signals (600 kHz and 400 kHz) are required for modulation in the drive and sense modes, respectively, which will increase the circuit complexity. Only a PI (Proportion-integral) controller is designed to control the structure rather than a PID (Proportion-integral-derivative) module, which may result in incomplete control, especially while the microgyroscope works in a complex dynamic environment. Moreover, the adjustable phase reference for modulation is very simple such as 0° and 90°. As for the actual gyroscope, the phase shift is actually very difficult to anticipate because the signal phase will be effected by both circuitry and two unmatched modes of the structure, which needs an adjustable phase shifter.

In this design, only one carrier signal (1 MHz) is input to the proof mass port for simultaneous modulation of both drive and sense modes. On the other hand, the open-loop system bandwidth is approximatively decided by the difference between resonant frequencies in two modes, and it is quite difficult to adjust them once the structure is produced. In this paper, an effective resolution is found to regulate the bandwidth using PID compensator module in the closed-loop system.

Above all, in order to develop a miniature silicon microgyroscope, a newly improved type of structure with vacuum packaging was investigated first. In drive mode, a self-oscillating drive circuit will be designed to ensure a constant motion, and in sense mode, a dual-channel closed-loop detection method using carrier signal modulation and demodulation in pure separate analogy components is presented and implemented successfully. Meanwhile, since there always exists zero bias drift with temperature [13,14], a practical digital compensation system using microprocessor should be designed to attenuate the temperature influence.

2. DESIGN OF THE SMG STRUCTURE

In terms of structure design, decoupling, sensitivity, and linearity there are three main important rules. Most conventional decoupled microgyroscopes have two or more degrees-of-freedom (DOF) to ensure their decoupling but ignoring their linearity. Though the combination of linear-linear structure and rotary-rotary structure achieves a good decoupling performance [15], the nonlinearity becomes prominent because of its arched capacitor plates. Another problem is that most decoupled gyroscopes adopt varying-distance detection in sense mode, despite having a good symmetrical style, which will also cause a serious nonlinearity while working at a comparatively large angular rate.

Based on the structure in [16], which cannot be a completely symmetric structure because of its varying-distance style in sense modes, a newly improved type of structure is designed. What is special is that drive mode and sense mode are completely symmetric because they are both varying-area style. A simplified model of the SMG is shown in Figure 1.

In this design, a varying-area detection method is applied in sense mode, which ensures a better linearity than the former mentioned gyroscopes. In addition, this gyroscope, with slide-film damping, has a quality factor nearly 10 times higher than squeeze-film damping at the same ambient pressure, in spite of about 1/10 conversion coefficient from displacement to capacitor, so the general sensitivity to Coriolis force is at the same level for both varying-distance and varying-area styles. In order to improve its sensitivity, multi rows of interdigital comb-finger

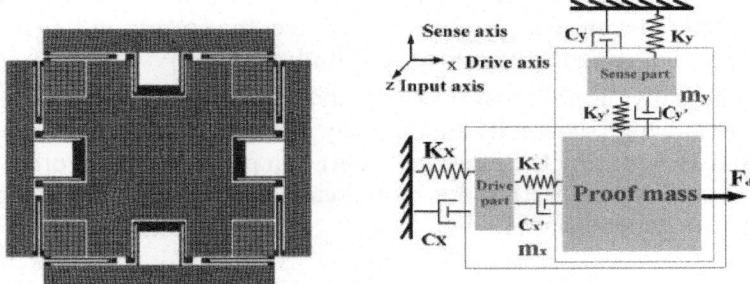

Figure 1. The structure and model of SMG.

capacitors are arranged orderly to amplify the displacement-capacitive-ratio conversion. Folded beams are anchored toward the inner direction of the proof mass and have an excellent symmetry in both drive and sense mode. In the actual structure, an added stop block is designed in the center of the proof mass to implement the over shock preventing function.

The symmetrical and decoupled microgyroscope designed in this paper is essentially a linear vibratory gyroscope to detect rotation angular rate by the oscillating components. Its simple model can be described as a total four DOFS system including proof mass, drive component, sense component, and damping elements. Specifically, the drive component has one DOF in the drive direction, the sense component has one DOF in the sense direction, and the common proof mass has two DOFs in two directions. In this model, m_x is the general mass with the sum of drive part and proof mass in drive direction, K_x and C_x are respectively the general stiffness coefficient and damping coefficient in drive direction, $K_{x'}$ and $C_{x'}$ are the inner stiffness coefficient and damping coefficient between drive part and proof mass respectively. Similarly, m_y is the general mass with the sum of sense part and proof mass in sense direction, K_y and C_y are respectively the general stiffness coefficient and damping coefficient in sense direction, $K_{y'}$ and $C_{y'}$ are the inner stiff coefficient and damping coefficient between sense part and proof mass respectively. F_d is the electrostatic force to drive the structure along x-axis.

According to the design rule of all the folded beam stiffness (the stiffness in the required direction is greater than that in its orthogonal direction), it is obvious that both the drive and sense components primarily vibrate in its corresponding DOF. In other words, the sense component almost vibrates in sense direction even if the drive component sufficiently vibrates in drive direction. So the coupling between the two modes is enormously reduced. Its working mechanism includes that:

(1) The drive component is driven as steadily as possible to ensure the drive frequency and vibrating amplitude are both stable and constant.

(2) Due to the Coriolis effect, the Coroilis force along the sense direction is generated when a rotation happens in the perpendicular direction.

(3) Because the vibratory amplitude in sense mode is in direct proportion to the outer angular rate, the gyroscope can detect the corresponding angular rate through detecting the vibratory amplitude.

Additionally, use of a varying-area style in the sense part and in the drive part can ensure high sensitivity and linearity in the detection of the Coriolis signal. For simplicity, ignoring the inner force between proof mass and other two parts (drive part and sense part), the symmetrical and decoupled microgyro motion equations are generalized [17] by:

$$m_x \ddot{x}(\omega t) + C_x \dot{x}(\omega t) + K_x x(\omega t) = m_x \left[\ddot{x}(\omega t) + \frac{\omega_{nx}}{Q_x} \dot{x}(\omega t) + \omega_{nx}^2 x(\omega t) \right] = F_d + k_{xy} y(\omega t) \tag{1}$$

$$m_y \ddot{y}(\omega t) + C_y \dot{y}(\omega t) + K_y y(\omega t) = m_y \left[\ddot{y}(\omega t) + \frac{\omega_{ny}}{Q_y} \dot{y}(\omega t) + \omega_{ny}^2 y(\omega t) \right] = -2m_y \Omega_z \dot{x}(\omega t) + k_{yx} x(\omega t) + \varepsilon x(\omega t + \varphi) \tag{2}$$

where x and y are displacements in drive and sense modes, ω_{nx} and ω_{ny} are resonant frequencies in two modes, Q_x and Q_y are quality factors in two modes, F_d is the electrostatic driving force in drive mode, Ω_z is the input angular rate in z direction, $-2m_y \Omega_z x$ is the Coriolis force component, $k_{xy} y$ and $k_{yx} x$ denote the forces generating orthogonal coupling signal in two modes, $\varepsilon x(\omega t + \varphi)$ is the force generating offset error signal, which will cause offset output without the rotation of gyroscope, ε is a proportional coefficient between the force and the displacement, φ is a actual phase shift which will decide the phase relationship among the offset error signal, Coriolis signal and quadrature signal. In [18], the researchers have intensively studied its mathematic model and recognized its sources, which are different from the Coriolis and quadrature signals. Actually, their phase relationship is such that the Coriolis and quadrature signals are orthogonal, the offset error signal is not in phase with them when the modes are unmatched, but the offset error signal is in same phase as the Coriolis signal if the modes are matched.

Structure design and fabrication imperfections are always major factors resulting in quadrature and offset errors [18,19]. In theory, these imperfections are reflected in asymmetry and anisoelasticity of the structure, and can be captured as misalignment of principal axes of elasticity from intended axes of symmetry of the structure. There would potentially be a solution to reduce stiffness and displacement coupling coefficients mentioned previously by using this special symmetrical and decoupled style. Figure 2 shows the package and SEM photos of a SMG structure.

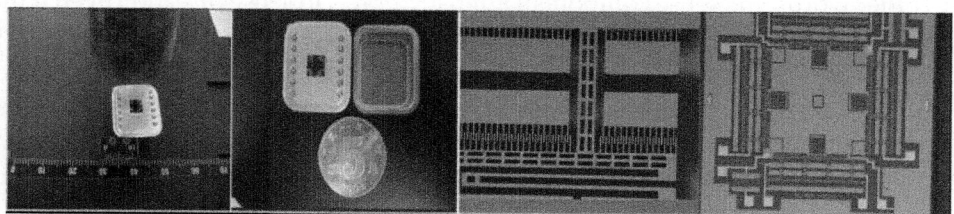

Figure 2. The package and SEM photos of a SMG structure.

3. DESIGN SCHEMES OF THE SMG IN DRIVE AND SENSE MODES

3.1. Self-Oscillating Scheme in Drive Mode

Currently most SMGs utilize capacitive electrostatic excitation, which can guarantee that the driving frequency ω_d approaches the natural frequency ω_{nx} as close as possible in drive mode, and the amplitude is as stable as possible. To satisfy these requirements, closed-loop control must be achieved in the actuation of the SMG. In order to achieve the self-oscillating startup of an oscillator, the closed-loop control must meet two requirements, *i.e.*, the phase angle of the whole loop $\theta = 2n\pi$ (n is an integer) and the gain of the whole loop A > 1 [20]. The closed-loop actuation of an SMG commonly adopts AGC, which implements the closed-loop driving with a non-linearity dynamics characteristic, reducing the closed-loop gains of the entire loop to 1 [21]. In most of previously mentioned literature [4-6], AGC and PLL modules have been used successfully. However, it is difficult to simplify the electronics when both these modules are used. On the other hand, tracking accuracy is also affected by the presence of Brownian noise and capacitive position sensing noise [22]. Due to the complicated Micro-Electro-Mechanical interface between all kinds of plate electrodes of the gyroscope, the accurate detection of the driving motion using a large bias voltage Vp (even 40 V) is difficult to carry out because of a variety of unknown parasitic capacitances and low SNR in actual separate components, which will simultaneously cause a large power consumption.

Considering the above reasons, a high SNR detection of driving motion using high-frequency (1 MHz) carrier-signal modulation and demodulation, and a wide flexible phase adjustment by a precision phase shifter are both adopted to improve the scheme of self-oscillation-driven of the SMG. The AGC goal is achieved through adding an automatic varying DC (direct current) voltage V_{dc} with the AC (Alternating current) voltage signal to produce a servo driving force, which can make the drive mode vibration amplitude constant.

The scheme succeeds in decoupling the phase angle and gain of the self-oscillation-driven, where the phase angle and the gain can be irrelevant and adjusted separately, which thereby increasing the stability of drive frequency and the amplitude. From the analysis above, the role of closed-loop control of driving circuit is to reduce the resonant frequency deviation, *i.e.*, $|\Delta\omega/\omega_d|$ as small as possible and increase the stability of the driving frequency and amplitude.

As shown in Figure 3, a whole scheme of driving circuit method is realized by orderly composition of the SMG's transfer function in drive mode, modulation module, K_{xc} module (the transfer coefficient from combo finger displace to its equivalent capacitance), demodulation module, K_{cv} module (the transfer coefficient from capacitor to voltage), AGC loop, phase shifter module, and force feedback generator module F_v. Especially, a high frequency precision carrier (1 MHz) is provided for modulation and demodulation. In the AGC loop, a PID controller module can ensure that the oscillating amplitude of the gyroscope in drive mode is equal to the set value by adjusting V_{ref}, and the absolute value circuit can complete the rectification of the demodulated AC signal.

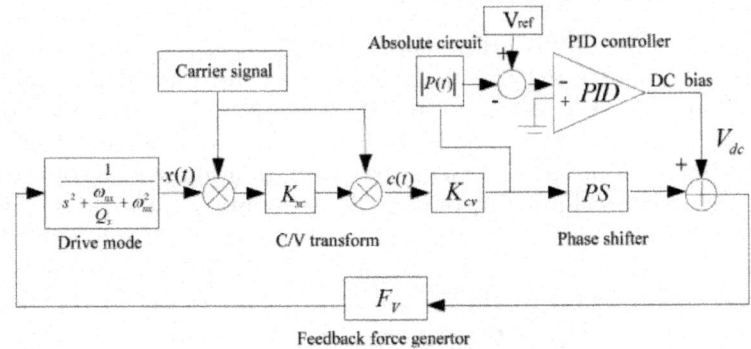

Figure 3. Drive mode circuitry method.

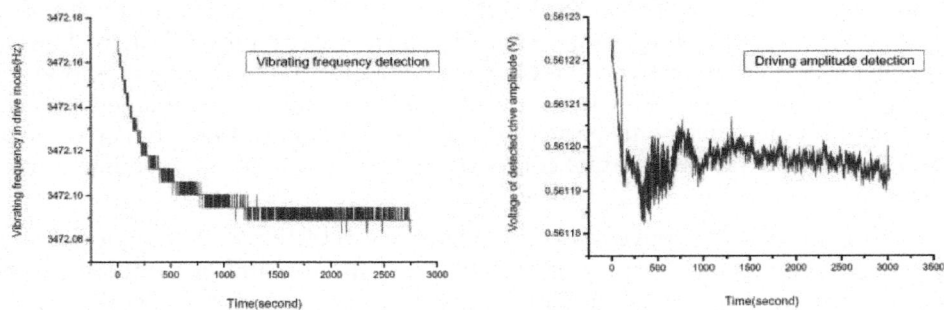

Figure 4. Vibrating frequency and amplitude in drive modes.

One hour detection results of vibrating frequency and amplitude in drive mode are shown in Figure 4. Since the power startup, the mean value of the vibrating frequency in drive mode is near 3,472.1 Hz, and its variance is 0.01538 Hz. Similarly, the mean detection voltage value of vibrating amplitude in drive mode is near 0.5612 V, and its variance is 4.49182×10^{-6} V, so the stability of both vibrating frequency and amplitude in drive mode have achieved a high precision (the relative error is lower than 1×10^{-5}).

3.2. Dual-Channel Closed-Loop Detection of SMG

Due to a variety of advantages over open-loop such as reduced noise and better stability, all kinds of digital or analog closed-loops are adopted in the gyroscope detection circuits in some literature [6-12,23]. In digital closed-loop [9-11], a sigma-delta interface can be functionally implemented to detect the minor motion signal, but the switch noise from sequence control clock probably becomes a noise source if used in the separate circuits here without careful treatment. In addition, a signal-channel closed-loop control used in sense mode cannot completely counteract the useful signal, the quadrature signal and offset error, because the force feedback is phase insensitive before performed synchronous demodulation to Coriolis plus offset and quadrature plus offset components.

In [7] even three loops in analog closed-loop were applied without consideration of the concrete phase relationship among the above signals, Especially, when there always exits mismatch (resonant frequency difference) in two modes of gyroscope, the phase relationship of them is complex to need modification and demodulation to reach a better sensitivity. In all designs mentioned above, the motion signal detection is difficult to reach high SNR because of all kinds of the parasite signals, which mainly arise from the complex Micro-Electro-Mechanical interface and appear in low frequency band near resonant frequency. In [24,25] two-channel decomposition and reconstruction closed loops are applied, the testing results demonstrate that the useful and quardrature signal will not interact because of their phase decoupling. However, the controlled gyroscope named szg3 is varying-distance style, and here the newly designed varying-area style gyroscope named E17 is investigated with the similar control strategy and normalized control parameters. Though the simulated control model is the same as before, due to using improved capacitance detection converter and other high performance circuitry, the actual testing results in this work will be verified to show the obvious improvement for this overall system. In this case, the whole control method will be further analyzed in detail. Based on our previous work, an improved dual-channel closed-loop detection block diagram of this proposed kind of SMG mentioned above is given in Figure 5, which will take full advantage of modulation and demodulation method of high frequency carrier, three adjustable precision phase shifters, and optimal PID compensators.

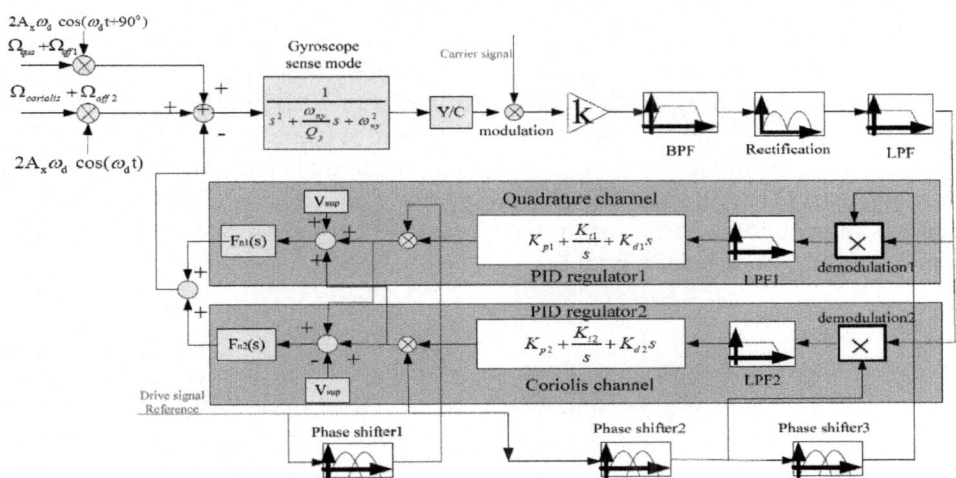

Figure 5. Sense mode circuitry method.

The output signals of the SMG are appropriately complicated; they principally contain two parts which are orthogonal to each other: quadrature part and

Coriolis part. The quadrature part contains the quadrature coupling signal and one part of the offset signal. The Coriolis part contains the Coriolis signal and the other part of the offset signal. For this gyroscope with a better decoupled structure, the offset error signal is relatively so weaker than the Coriolis signal that it is easy to cut off the interfering signal from the Coriolis signal. According to the theory of signal processing, the optimal method of separating the interfering signal from the Coriolis signal is the precision phase modulation and demodulation. In Figure 5, the input signals in the Coriolis and quadrature channels are expressed respectively as:

$$S_{qua}(t) = 2A_x \omega_d \cos(\omega_d t + 90^o) \cdot (\Omega_{qua} + \Omega_{off1}) \tag{3}$$

$$S_{coriolis}(t) = 2A_x \omega_d \cos(\omega_d t) \cdot (\Omega_{coriolis} + \Omega_{off2}) \tag{4}$$

where A_x is the oscillation amplitude in drive mode, ω_d is driving frequency same as its natural frequency ω_{nx} according to the designed structure parameters, $\Omega_{coriolis}$ and Ω_{qua} are the input angular rates of the Coriolis signal and equivalent quadrature signal, respectively, and they are 90 degree out-of-phase. Ω_{off1} and Ω_{off2} are two decomposition components of offset error in phase of $\Omega_{coriolis}$ and Ω_{qua}. According to superposition principle in control theory, both of them can be analyzed separately. The common forward part, including transfer function of gyroscope in sense mode, Y/C module (the conversion coefficient from displacement in sense direction to variable capacitor), K (the total gain of amplifiers), BPF (band-pass filter), Rectification module (for first demodulation of both two signals), and LPF (low-pass filter), can pick off these mixed signals. There are two channels including quadrature channel and Coriolis channel, which are similarly composed of the secondary demodulation module, LPF, PID compensator, and electrostatic force feedback generator $F_n(s)$ by adjusting the preloaded voltage V_{sup}, so the whole block diagram constitutes the dual-loop detection of the gyroscope in sense mode. According to the block diagram of dual-channel closed-loop control, Figure 6 illustrates the simplified control scheme of closed-loop detection only in the Coriolis signal channel because these two channels have the same systemic framework. Through this control strategy, the final effect is that the quadrature part is almost completely suppressed and the output signal can automatically follow the input angular signal $\Omega_{coriolis}$.

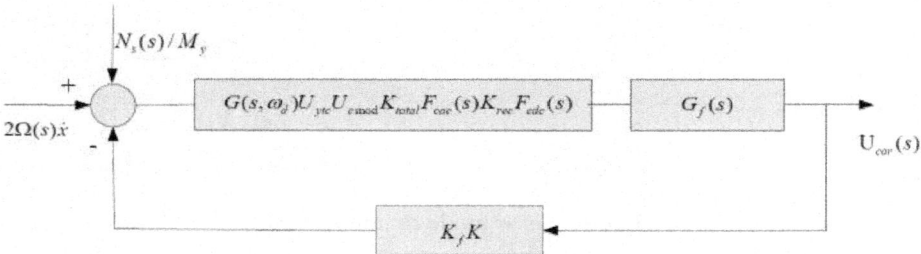

Figure 6. The simplified block diagram of coriolis signal closed-loop detection.

As shown in Figure 6, $2\Omega(s)\dot{x}$ is the Coriolis acceleration, $N_s(s)/M_y$ is equivalent to the input angular rate noise. K_fK is the actuator and adjustable feedback gain. $G(s, \omega_d)K_{ytc}K_{cmod}K_{total}F_{cac}(s)\,K_{rec}F_{cdc}\,(s)$ is the total forward path transfer function, where $G(s, \omega_d)$ is the transfer function of gyroscope in sense mode, K_{ytc} and K_{cmod} are gain coefficients from Y/C module and carrier signal modulation respectively, K_{total} and K_{rec} are gain coefficients from the total amplifiers and Rectification module respectively. $F_{cac}(s)$ and $F_{cdc}(s)$ are transfer functions of BPF and LPF, respectively. $Gf(s)$ is PID compensator. $U_{cor}(s)$ is the output signal.

In fact, the principle of the dual-channel closed-loop detection achieved by the quadrature channel is as same as which achieved by the Coriolis channel. Both signals processed have the same frequency that equal the natural frequency in the SMG's drive mode, the difference between both of them are phase angle and the varying trend, so the closed-loop detection with these two channels, essentially, is an AC feedback servo system which is rapidly controlled to keep the balance of AC force. It means that there would be almost a null displacement of the proof mass in the sense direction. Considering a special AC force feedback, these two channels need both modulation and demodulation, which can achieve a quick closed-loop servo control. These two orthogonal channels are decoupled and do not interfere with each other, so the closed-loop control performance of the two channels can be analyzed separately. For the useful signal channel, ie. the Coriolis channel, the closed-loop transfer function of the closed-loop system is calculated as follows:

$$\text{Let } G_0(s) = G(s, \omega_d)K_{ytc}K_{cmod}K_{total}F_{cac}(s)K_{rec}F_{cdc}(s) \text{ and } G(s, \omega_d) = 1/s^2 + \frac{\omega_{ny}}{Q_y}s + \omega_{ny}^2, \quad (5)$$

then:

$$G_c(s) = \frac{G_o(s)G_f(s)}{1 + G_o(s)K_fG_f(s)}$$

$$= \frac{\dfrac{K_{ytc}K_{cmod}K_{total}F_{cac}(s)K_{rec}F_{cdc}(s)}{s^2 + \dfrac{\omega_{ny}}{Q_y}s + \omega_{ny}^2}(K_p + \dfrac{K_p}{T_i}\cdot\dfrac{1}{s} + K_p\tau s)}{1 + \dfrac{K_{ytc}K_{cmod}K_{total}F_{cac}(s)K_{rec}F_{cdc}(s)}{s^2 + \dfrac{\omega_{ny}}{Q_y}s + \omega_{ny}^2}K_f(K_p + \dfrac{K_p}{T_i}\cdot\dfrac{1}{s} + K_p\tau s)} \quad (6)$$

where the PID compensator is:

$$G_f(s) = K_p + \frac{K_p}{T_i}\cdot\frac{1}{s} + K_p\tau s \quad (7)$$

Obviously, the closed-loop transfer function is a high-order system. It can be computed through Matlab tools. The effect of the PID compensator's parameters on the system bandwidth is shown in Table 1, and Figure 7 shows the result of simulation.

Table 1. Effect of the PID compensator's parameters on the system bandwidth.

PID compensator parameters	$\dfrac{0.002s^2 + 0.5s + 1}{s}$	$\dfrac{0.0015s^2 + 2s + 1}{s}$	$\dfrac{0.0025s^2 + 3s + 1}{s}$	$\dfrac{0.005s^2 + 3s + 2}{s}$
K_p	0.5	2	3	3
T_i (s)	0.5	2	0.3	1.5
τ (s)	0.004	0.0075	0.0025	0.0025
Simulated bandwidth (Hz)	60	99.8	146	205

As shown in Table 1, changing the transfer function of the compensator module (just changing the PID parameters: the proportional coefficient K_p, time integral constant T_i, time differential constant τ) can influence the closed-loop system bandwidth of gyroscope. Meanwhile, Figures 7(a), (b) illustrate the changes of closed-loop system bandwidth with different compensator models, and the system Bode graph shows the changes of gain margin and phase margin. Therefore the chosen parameters of the designed compensator should not only meet the system bandwidth requirements, but also guarantee the stability of the system. From the above analysis, there exists a complex nonlinear relationship between the bandwidth and compensator parameters. However the tradeoff among these parameters can be found to reach a certain bandwidth together with a better stability by Matlab simulation tools. In this case, when K_p approaches near 2, time integral constant T_i approaches near 2s, and time differential constant τ approaches near 0.0075 s, the closed-loop system shown in the Figure 7(b) displays its stability and reliability with the phase margin greater than 45 degree, the amplitude margin more than 30 dB, and the bandwidth approaching 100 Hz. Though the bandwidth can be further increased by two later compensator parameters, the corresponding phase and amplitude margins tend to drop almost near to 45° and 30 dB, which demonstrates that the instability of closed-loop. The step response in Figure 7(c) displays a small overshoots less than 30 percent and quick response time near 5 ms when input angular velocity of rotation is 1°/s.

(a) Bode graph.

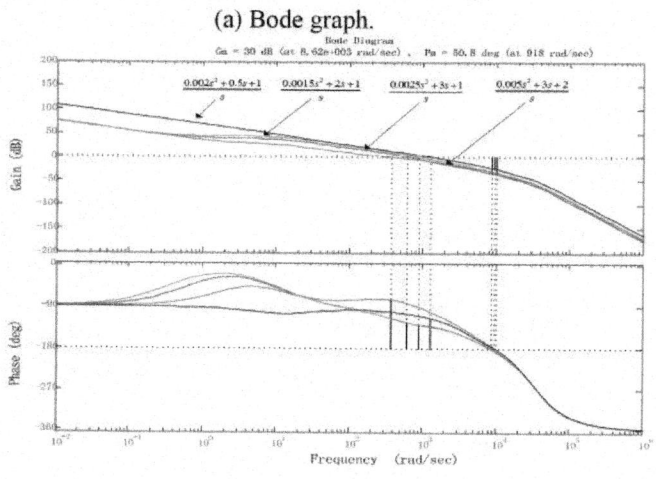

(b) Closed-loop bandwidth.

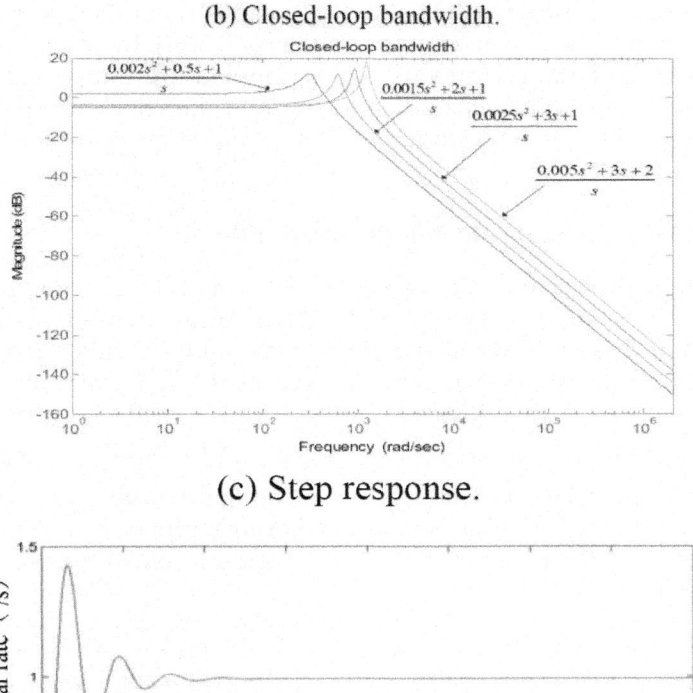

(c) Step response.

Figure 7. The simulated performances of the whole closed-loop system.

4. DIGITAL TEMPERATURE COMPENSATION

The proposed silicon micro-machined gyroscope is made of silicon, which would be influenced by temperature variation, resulting in the distortion of the structure, the change of the Young's modulus, the resonant frequency, and the Q factor etc. The effects of the environmental temperature on the performance is of great interest, since they can be an important source of error in gyroscopes. Essentially, temperature variation will affect the Young's modulus of the material and result in deformation of the structure, deviations in resonant frequencies and change of Q factor. Therefore the output of the system has a complicated changing process, with the result that the bias will drift greatly owing to the temperature variation. The temperature dependent characteristics of the gyroscope were investigated without compensation resolution in [13,14]. In order to eliminate the impact from the temperature variation, a temperature compensation system is designed which can calculate the compensation values at each temperature point ranging from -40 °C to 80 °C in terms of the compensation model previously set up by experiments .

4.1. The Mechanism of the Temperature Effect

Because L is the parameter of both Young's modulus and the spring stiffness K, all of them would be changed with temperature variation. Suppose that the coefficient of thermal expansion of silicon is α, and the initial length of silicon structure is l_0 . The temperature change from initial T_0 to T will result in the change of the three physical quantities as follows:

$$L = l_0[1 + \alpha(T\text{-}T_0)], \; E(T) = E_0\text{-}E_0k_{ET}(T\text{-}T_0), \; K(T) = K_0\text{-}K_0k_{ET}(T\text{-}T_0) \tag{8}$$

where k_ET is the variation coefficient of the Young's modulus E over temperature. E_0 and K_0 are initial Young's modulus and spring stiffness, respectively. Then the natural resonant frequency of the silicon micro-machined gyroscope would be changed as follows:

$$f(T) = \sqrt{K(T)/m} = \sqrt{K_0\left(1 - k_{ET}\left(T - T_0\right)\right)/m} = f_0\sqrt{1 - k_{ET}\left(T - T_0\right)} \tag{9}$$

where f is the resonant frequency of proof mass m. Meanwhile, the Q factor of any mode can be expressed as:

$$Q = \frac{d}{\mu A}\left(mK\right)^{1/2}, \mu = \frac{\sqrt{2}\pi d^2 \mu_0 pd}{k_b T} \tag{10}$$

where d is the fluid thickness, A is the plate area parallel with the substrate, p is the ambient pressure, k_b is Boltzmann constant, u_0 and u are original and changed viscosity coefficient of the gas respectively.

From above analysis, the natural resonant frequency and Q factor are both directly correlated with the sensitivity, stability, and other static and dynamic performances, so these factors would all be affected by temperature variation. Besides, the performance of the electronic components in drive and sense modes would also be influenced by temperature variation.

4.2. Establishment of the Compensation Model

From the above analysis, the complication of the influence on the zero bias of silicon micro-machined gyroscope is obvious, and it is different to deduce the precise math expressions theoretically. Therefore, it is relatively easier to derive the relationship expression between zero bias and temperature through lots of temperature experiments. The silicon micro-machined gyroscope was used in temperature experiment. The experimental steps are as follows: Firstly, the gyroscope is put into the temperature controlling casing, and the temperature is dropped down to -40 °C. Secondly, the zero bias of gyroscope is measured every 5 °C and sampled for one minute at each temperature point. Thirdly, the average zero bias value is calculated at each point, and the processed data is plotted in a graph. Lastly, the precise math expressions with least square fitting method is figured out from -40 °C to 80 °C. In order to reduce the fitting error and corresponding the polynomial orders, the fitting expressions of three orders term could be divided into three segments as follows:

When $- 40°C \leq T < -10°C$, then:

$$V_{bias} = 0.4813 + 0.00719T - 0.000735T^2 + 0.0000416T^3 \tag{11}$$

When $- 10°C \leq T < 50°C$, then:

$$V_{bias} = 0.4313 - 0.00629T + 0.00121T^2 + 0.00000642T^3 \tag{12}$$

When $50°C \leq T < 80°C$, then:

$$V_{bias} = -673.07 + 42.879T - 1.0146T^2 + 0.01053T^3 \tag{13}$$

The polynomials in three temperature segments are stored in the memory of microprocessor, and in the compensation process, the microprocessor would calculate the compensation values according to the real-time temperature value measured by digital temperature sensor inside the gyroscope box, and at the same time the microprocessor calculates the desired signal by subtracting the compensation value from the sampled gyroscope signal.

4.3. The Design of Temperature Compensation System

Due to the accurate relationship between the zero bias and the ambient temperature of the gyroscope, the actual temperature compensation can be realized by designing a hardware platform. Figure 8 shows the block diagram of the main hardware modules: the detected uncompensated analog signal of the gyroscope is converted into digital signal by an ADS1251 AD converter (Texas Instruments Company). Meantime, the temperature value is measured by a DS18B20 digital temperature sensor (Dallas Semiconductor) and transmitted to a C8051F360 microprocessor (Silicon Labs Company) which is the core of whole system. The fitting polynomials are previously stored in the flash memory of the C8051F360 and used to calculate the compensation value in terms of the measured temperature, and then the compensated gyroscope signal is converted into analog signal by AD5060 (Analog Device Company) and transferred out. To facilitate debugging, the PC can communicate with the C8051F360 through UART and download successfully complied program into the C8051F360 by C2 interface. These components are all low power with high performance.

(1) DS18B20 is a kind of 1-wire digital temperature sensor, with a wide measuring range from -55 °C to +125 °C, 9 - 12 bit digital temperature readout, and accuracy of ±0.5 °C. In the design, 12 bit resolution is chosen, and 3.3 V power supply is applied. Data acquision (DQ) pin is connected to pin P1.6 of the C8051F360.

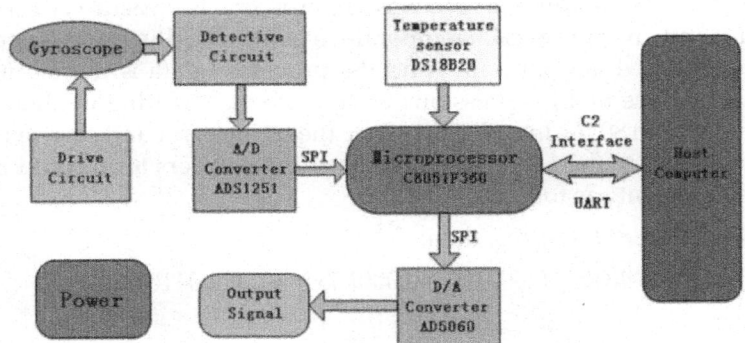

Figure 8. The hardware setup of the temperature compensation system.

(2) Because the gyroscope signal is quite weak, a high-accuracy, wide dynamic range of 24-bit Δ-Σ structure ADS1251 AD converter, which can achieve 19 bits of effective accuracy when its readout rate is up to 20 kHz, is ideal for this requirement. It is a single channel converter, with SO-8 seal, very small volume, and 2-wire serial interface, which can be directly connected to the microprocessor through a Serial Peripheral Interface (SPI).

(3) According to the analog output requirement of the system, after converted by A/D and then compensated in the microprocessor, the compensated digital signal should be transmitted out again through a D/A converter. In this design, AD5060 with a 3-wire serial interface, which can communicate with microprocessor by SPI, is used to achieve outputting gyroscope angular rate signal compensated. C8051F360 has only one SPI port, which has to be connected with both ADS1251 and AD5060. Apart from SCLK, different data pins of SPI are connected with each separate pin SDATA for data transmission respectively, even at the same time. Though ADS1251 and AD5060 use the same SPI interface of the microprocessor, there is no conflict between them because an optimal program will enable them to work orderly based on time-division-multiplexing. In addition, a special pin of AD5060 will be connected to the microprocessor, which is used to generate an interrupt request to inform that the data conversion is over. The AD5060 can achieve 16 bits of effective accuracy for D/A conversion.

(4) In order to facilitate the observation when debugging and recording data during the temperature experiments, the UART serial port is used to communicate with host computer in this design, port Tx and port Rx are configured respectively to the relevant pins of the microprocessor by using the overlapping switch function. The UC-5 debugging adapter (New Hualong Corporation) is connected to the debugging interface C2 of C8051F360 and an USB port connecter with the host computer is used to download the program to the internal flash memory of the C8051F360. A tiny-6 pin interface embedded in the printed circuit board (PCB) circuitry is connected to the standard 5 mm × 2 mm interface through a connector designed, which will realize not

only downloading program conveniently in the debugging process, but also a reduction of area of the PCB and decreased volume in favor of miniaturization.

A temperature compensation system using a C8051F360 microprocessor is thus designed. The gyroscope signal would be compensated by subtracting the compensation values which can be calculated through the expressions in terms of the temperature values measured. The PCB photos of the temperature compensation system are shown in Figure 9. In this paper, A/D and D/A are only used in digital compensation system. In Figure 5, the gyroscope's circuit in sense mode is pure analog using many discrete amplifiers which is independent of Figure 9. The compensated signal is processed by the circuit (see Figure 9) according to the detected temperature outside, and retransferred to analog signal with sum of gyroscope's output analog signal. In this implementation, Figure 9 shows only the circuit of digital temperature part while the packaged gyroscope is shown in later Figure 11.

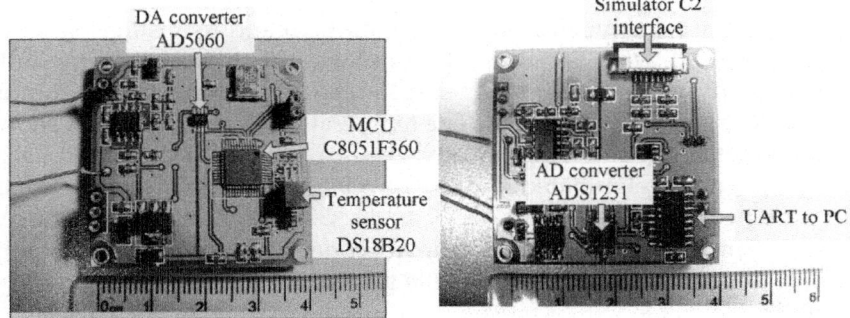

Figure 9. The PCB photos of the temperature compensation system.

5. EXPERIMENTAL RESULTS

A SMG with the structure of changing-area capacitor detection is adopted in this paper, and the quality factor of a SMG (named as E17 sealed in metal can package working at an ambient pressure of 10 Pa) in two modes is above 10,000, so its sense mode detection essentially attains better linear performance and high sensitivity. Its responses in drive and sense modes (at atmospheric pressure and in vacuum) are shown in Figure 10. Single side driving and single side detection method is applied to make both vibrating amplitude and frequency highly stable in drive mode circuit. Meanwhile the dual-channel closed-loop detection method is utilized to achieve high precision detection in sense mode circuit. A miniature prototype scheme based on PCB technology has been realized with the dimention of 40 mm × 40 mm × 30 mm and power consumption less than 200 mW. The testing results demonstrate that the useful signal and quadrature signal would not interact with each other because of their phase decoupling. Under the condition of the scale factor of 9.6 mV/°/s, in full measurement range of ± 300deg/s, the zero bias stability attains 15 °/h with worse-case nonlinearity of 400 ppm, the minimum measurable angular rate is 0.02°/s, the noise equivalent rate is 0.01°/s, and the noise is 0.0024°/s/√Hz by ALLAN variance analysis.

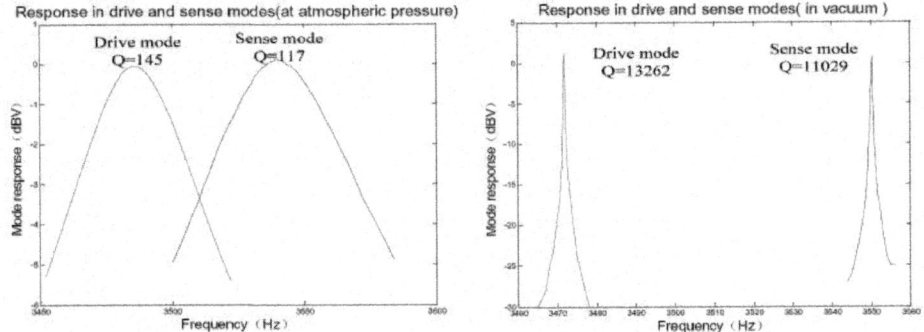

Figure 10. Response in drive and sense modes.

A miniature prototype of SMG based on PCB and its testing platform are shown in Figure 11. Some important performances such as scale factor and zero bias stability are shown in Figures 12 and 13. The digital temperature compensation results are shown in Figure 14. The temperature experiment tests show that the influence of temperature variation on the zero bias of silicon micro-machined gyroscope is greatly eliminated and the zero bias is reduced to one tenth of the original after compensation from -40 °C to 80 °C. Table 2 generalizes the overall performance index of a miniature microgyroscope prototype at different temperature. In summary, the experiment results testify that the miniature prototype of SMG can be realizable with a satisfied performance.

Figure 11. Miniature microgyro prototype.

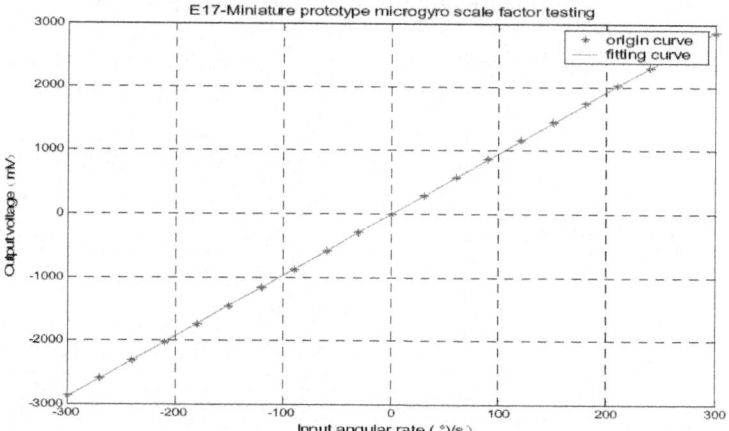

Figure 12. Scale factor testing.

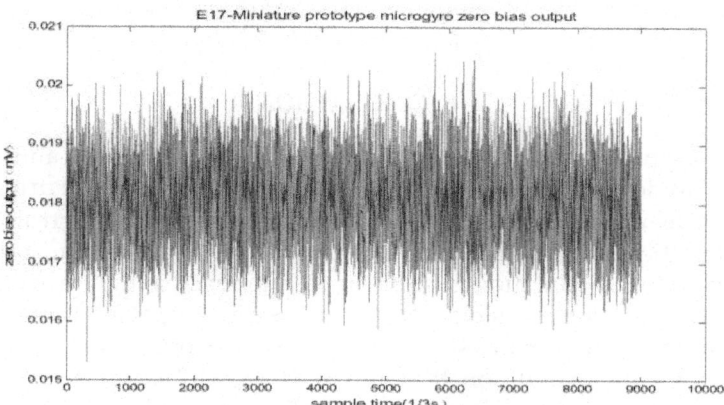

Figure 13. Zero bias stability testing.

Figure 14. Efforts of temperature on ZRO before and after compensation.

(a) Before compensation. (b) After compensation.

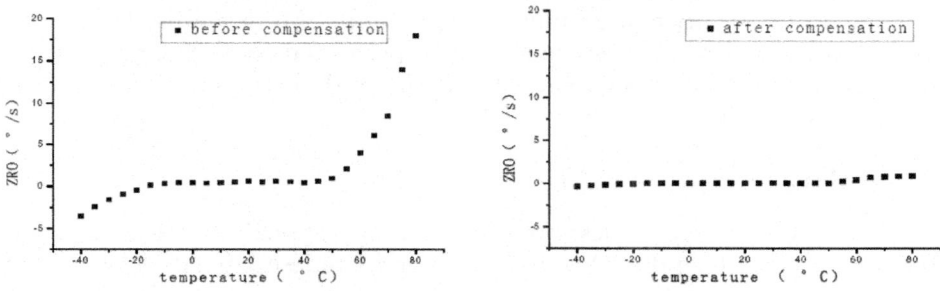

Table 2. Performance index of miniature microgyro prototype.

Technical data		Value
Performance(+25 °C)	Scale factor	9.6 mV/°/s
	Bias stability	15 °/h
	Noise	0.0024°/s/√Hz
	Noise equivalent rate	0.01°/s
	Dynamic range	± 300°/s
	The minimum measurable angular rate	0.02°/s
	Linearity	≤ 400 ppm
	Bandwidth	> 80 Hz
Power supply	Supply voltage	± 5 V
	Current dissipation	20 mA
Environment(-40 °C~ + 80 °C)	Bias stability	3°/s
	Shock survival	1000 g
	Temperature drift	< 0.03°/s/°C

6. CONCLUSIONS

A miniature vacuum-packaged SMG with symmetrical and decoupled structure was designed to prevent unintended coupling between drive and sense modes. To ensure high resonant stability and strong capacity of resisting distur-bance, simplified self-oscillating closed-loop circuit include AGC loop and high frequency carrier modulation is implemented to reach a high precision in drive mode. The application occasion of the gyroscope is complicated, usually affected by many kinds of disturbances such as interference, over shock, vibration, tem-perature and so on. In addition, the bandwidth of SMG is approximately decided by the designed frequency difference between drive and sense modes. In order to improve the accuracy and reliability and broaden the system bandwidth, an improved dual-channel closed-loop detection method is adopted in this paper, which can ensure the stability and reliability of silicon micro-gyroscope's output by testing all kinds of specifications of SMG. Finally a miniature prototype of SMG with temperature compensation based on PCB is designed successfully. The ex-periment results testify that the miniature prototype SMG is realizable with good performance. Nevertheless, some important performances such as startup time of SMG and intelligent function of self detection and calibration will be perfected through further efforts.

Acknowledgements

The authors gratefully acknowledge Chinese National Science Foundation's financial support (Contract No. 60575010) and Chinese Hi-Tech Research and Development Program's financial support (Contract No. 2002AA812038).

REFERENCES AND NOTES

1. Sharma, A.; Zaman, M.F.; Ayazi, F. A 0.2o/hr micro-gyroscope with automatic CMOS mode matching. In *Proceedings of IEEE International Solid-State Circuits Conference, Digest of Technical Papers*, San Francisco, CA, USA, December 2007; pp. 386-387.

2. Kulygin, A.; Schmid, U.; Seidel, H. Characterization of a novel micromachined gyroscope under varying ambient pressure conditions. *Sens. Actuat. A* 2008, 145-146, 52-58.

3. Li, Z.; Yang, Z.; Xiao, Z. A bulk micromachined vibratory lateral gyroscope fabricated with wafer bonding and deep trench etching. *Sens. Actuat.* 2000, 83, 24-29.

4. Chang, B.S.; Kim, S.; Lee, J.G.; Kang, D.T.; Sung, W. Design and performance analysis of PLL based self oscillation loop in vibrating gyroscope. In *Proceedings of SPIE*, San Jose, CA, USA February 2005; pp.1-6.

5. Sharma, A.; Zaman, M.F.; Ayazi, F. A 104-dB dynamic range transimpedance-based CMOS ASIC for tuning fork microgyroscopes. *IEEE J. Solid-State Circuits.* 2007, 42, 1790-1802.

6. Phani, A.S.; Seshia, A.A.; Palaniapan, M.; Howe, R.T.; Yasaitis, J.A. Modal coupling in micromechanical vibratory rate gyroscopes. *IEEE Sensors J.* 2006, 6, 1144-1152.

7. Clark, W.A.; Howe, R.T. Surface micromachined Z-axis vibratory rate gyroscope. In *Proceedings of Solid-State Sensor and Actuator Workshop*, Hilton Head Island, SC, USA, June 1996; pp. 283-287.

8. Loveday, P.W.; Rogers, CA. The influence of control system design on the performance of vibratory gyroscopes. *J. Sound. Vib.* 2002, 255, 417-432.

9. Jiang, X. *Capacitive position-sensing interface for micromachined inertial sensors.* Ph.D. Dissertation, University of California, Berkeley, CA, USA, 2003.

10. Petkov, V.P.; Boser, B.E. A fourth-order sigma-delta interface for micromachined inertial sensors. *IEEE J. Solid-State Circuits* 2005, 40, 1602-1609.

11. Ezekwe, CD.; Boser, B.E. A mode-matching EA closed-loop vibratory gyroscope readout interface with a 0.004 °/s/VHz noise floor over a 50 Hz band. *IEEE J. Solid-State Circuits* 2008, 43, 3039-3048.

12. Gaißer, A.; Geiger, W.; Link, T.; Merz, J.; Steigmajer, S.; Hauser, A.; Sandmaier, H.; Lang, W.; Niklasch, N. New digital readout electronics for capacitive sensors by the example of micromachined gyroscopes. *Sens. Actuat. A-Phys.* 2002, 97-98, 557-562.

13. Ferguson, M.I.; Keymenlen, D.; Peay, C; Yee, K. Effect of temperture on vibratory rate gyroscope. In *Proceedings of IEEE Aerospace Conference*, Big Sky, MT, USA, March 2005; pp.1-6.

14. Shcheglov, K.; Evans, C; Gutierrez, R.; Tang, T.K. Temperature dependent characteristics of the JPL silicon MEMS gyroscope. In *Proceedings of IEEE Aerospace Conference*, Big Sky, MT, USA, March 2000; pp.403-411.

15. Geiger, W.; Butt, W.U.; Gaißer, A.; French, J., Braxmaier, M.; Link, T.; Kohne, A.; Nommensen, P.; Sandmaier, H.; Lang, W.; Sandmaier, H. Decoupled microgyros and the design principle DAVED. *Sens. Actuat. A-Phys.* 2002, 95, 239-249.

16. Alper, S.E.; Emre, S.; Yuksel, T.; Tayfun, A. A compact angular rate sensor system using a fully decoupled silicon-on-glass MEMS gyroscope. *J. Microelectromech. Syst.* 2008, 17, 1418-1429.

17. Park, S.; Horowitz, R. Adaptive control of the conventional mode of operation of MEMS gyroscopes. *J. Microelectromech. Syst.* 2003, 12, 101-108.

18. Saukoski, M.; Aaltonen, L.; Halonen, K.A.I. Zero-rate output and quadrature compensation in vibratory MEMS gyroscopes. *IEEE Sens. J.* 2007, 7, 1639-1652.

19. Yeh, B.Y.; Liang, Y.C. Mathematical modelling on the quadrature error of low-rate microgyroscope for aerospace applications. *Analog. Integr. Circuit. Signal.* 2001, 29, 85-94.

20. Nguyen, C.T.-C.; Howe, R.T. An integrated CMOS micromechanical resonator high-Q oscillator. *IEEE J. Solid-State Circuits* 1999, 34, 440-455.

21. M'Closkey, R.T.; Vakakis, A. Analysis of a microsensor automatic gain control loop.In *Proceedings of the American Control Conference*, San Diego, CA, USA, June 1999; pp. 3307-3311.

22. Sun, X.; Horowitz, R.; Komvopoulos, K. Stability and resolution analysis of a phase-locked loop natural frequency tracking system for MEMS fatigue testing. *J. Dyn. Syst. Meas. Contr.* **2002**, *124*, 599-605.

23. Saukoski, M.; Aaltonen, L.; Salo, T.; Halonen, K. Integrated readout and control electronics for a microelectromechanical angular velocity sensor. In *Proceedings of the 32nd European Solid-State Circuits Conference*, Montreux, Switzerland, September 2006; pp. 243-246.

24. Xia, D.Z.; Zhou, B.L.; Wang, S.R. Design and testing analysis of double closed loop vacuum silicon microgyroscope. *Chin. J. Sens. Actuat.* **2008**, *21*, 241-243.

25. Xia, D.Z.; Zhou, B.L. The design of silicon micromechanical Z–axis gyroscope closed-loop servo control System. *Chin. J. Meas. Contr. Tech.* **2007**, *26*, 1-6.

INDEX